A PRACTICAL APPROACH TO MICROARRAY DATA ANALYSIS

A PRACTICAL APPROACH TO MICROARRAY DATA ANALYSIS

edited by

Daniel P. Berrar
School of Biomedical Sciences
University of Ulster at Coleraine, Northern Ireland

Werner Dubitzky
Faculty of Life and Health Science
and Faculty of Informatics
University of Ulster at Coleraine, Northern Ireland

Martin Granzow
4T2consulting
Weingarten, Germany

KLUWER ACADEMIC PUBLISHERS
Boston / Dordrecht / London

Distributors for North, Central and South America:
Kluwer Academic Publishers
101 Philip Drive
Assinippi Park
Norwell, Massachusetts 02061 USA
Telephone (781) 871-6600
Fax (781) 681-9045
E-Mail: kluwer@wkap.com

Distributors for all other countries:
Kluwer Academic Publishers Group
Post Office Box 322
3300 AH Dordrecht, THE NETHERLANDS
Telephone 31 78 6576 000
Fax 31 78 6576 254
E-Mail: services@wkap.nl

 Electronic Services < http://www.wkap.nl>

Library of Congress Cataloging-in-Publication Data

A C.I.P. Catalogue record for this book is available
from the Library of Congress.

A Practical Approach to Microarray Data Analysis edited by Daniel P. Berrar, Werner
Dubitzky and Martin Granzow
ISBN 1-4020-7260-0

Printed on acid-free paper.

Printed in Great Britain by IBT Global, London

The Publisher offers discounts on this book for course use and bulk purchases.
For further information, send email to <joanne.tracy@wkap.com> .

Contents

Acknowledgements

The editors would like to thank the contributing authors for their excellent work. Furthermore, the editors would like to thank Joanne Tracy and Dianne Wuori from Kluwer Academic Publishers for their help and support in editing this volume.

Preface

In the past several years, DNA microarray technology has attracted tremendous interest in both the scientific community and in industry. With its ability to simultaneously measure the activity and interactions of thousands of genes, this modern technology promises unprecedented new insights into mechanisms of living systems. Currently, the primary applications of microarrays include gene discovery, disease diagnosis and prognosis, drug discovery (pharmacogenomics), and toxicological research (toxicogenomics).

Typical *scientific tasks* addressed by microarray experiments include the identification of coexpressed genes, discovery of sample or gene groups with similar expression patterns, identification of genes whose expression patterns are highly differentiating with respect to a set of discerned biological entities (e.g., tumor types), and the study of gene activity patterns under various stress conditions (e.g., chemical treatment). More recently, the discovery, modeling, and simulation of regulatory gene networks, and the mapping of expression data to metabolic pathways and chromosome locations have been added to the list of scientific tasks that are being tackled by microarray technology.

Each scientific task corresponds to one or more so-called *data analysis tasks*. Different types of scientific questions require different sets of data analytical techniques. Broadly speaking, there are two classes of elementary data analysis tasks, *predictive modeling* and *pattern-detection*. Predictive modeling tasks are concerned with learning a classification or estimation function, whereas pattern-detection methods screen the available data for interesting, previously unknown regularities or relationships.

A plethora of sophisticated methods and tools have been developed to address these tasks. However, each of these methods is characterized by a set of idiosyncratic requirements in terms of data pre-processing, parameter configuration, and result evaluation and interpretation. To optimally design and analyze microarray experiments, researchers and developers need a sufficient overview of existing methodologies and tools and a basic understanding of how to apply them.

We believe that one significant barrier to the widespread effective and efficient use of microarray analysis methods and tools is a lack of a clear understanding of how such techniques are used, what their merits and limitations are, and what obstacles are involved in deploying them. Our goal in developing this book was to address this issue, by providing what is simultaneously a *design blueprint, user guide,* and *research agenda* for current and future developments in the field.

As design blueprint, the book is intended for life scientists, statisticians, computer experts, technology developers, managers, and other professionals who will be tasked with developing, deploying, and using microarray technology including the necessary computational infrastructure and analytical tools.

As a user guide, the book seeks to address the requirement of scientists and researchers to gain a basic understanding of microarray analysis methodologies and tools. For these users, we seek to explain the key concepts and assumptions of the various techniques, their conceptual and computational merits and limitations, and give guidelines for choosing the methods and tools most appropriate for the analytical task at hand. Our emphasis is not on a complete and intricate mathematical treatment of the presented analysis methodologies. Instead, we aim at providing the users with a clear understanding and practical know-how of the relevant methods so that they are able to make informed and effective choices for data preparation, parameter setting, output post-processing, and result interpretation and validation. For methodologies where free software exists we will also provide practical tips for obtaining and using the tools.

As a research agenda, this volume is intended for students, teachers, researchers, and research managers who want to understand the state of the art of the presented methodologies and the areas in which gaps in our knowledge demand further research and development. To this end, our aim was to maintain the readability and accessibility of a textbook throughout the chapters, rather than compiling a mere reference manual. Therefore, considerable effort was made to ensure that the presented material, which stresses the applied aspects of microarray analysis, is supplemented by rich literature cross-references to more foundational work.

Clearly, we cannot expect to do justice to all three goals in a single book. However, we do believe that we have succeeded in taking useful steps toward each goal. In doing so, we hope to advance the understanding of both the methodologies and tools needed to analyze microarray data, and the implications for future developments of microarray technology and its support technologies.

The design and subsequent analytical examination of microarray experiments rests on the scientific expertise of the experimenters, their knowledge of the relevant microarray technology and experimental protocols, and their understanding of analysis methods and tools. The available machinery of microarray analysis methods ranges from classical statistical approaches, to machine learning techniques and to methods from artificial intelligence. Hence, the preparation of this book must draw upon the experts from many diverse subfields in mathematics and computer science. In developing this volume, we have assembled a distinguished set of authors, each recognized as an authority in one or more of these fields. We have asked these authors to present a selected set of state-of-the-art methodologies and tools for analyzing microarrays from a highly practical, user-oriented perspective, emphasizing the how-to aspects of the presented techniques. To support the research agenda of this book, we have also asked the authors to identify where future developments are likely to take place and to provide a rich set of pointers to theoretical works underpinning the presented methods. The result, we hope, is a book that will be valuable for a long time, as summary of where we are, as a practical user guide for making informed choices on actual microarray analysis projects, and as roadmap for where we need to go in order to improve and further develop future microarray analysis technology.

This book contains one introductory chapter and 19 technical chapters, dealing with specific methods or class of methods. As illustrated in Table 1, the technical chapters are roughly grouped into two broad categories, namely *data preparation* and *exploratory data analysis* respectively. Partitioning the chapters into these areas largely mirrors the current state of the art in the field. Different protocols, experimental conditions, analysis goals, data complexity, and sources of systematic variation in microarray experiments normally require different ways for selecting and preparing the raw data obtained from the detection devices. These methods range from missing value imputation and normalization to feature subset selection and data integration. Collectively, we refer to these methods as data preparation or *pre-processing* techniques. Once the final format for the data is achieved, data exploration or analysis can commence. This part of the data analysis process is referred to as *exploratory data analysis*. Typical exploratory analysis tasks include *classification* (or class prediction), *clustering* (or automatic classification), correlation and association analysis, and others.

As much as possible, the chapters are presented in an order that reflects the overall data analysis process. In Chapter 1, we provide an introduction to microarray analysis with the aim of (a) providing an easy-to-understand description of the entire process, and (b) establishing a common terminology. First, we recapitulate the biological and technological background of microarray hybridization experiments. This includes the main types of arrays that exist, aspects of their protocols, and what kind of quantities they are measuring. Second, we categorize the classes of questions life scientists hope to answer with microarray experiments, and what kind of analytical tasks they imply. Third, we describe the entire process from the inception of a scientific question or hypothesis, to the design and execution of a microarray experiment, and finally to data preparation, analysis, and interpretation. We then discuss some of the conceptual and practical difficulties the experimenter faces when choosing and applying specific data analysis techniques. The remaining chapters are intended to shed more detailed light on these issues.

Table 1. Roadmap to the content of the book.

Part	Topic	Chapter#
	Introduction: Overview of microarray analysis process	1
Pre-processing	Foundations, issues, and methods	2
	Missing value imputation	3
	Error handling and normalization	4
	Singular value decomposition, principal component analysis	5
	Feature selection: established and recent techniques	6
Classification	Statistical foundations and methods	7
	Bayesian networks	8
	Support vector machines	9
	Weighted flexible compound covariate method and decision trees	10
	Artificial neural networks	11
	k-nearest neighbor and genetic algorithms	12
Cluster Analysis	Overview and review of some methods	13
	Hierarchical clustering methods	14
	Self-organizing maps	15
	Other non-hierarchical methods	16
Other	Correlation and association analysis methods	17
	Functional interpretation analytical results	18
Tools	Systematic review of free and commercial software	19
	Managing microarray data analysis: workflow and process	20

Chapter 2 addresses the issue of data pre-processing in microarray analysis in general. It is written for the newcomer to this field and explains the basic concepts and provides a useful vocabulary. It discusses the motivation for normalization, data centralization, data re-scaling, and missing value imputation. This chapter represents an introduction to the Chapters 2 to 5.

Chapter 3 presents three different methods for missing value imputation in microarray data. This includes a *k-nearest-neighbor* approach, a method based on *singular value decomposition,* and *row averaging.* Practical guidelines are presented for using publicly available free software tools.

Chapter 4 discusses various sources of errors in microarray data, and then proceeds with a detailed discourse on normalization. In contrast to Chapter 2, the focuses is on mathematical considerations.

Chapter 5 is concerned with a major problem in microarray data analysis – the so-called *large-p-small-n problem* also known as the *curse of dimensionality.* This refers to the fact that for many microarray experiments the number of variables (genes) exceeds the number of observations (samples) by a factor of 10, 1000, or more. Feature selection and dimension reduction methods refer to techniques designed to deal with this "curse". The chapter discusses the use – and misuse – of *singular value decomposition* and *principal component analysis.*

Chapter 6 is a survey of several important feature selection techniques used to ward off the curse of dimensionality. First, it presents classic *filter* and *wrapper* approaches and some recent variants of explicit feature selection. Second, it outlines several feature weighting techniques including WINNOW and Bayesian feature selection. Third, towards the end, the chapter describes some recent work on feature selection for clustering tasks, a subject that has been largely neglected.

Chapter 7 discusses statistical issues arising in the classification of gene expression data. This chapter introduces the statistical foundations of classification. It provides an overview of traditional classifiers, such as linear discriminant analysis and nearest neighbor classifiers, in the context of microarray analyses. The general issues of feature selection and classifier performance assessment are discussed in detail.

Chapter 8 looks at Bayesian networks for the classification of microarray data. It introduces the basic concept of this approach, and reports on a study where the performance of Bayesian networks was compared with other state-of-the-art classifiers.

Chapter 9 describes a classification method that has been gaining increasing popularity in the microarray arena – *support vector machines* (SVMs). It provides an informal theoretical motivation of SVMs, both from a geometric and algorithmic perspective. Instead of focusing on mathematical completeness, the intention of this chapter is to provide the practitioner with some "rules of thumb" for using SVMs in the context of microarray data. Finally, pointers to relevant, publicly available free software resources are given.

Chapter 10 reports on a recent case study of gene expression analysis in lung cancer. The authors describe the *weighted flexible compound covariate* method for classifying the microarray data. They also demonstrate how this relatively new method is related to decision trees.

Chapter 11 deals with a widely used machine learning technique called *artificial neural networks* (ANNs). The authors describe the application of ANNs to micrarray classification task. They discuss how a principal component analysis, *cross-validation* and *random permutation tests* can be employed to improve and evaluate the predictive performance of ANNs. The problem of extracting important genes from a constructed ANN is also addressed.

Chapter 12 represents the last chapter on classification. It presents the *k-nearest-neighbor* strategy and *genetic algorithms* for classifying microarray data. It discusses the general motivation and the concepts of these methods, and demonstrates their performance on microarray data sets. The authors provide references to publicly available free software resources.

Chapter 13 provides an overview of the major types of clustering problems and techniques for microarray data. It focuses on crucial design and analytical aspects of the clustering process. The authors provide some important criteria for selecting clustering methods. Furthermore, the chapter describes a scheme for evaluating clustering results based on their relevance and validity (both computational and biological).

Chapter 14 addresses hierarchical clustering methods in the context of microarray data. The discussed methods include hierarchical clustering methods, including *adaptive single linkage clustering,* a new method designed to provide adaptive cluster detection while maintaining scalability. Furthermore, the chapter provides examples using both simulated and real data.

Chapter 15 presents *self-organizing maps* (SOMs) for clustering microarray data. It discusses question such as: How do these models work? Which are their advantages and limitations? Which are the alternatives? In answering these questions, this chapter constitutes a rich source of practical guidelines for using SOMs to analyze microarray data.

Chapter 16 examines a number of non-hierarchical clustering algorithms for microarray analysis, namely *cluster affinity search technique*, the famous *k-means* technique, *partitioning around medoids*, and *model-based clustering*. The chapter puts emphasis on the practical aspects of these algorithms, such as guidelines for parameter setting, the specific algorithmic properties, and practical tips for implementation.

Chapter 17 addresses correlation and association analysis methods. It addresses questions that should help the user to assess the limitations and merits of these methods, such as: How to statistically measure the strength between two variables and test their significance? What is correlation, what is association? Which conclusions do correlation and association analysis allow in the context of microarray data?

Chapter 18 discusses the global functional interpretation of gene expression experiments. After a researcher has found differentially expressed genes using one of the above described methods, he must face the challenge of translating his results into a better understanding of the underlying biological phenomena. This chapter shows how this can be achieved.

Chapter 19 provides an overview of both publicly available, free software and commercial software packages for analyzing microarray data. The aim of this review is to provide an overview of various microarray software categorized by their function and characteristics. This review should be a great help for those who are currently consider obtaining such software.

Finally, Chapter 20 describes the microarray data analysis from process perspective, highlighting practical issues such as project management and workflow considerations.

The book is designed to be used by the practicing professional tasked with the design and analysis of microarray experiments or as a text for a senior undergraduate- or graduate level course in analytical genetics, biology, bioinformatics, computational biology, statistics and data mining, or applied computer science. In a quarter-length course, one lecture can be spent on each chapter, and a project may be assigned based on one of the topics or techniques discussed in a chapter. In a semester-length course, some topics can be covered in greater depth, covering more of the formal background of the discussed methods. Each chapter includes recommendations for further reading. Questions or comments about the book should be directed to the editors by e-mail under *dp.berrar@ulster.ac.uk*, *w.dubitzky@ulster.ac.uk*, or *granzow@4T2consulting.de*. For further details on the editors, please refer to the following URL:

http://www.infj.ulst.ac.uk/~cbbg23/interests.html.

Daniel Berrar

Werner Dubitzky

Martin Granzow

Chapter 1

INTRODUCTION TO MICROARRAY DATA ANALYSIS

Werner Dubitzky[1], Martin Granzow[2], C. Stephen Downes[1], Daniel Berrar[1]

[1]*University of Ulster, School of Biomedical Sciences, Cromore Rd., Coleraine BT52 1SA, Northern Ireland,*
e-mail: {w.dubitzky, cs downes, dp.berrar}@ulster.ac.uk

[2]*4T2consulting, Ringstrasse 61, D-76356 Weingarten, Germany,*
e-mail: granzow@4T2consulting.de

1. INTRODUCTION

DNA microarray technology is attracting tremendous interest both among the scientific community and in industry. With its ability to measure simultaneously the activities and interactions of thousands of genes, this modern technology promises new insights into the mechanisms of living systems. Typical scientific questions addressed by microarray experiments include the identification of coexpressed genes, either as genes expressed throughout a sub-population or as genes always expressed together (sample or gene groups), identification of genes whose expression patterns make it possible to differentiate between biological entities that are otherwise indistinguishable (e.g., tumour samples that are clinically grouped together despite differences in molecular defects), and the study of gene activity patterns under various stress conditions (e.g., chemical treatment).

Although microarrays have been applied in many biological studies, the handling and analysis of the large volumes of data generated is not trivial. Different types of scientific questions require different data analytical techniques. Broadly speaking, there are two classes of elementary data analysis tasks, *predictive modeling* and *pattern-detection*. Predictive modeling tasks are concerned with learning a classification or estimation function, whereas pattern-detection methods screen the available data for interesting and previously unknown regularities or relationships. A wide range of sophisticated methods and tools have been developed to address these tasks. To facilitate predictive modeling and pattern-detection and to address data errors arising from the processes involved in microarray hybridization experiments, a great portion of this discussion will focus on data preparation and pre-processing.

Instead of looking at particular methods in detail, this article concentrates on the overall microarray data analysis process, stretching from hypothesis conception to array design to model construction and validation. The basic aim is to describe the various steps of this process in an illustrative manner, rather than analyzing each and every minute detail. This article is intended, therefore, to appeal to a wide readership interested in microarray data analysis.

2.   BASIC BIOLOGY AND ARRAY PROTOCOL

2.1   Genes"R"us

Understanding and using microarray analysis techniques requires a basic understanding of the fundamental mechanisms of gene expression itself. This section reviews some of the most important aspects of the underlying concepts. Readers familiar with the fundamentals of gene expression may safely skip this section.

Describing gene expression starts necessarily with *deoxyribonucleic acid* (DNA), the very stuff genes are made from, and *ribonucleic acid* (RNA). Both DNA and RNA are *polymers*, that is, molecules that are constructed by sequentially binding members of a small set of subunits called *nucleotides* into a linear strand or sequence. Each nucleotide consists of a base, attached to a sugar, which is attached to a phosphate group. The linear strand consists of alternate sugars and phosphates, with the bases protruding from the sugars. In DNA, the sugar is deoxyribose and the bases are named guanine, adenine, thymine, and cytosine; in RNA the sugar is ribose and the bases are guanine, adenine, uracil, and cytosine (Alberts et al., 1989). The sugar-phosphate backbone can, for the purposes of informatics, be considered as straight (though actually it has all sorts of twists, kinks and loops – higher-order structures – that are of interest to those who care about such things). The bases that protrude from the backbone are far more informative. They can form pairs, via hydrogen bonds, with bases in other nucleic acid strands: adenine binds to thymine (or uracil) and guanine to cytosine, by the formation of two and three hydrogen bonds respectively. Such base pairing allows DNA to be organized as a double-stranded polymer whose characteristic three-dimensional helix structure has become famous. The two DNA strands are complementary to each other, meaning that every guanine in one strand corresponds to a cytosine in the other (complementary) strand. The same mechanisms apply to the complementary DNA nucleotides adenine and thymine.

It is this sequence of paired bases which allows DNA to encode information, and to replicate it by using each strand as a template against which to assemble a new complementary strand. As an encoding device, DNA has an extraordinary storage density. One cubic micrometre of DNA can encode around 150 Mbytes of information; this is about ten orders of magnitude better than a CD-ROM optical memory, and twelve more than most computer hard disks.

Within the DNA, *genes* are unique sequences of variable length. The genes within a cell comprise the cell's *genome*: it contains the information necessary for synthesizing (constructing) proteins, which do all that a cell needs. The genome also contains the information that controls which proteins are synthesized in a given cell under particular circumstances.

Implicit in the structure of a cell's genome are mechanisms for self-replication and for transforming gene information to proteins. The gene-to-protein transformation constitutes the "central dogma of molecular biology"; it is described by a two-step process:

Step 1: Transcription: *Gene (DNA) makes RNA*
Step 2: Translation: *RNA makes protein* $\left.\vphantom{\begin{array}{c}a\\b\end{array}}\right\}$ Expression

That is, the information represented by the DNA sequence of genes is transferred into an intermediate molecular representation, an RNA sequence, using part of a DNA strand as a template for assembling the RNA. The information represented by the RNA is then used as a template for constructing proteins, according to a code in which each amino acid is represented by three bases in the RNA. The RNA occurring as intermediate structure is referred to as *messenger* RNA (mRNA). The term *transcription* is commonly used to describe Step 1 and the term *translation* for Step 2. Collectively, the overall process consisting of transcription and translation is known as *gene expression*. Notice, in most organisms only a small subset of genomic DNA is capable of being transcribed to mRNA or expressed as proteins. Some regions of the genome are devoted to control mechanisms, and a substantial amount of the genomes of higher-level organisms appears to serve no informational function at all. These DNA sections are also known as *junk* DNA.

With enormous effort and expense, almost the whole human genome ($\sim 2.8 \times 10^9$ basepairs) has been sequenced, junk and all; there are a few refractory regions which will succumb soon. The human genome encodes at least 40,000 genes.

A natural error is to suppose that, once the genome is known, everything important in human biology is understood. This is far from the truth. As an analogy: suppose we have obtained a catalogue of all the 40,000-odd parts and tools needed to make an automobile; do we then understand its design, can we improve it? Not if the catalogue is like the human genome, with entries mostly in random order, with no indication of which tool is needed for fitting which part, or where the parts go, or which has to be connected to what, or how many of each we need. And imagine that (as experience teaches us is true for the genome) that many catalogue entries may be ambiguous, so that the same part number can describe a range of more or less related parts. Still worse; many parts, as specified in the catalogue, don't actually fit, and have to be machined down or have holes drilled through them or small attachments stuck onto them before they can be used at all. Add that some entries in the catalogue are for parts that aren't used any longer, or are unusable...

All that analogy is true for the proteins encoded by the genome. Proteins are the ultimate product of the gene expression process. All the proteins synthesized from a cell's genome constitute its *proteome*. Chemically, proteins are polymers that are formed from 20 different subunits called *amino acids*. The linear chain of amino acids making up a protein, dictated by the sequence of bases in the mRNA, is known as its *primary structure* or *sequence*. Within its normal physiological environment, an amino acid sequence assumes a three-dimensional conformation, which is the major determinant of the protein's biological function. For each amino acid sequence, there is a stable three-dimensional structure sometimes referred to as the protein's *native state*; many proteins have a range of possible native states, and can switch between them according to their interactions with other molecules. The native state of a protein and the folding process involved in reaching this state from its initial linear orientation are dictated by the primary sequence of the protein. Despite the strong deterministic correspondence between the primary sequence and the native state of a protein, the processes involved in protein folding are highly complex and difficult to capture and describe logically. Protein folding and structure prediction have been the subjects of ongoing research for some time.

In many ways, proteins can be considered as the biochemical "workhorses" of an organism. Proteins play a variety of roles in life processes, ranging from structural (e.g., skin, cytoskeleton) to catalytic (enzymes) proteins, to proteins involved in transport (e.g., haemoglobin), and regulatory processes (e.g., hormones, receptor/signal transduction), and to proteins controlling genetic transcription and the proteins of the immune system. Given their importance in terms of biological function, it is no surprise that many applications in biotechnology are directly related to the understanding of protein structure and function. This is perhaps most impressively demonstrated by modern drug development. The principal working mechanism of most known drugs is based on the idea of selectively modifying (by interaction) the function of a protein to affect the symptoms or underlying causes of a disease. Typical target proteins of most existing drugs include receptors, enzymes, and hormones. Therefore, many experts foresee *proteomics*, the study of structure and function of proteins, as the next big step in biomolecular research.

In many biomolecular studies, the most important issue is to measure *real* gene expression, that is the abundance of proteins. However, as we will see in more detail below, DNA microarray experiments do measure the abundance of mRNA, but not protein abundance. According to a simple, traditional view of gene expression, there is a direct one-to-one mapping from DNA to mRNA to protein. To put it another way, a specific gene (i.e. genomic DNA sequence) will always produce one and the same amino acid sequence of the corresponding protein, which will then fold to assume its native state. Given this simplified scheme, measuring mRNA abundance would provide us with highly accurate information on protein abundance, as protein and mRNA abundance are proportional due to the direct mapping. We would also know the primary structure of the proteins corresponding to the measured mRNA, since the genetic code allows us to deduce the amino acid sequence from a given DNA or RNA sequence. Unfortunately, gene expression is in reality more complex.

The modern view of gene expression paints a more intricate picture, which suggests a highly dynamic gene expression scenario. There are various ways in which proteins are formed and modified; and indeed, the genome itself is subject to alterations.

First, genomic DNA itself may undergo changes, as a result of the replication machinery making mistakes, or attempting to copy damaged DNA and getting it wrong. DNA bases can be changed; small or large regions can be inserted or deleted or duplicated, pieces of one DNA molecule can be joined onto pieces of another, RNA can be copied backwards into DNA (reverse transcription), regions of junk DNA derived from viruses can jump about the genome.

Second, the (forward) transcription process is more complicated. One gene may be transcribed to give a range of possible products. When composing the final mRNA sequence, the transcription process uses a mechanism called *splicing* to cut out some regions (*introns*), so that only the remaining segments (*exons*) survive to be translated to protein. This process may yield alternative versions of mRNA sequences due to *alternative splicing*. A region of mRNA that in one case is treated as intron/exon/intron may in another be treated as one large intron; a gene coding for 3 possible introns, A, B and C can thus produce a mRNA with an ABC sequence, or alternatively just AC. (With multiple introns, much greater permutations are of course possible. The current record is held by a gene expressed in the brain of the fruitfly Drosophila, which could in principle be alternatively spliced in over 40,000 different ways)

Another source of variation comes from promoter choice. A *promoter* is a specific location or site where the transcription of a gene from DNA to mRNA begins. Genes may have multiple promoters, thus giving rise to different transcript versions and hence different proteins even without alternative splicing

It is also possible – some cases are known, and we do not know how many there are – for an mRNA to be "edited" before it is translated. One base in the mRNA can be replaced by another, altering one amino acid (or, drastically, converting a signal for an amino acid into a signal to stop translating early).

Furthermore, there is no necessary connection between the amount of an mRNA present (in whatever spliced or edited form) and the amount of protein translated from it. There is, in a general sort of way, a correlation between mRNA abundance and the corresponding protein abundance, but there is no doubt that rates of translation can be differentially controlled.

Lastly, there are post-translational modifications. These are structure-modifying alterations occurring after the translation process. Proteins may be split up into smaller fragments, or have their ends or internal regions removed; amino acids may be altered by adding (temporarily or permanently) other chemical groups to them, sometimes with drastic effects on the protein's catalytic or structural properties.

With all these caveats, we should stress that although DNA gene microarrays are often thought of as instruments for measuring gene expression, what they really measure is mRNA transcript abundances. They do not always distinguish between different forms of processed mRNA, and they can give no information about differential translation rates, nor about post-translational modification. But they do give some valuable information, quickly and fairly easily. One of the main reasons why researchers are pursuing DNA microarray studies with such intensity, in the full knowledge of their limitations, is the fact that protein expression and modification studies are still very expensive, and often involve highly specialized and delicate techniques, (e.g., 2D-gel electrophoresis, mass spectrometry). High-throughput protein-detecting arrays or chips are beginning to emerge; however, there are still a number of issues to be resolved before this technology is mature. Critical issues involve efficient methods for large-scale separation and purification of proteins, and the maintenance of the active biological configuration of proteins while they are attached to the surface of a chip. So at the present time, and for the immediately foreseeable future, DNA microarray technology constitutes a useful compromise for carrying out explorative high-throughput experiments. Because of the inherent limitations, one should exercise extreme caution when interpreting the results of such microarray experiments, even if their analysis is perfect.

The biochemical details of gene expression are highly intricate. Many good texts exist on this subject, for example, Schena et al., 1995; Raskó and Downes, 1995.

2.2 Brewing up the Hybridization Soup

To summarize the last section: a great deal of modern molecular biology revolves around nucleotide polymers – DNA and RNA – and amino acid polymers, proteins. DNA microarrays measure a cell's transcript via the abundance of mRNA molecules, but not protein concentrations. This section will illustrate the biochemical principles involved in measuring transcripts with microarrays.

A number of techniques have been developed for measuring gene expression levels, including *northern blots*, *differential display*, and *serial analysis of gene expression*. DNA and oligonucleotide microarrays are the latest in this line of methods. They facilitate the study of expression levels in parallel (Duggan et al., 1999). All these techniques exploit a potent feature of the DNA duplex – the sequence complementarity of the two strands. This feature makes *hybridization* possible. Hybridization is a chemical reaction in which single-stranded DNA or RNA molecules combine to form double-stranded complexes (see schematic illustration in Figure 1.1). The famous DNA double helix is an example of such a molecular structure.

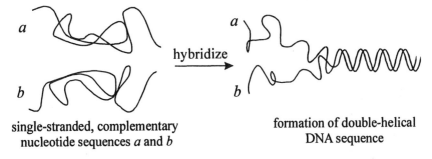

single-stranded, complementary
nucleotide sequences *a* and *b*

formation of double-helical
DNA sequence

Figure 1.1. Hybridization of two single-stranded nucleic acid sequences to a double-stranded, helical complex.

The hybridization process is governed by the base-pairing rules: specific bases in different strands form hydrogen bonds with each other. For DNA the matching pairs are adenine-thymine and cytosine-guanine. Hybridization is a nonlinear reaction. *Yield* – the number or concentration of nucleic acid elements binding with each other in the resulting double-stranded molecule – depends critically on the concentration of the original single-stranded polymers and on how well their sequences align or match. It is this yield that is measured in a microarray experiment.

Before we proceed to an actual hypothetical microarray experiment, let us look at color-coded genes (green and red ones) in *competitive* and *comparative hybridization*. In order to selectively detect and measure the amount of mRNA that is contained in an investigated sample, we must label the mRNA with *reporter* molecules. The reporters currently used in microarray experiments include fluorescent dyes (fluors), for example, cyanine 3 (Cy3) and cyanine 5 (Cy5).

Let us assume we have two samples of transcribed mRNA from two different sources, *sample* 1 and *sample* 2. Both samples may consist of multiple copies of many genes. We have also a *probe*, which is a specific nucleic acid sequence, perhaps a gene, or a characteristic subsequence of a gene, or a short, artificially composed nucleotide sequence. Like the two samples, the probe will contain many copies of the sequence in question. This is important because sufficient amounts are needed to get the hybridization reaction going, and to be able to detect and measure the various concentrations. What we want to find out is the relative abundance of the mRNA complementary to the probe sequence within sample 1 and sample 2. Sample 1, for example, may contain three times as many copies of sequences complementary to the probe as sample 2, or they may not be contained in either sample at all. To find out the exact answer, we proceed as follows (see also Figure 1.2):

1. Prepare a mixture consisting of identical probe sequences. In this scheme of things the probe is a kind of "sitting duck", awaiting hybridization.
2. Label sample 1 with green-dyed reporter.
3. Label sample 2 with red-dyed reporter.
4. Simultaneously give both sample mixtures the chance to hybridize with the probe mixture. Here, sample 1 and sample 2 are said to *compete* with each other in an attempt to hybridize with the probe.
5. Gently stir for five minutes.
6. Filter the mixture to retain only those probe sequences that have hybridized, that is, formed a double-stranded polymer.

7. Measure the amount or intensity of green and red in the filtered mixture, and *compare* the amounts to determine the relative abundance of the probe sequence.

8. Jot down the result, add a little salt, and enjoy.

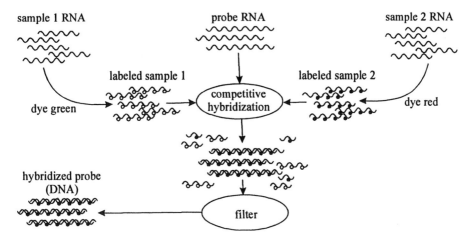

Figure 1.2. Competitive/comparative hybridization.

2.3 What You Always Wanted to know About Gene X

This section briefly looks at a hypothetical DNA microarray experiment and the various steps involved. However, for reasons of illustration this experiment is dealing with a rudimentary array consisting of only four genes rather than hundreds or thousands commonly used in such experiments. Notice, that because of RNA's inherent chemical instability, it is often useful to work with a more stable *complementary* DNA (cDNA) made by reverse transcription, rather than with mRNA, at intermediate steps. However, before the array is made, the cDNA is *denatured* (broken up into its individual strands) to allow the hybridization reaction.

Typical goals of DNA microarray experiments involve the comparison of gene transcription (expression) in two or more kinds of cells (e.g., cardiac muscle versus prostate epithelium), and in cells exposed to different conditions, for example, physical (e.g., temperature, radiation), chemical (e.g., environmental toxins), and biological conditions (e.g., normal versus disease, changing nutrient availability, cell cycle variations, drug response). Genetic diseases like cancer are characterized by genes being inappropriately transcribed, or missing altogether. A cDNA microarray study can pinpoint the transcription differences between normal and diseased, or it can reveal different patterns of abnormal transcription to identify different disease variations or stages.

Let us imagine a study involving ten human patients with two different forms of the same type of cancer; six patients suffer from form A and four from form B. Further, let us assume that we want to investigate four different genes, *a, b, c,* and *d,* and their roles in the disease. To help us answer this question we set up a cDNA microarray experiment. Initially, we would hope to find characteristic expression patterns, which would help us to formulate more specific hypotheses. Clearly, in a more realistic scenario, we would like to analyze many more genes at the same time. But for the sake of this illustration, we are just looking at four genes. The following outlines the various steps we need to take care of for this experiments (see also Figure 1.3).

1. *Probe preparation.* Prepare one DNA microarray per patient, using a standard DNA (possibly cDNA).

2. *Target sample preparation.* Obtain, purify, and dye target mRNA samples.

3. *Reference sample preparation.* Obtain and prepare *reference* or *control* mRNA and label it.

4. *Competitive hybridization.* Hybridize target and reference mRNA with the cDNA on the array.

5. Wash up the dishes.

6. *Detect red-green intensities.* Scan the array to determine how much target and reference mRNA is bound to each spot.

7. *Determine and record relative mRNA abundances.*

The probe preparation step involves the manufacturing of sufficient amounts of cDNA sequences that are identical to the sequence of the studied genes. In place of the full sequence, it is often more practical to use a characteristic subset with 500 to 2,500 nucleotides in length. The cDNA sequence mixtures representing the investigated genes are then affixed to the array (a kind of glass slide). Normally, the mixtures are placed as round spots on the array and arranged in a grid-like fashion, hence the name micro*array*. At least for larger experiments, we would like to record information about where on the array which gene is placed so that we can later track the right data. In addition, we would record any information relevant to the genes in question, for example, the precise nucleotide sequence, known biological function, pointers to relevant literature, and so on.

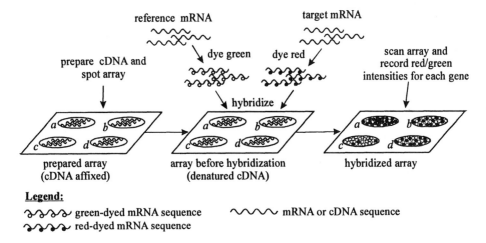

reference mRNA   target mRNA

prepare cDNA and spot array   dye green   dye red   scan array and record red/green intensities for each gene

hybridize

prepared array (cDNA affixed)   array before hybridization (denatured cDNA)   hybridized array

Legend:
green-dyed mRNA sequence   mRNA or cDNA sequence
red-dyed mRNA sequence

Figure 1.3. A Simplified 4-Gene Microarray Experiment.

Target refers to the actual entities or samples we want to measure: in this case, the transcribed mRNA from tissue or serum cells of our patients. First, we must obtain the tissue or serum cells. Once the cells are at hand, mRNA must be extracted from the cells, purified, dyed (i.e. labeled with reporter molecules), and converted into a suitable chemical RNA form. In the diagram of Figure 1.3 we label the target sample with "red" dye and the reference with green "dye". The colors red and green are arbitrary. It does not necessarily imply that the actual reporter molecules *are* indeed red or green. However, we use these colors here, since in the final digitized, false-color graphics image, the color red is chosen to represent target mRNA abundance and green for reference abundance.

The reference sample is used as a baseline relative to which we measure the abundance of target mRNA. There are two common choices for references samples: *standard reference* and *control*. Standard references (also called *universal references*) are often derived from mRNA pools unrelated to the target samples of the experiment. Standard references should contain sufficient amounts of mRNA from the genes studied with the array. So for our four-gene example, we could use a standard reference from which we know that it contains "standard" mRNA of the four genes *a*, *b*, *c*, and *d*. *Control* samples are also employed as references. In contrast to references, controls are somehow related to the experiment at hand. For example, in a normal-versus-disease study, the control may represent tissue from the normals. Once sufficient amounts of reference mRNA is produced, it is labeled with reporters. Clearly, the reporter molecules used here must be such that we can later distinguish between reference from target mRNA. In the four-gene example, we dye the reference sample green.

Now we are ready for hybridization between the labeled target and reference mixtures. We pour on the mixtures, and let them hybridize to the array.

Once the hybridization reaction is completed, we wash off any reference and target material that has not managed to find a probe partner. This leaves us with the array depicted in the right part of Figure 1.3. When you step back far enough from the book, you will recognize that the spot labeled *a* is dark, the one labeled *c* is bright, whereas the two other spots (labeled *b* and *d*) appear gray. Performing a suitable color conversion in your head, you will, after a while, be able to see a red, a green, and two orange spots on our four-gene array. Get the picture? Basically, what this tells us is that (a) gene *a* is much more active (highly expressed) in the target than in the standard or reference sample, as the red-dyed mRNA sequences have massively outperformed the green-dyed sequences in hybridizing to probe sequences representing gene *a*. By a similar argument, we can say (b) that gene *c* of the target sample is underexpressed when compared to the standard reference, and (c) that both the expression of gene *b* and *d* seems balanced, that is, the transcript mRNA abundance in the target is roughly the same as that in the reference. Hence, we get orange.

In real microarray experiments, it is impractical, if not impossible, to determine the relative mRNA abundances with the naked eye. To detect and measure the relative abundances of reference and target material, we use a device equipped with a laser and a microscope. The fluorescent reporter molecules with which we have labeled the samples emit detectable light of a particular wavelength when stimulated by a laser. The intensity of the emitted light allows us to estimate quantitatively the relative abundances of transcribed mRNA. After the scanning we are left with a high-resolution, false-color digital image. Image analysis software takes over here to derive the actual numerical estimates for the measured expression levels (Yang et al., 2000). These numbers may represent absolute levels or ratios, reflecting the target's expression level against that of the reference.

As far as our four-gene experiment is concerned, we stick with ratio measures. For example, for patient 1 of our study, we may have obtained the following expression ratios: $r(a) = 5.00$, $r(b) = 0.98$, $r(c) = 0.33$, and $r(d) = 0.89$. In this scheme, a value close to 1.00 means balanced expression, 2.00 means the target mRNA abundance is two times as high as that in the reference, and 0.50 means the reference abundance is twice as high as that of the target.

To complete our four-gene, ten-patient experiment, we must repeat the entire procedure ten times, and produce one array per patient. Once all arrays are done, we derive, record, and integrate the expression profiles of all patients in a single data matrix along with other information needed to analyze the data. Table 1.1 illustrates such data in the context of the four-gene example expression study. Notice, the data in Table 1.1 is not normalized and no specific data transformations have been carried out.

Table 1.1. Numerical expression data matrix obtained from our hypothetical four-gene, ten-patient, two-tumor microarray experiment.

Patient#	1	2	3	4	5	6	7	8	9	10
Tumor	A	A	A	A	A	A	B	B	B	B
r(a)	5.00	1.33	3.45	3.05	4.22	2.09	0.33	0.65	0.22	0.12
r(b)	0.98	0.87	1.04	1.10	?	2.11	1.23	1.32	0.85	0.77
r(c)	0.33	1.40	0.42	0.55	0.24	0.60	2.44	?	3.00	2.22
r(d)	0.89	0.90	1.00	0.92	0.66	1.05	1.32	1.01	0.97	0.87

Each numbered column in Table 1.1 holds the data related to a particular patient: patient identifier, tumor type, and the measured expression levels of the four studied genes. In a real experiment, the generated data matrix would of course be more complex (many more genes and more patients) and perhaps include also clinical data and information on the studied genes. However, the data table as depicted illustrates how the data from multiple arrays can be integrated along with other information so as to facilitate further processing. Owing to the small scale (four genes, ten patients) of our experiment and the deliberate choice of expression levels, we are able to analyze the data qualitatively by visual inspection.

First, we observe that there are no recorded expression values (depicted by question mark) for patient 5 and gene *b*, and for patient 8 and gene *c*. There may be many reasons why these values are missing. Numerous techniques exist to deal with *missing values* at the data analysis stage.

Second, for tumor A patients the expression levels of gene *a* seem to have a tendency to be by a factor 2 or more higher than the base line level of 1.00. At the same time, for tumor B patients, *a*'s expression levels tend to be a factor two or more lower than the reference level. This differential expression pattern of gene *a* suggests that the gene may be involved in the events deciding the destiny of the tumor cell in terms of developing into either of the two forms. If this difference is statistically significant on a particular confidence level remains to be seen.

Third, there seems to be also a differential expression pattern for gene *c*. However, here we observe the tendency to underexpressed levels for tumor A and overexpressed for tumor B.

Fourth, most expression values of gene *b* and *d* appear to be "hovering" about the base line of 1.00, suggesting that the two genes are not differentially expressed across the studied tumors.

Fifth, we observe that high expression levels of gene *a* are often matched by a low level of gene *c* for the same patient, and vice versa. This suggests that the two genes are (negatively) co-regulated.

We follow up on this example in subsequent sections. The reader may look at Figure 1.8 for a visualization of some of the expression patterns discussed above.

2.4   Into the Microarray Jungle

Different microarray technologies have been developed. These can be divided into three categories: spotted cDNA microarrays, spotted oligonucleotide microarrays, and Affymetrix chips. Spotted cDNA and oligonucleotide microarrays include both contact printing and the newer ink-jet technology (a wonderful spin-off; originally invented to deal with the delicate task of stamping dates on eggs without breaking the eggshells). They may be spotted onto glass slides, in which case laser fluorescence may be used to detect two-color hybridization from two samples at once. Or they may be spotted, rather more cheaply, onto filters, in which case radiolabelled material is used for hybridization, one sample at a time. Grandjeaud et al. (1999) provide an impartial guide to the advantages and disadvantages of either. One maker's slide arrays are generally compatible with another manufacturer's laser fluorescence analysis system; similarly, filters can be read by any scanner. There remain the very wonderful Affymetrix chips, manufactured in a unique way, and only readable with a special Affymetrix machine which cannot be used on any other maker's arrays; these can be regarded as a distinct subtype. Think of them as a sort of molecular Microsoft.

2.4.1   In more Detail – "Complementary" Sequences, Chip Spotting and Chop Splitting

Spotted microarrays consist of a solid surface (e.g., a microscope glass slide) onto which miniscule amounts (spots) of nucleotide sequences are placed in a grid-like arrangement. Each spot represents a specific gene, an *expressed sequence tag* (partial gene sequence providing a tag for a gene of which the full sequence or function may not be known), a *clone* (population of identical DNA sequences) derived from cDNA libraries (collections of DNA sequences made from mRNA by reverse transcription), or an *oligonucleotide* (short sequence specifically synthesized for experiment). The spots serve as probes against which target and reference mRNA is hybridized.

With spotted arrays the probes are deposited on the array by an automated process called *contact spotting* or *printing* (similar to ink-jet printer technology). The spotting machinery prints nucleotide spots (with a diameter of approximately 100μm) in close proximity on the array. In this way, 10,000 to 30,000 probes can be arranged on a single array. However, the number of probes does not necessarily match the number of genes. For reasons of reproducibility, a gene may be represented by more than one probes on the array.

Where does the probe DNA for the arrays come from? This is different for spotted cDNA and spotted oligonucleotide arrays. The cDNA approach relies on DNA from cDNA clones, which are often derived from DNA collections or libraries that were created for other purposes. An example collection is the I.M.A.G.E Consortium library. The typical length of cDNA probe sequence is in the range of 500 to 2,500 base pairs. In contrast to cDNA technology, whose probe sequences are "pre-determined", oligonucleotide arrays facilitate the *design* of probe sequences. The oligonucleotides represent short DNA probes synthesized on the basis of the sequences of existing or hypothetical genes. Typically, oligonucleotide probes are 50 to 70 bases in length. Oligonucleotide arrays allow more flexibility in the design of a microarray.

GeneChips® from Affymetrix are an excellent example of the increased flexibility that comes with oligonucleotide arrays. GeneChips use oligonucleotides of 25 bases per probe. The diameter of each probe spot is approximately 18μm, facilitating an impressive maximum of 500,000 probes per array. Affymetrix makes use of multiple probes to represent a gene. The most recent figure is 22 probes per gene, allowing for up to 23,000 genes per chip. This approach has its unique way to represent the genes on a chip. Each gene is characterized by a collection of probes called the *probe set*. In this set, multiple probe pairs make up the gene (currently 11 probe pairs per gene). Each probe pair consists of one probe called *perfect match* and another called *mismatch*. The former has a sequence identical to that in the gene of interest, the latter differs from the perfect match probe by a single base in the middle of the sequence. Multiple probe pairs are used to improve the specificity of the measurement.

2.5 Sounding out Life with Microarrays

The kinds of questions you are asking from microarrays are likely to affect array design and data analysis. Before we briefly discuss issues of array design and data analysis, we turn our attention to the classes of scientific questions that microarrays may help to answer. In trying to precisely formulate and answer a question, the scientist becomes engaged in what we call a *scientific task* or *study*. Each type of scientific task involves a set of characteristic concepts, subtasks, and steps that need to be carried out to accomplish the task. As we shall see later, one important subtask is the *analytical task*.

2.5.1   Differential Gene Expression Studies

Differential gene expression studies are searching for those genes that exhibit different expression levels under different experimental conditions, that is, in different tissue types, or in different developmental stages of the organism. Typical studies of this kind include normal-versus-diseased state investigations. The hypothetical four-gene experiment in Section 2.3 is an illustration of a differential-expression study for two tissues (tumor types). Essentially, differential expression studies look at a single *gene profile*.

2.5.2   Gene Co-regulation Studies

Gene co-regulation is somewhat similar to differential gene expression. However, instead of analyzing the expression variation of a single gene expression profile against the experimental conditions, it compares gene profiles with each other. Here the objective is to identify genes whose expression levels vary in a coordinated or correlated fashion across the studied experimental conditions or samples. Two basic patterns of gene co-regulation exist: *positive* and *negative*. There is a positive co-regulation between two genes if the expression level of one gene increases as that of the other increases. A negative co-regulation exists if the expression level of one gene *decreases* as that of the other increases. For example, the genes *a* and *c* in our four-gene example in Section 2.3 exhibit a negative co-regulation pattern across the ten studied samples or patients. Basically, gene co-regulation studies compare *gene profiles* of two or more genes.

2.5.3   Gene Function Identification Studies

Microarray experiments can help to reveal the function of novel genes. The principal process involves the comparison of the novel gene's expression profile under various conditions with the corresponding profiles of genes with known function. The functions of genes with highly similar expression profiles serve as candidates for inferring the function of the new gene.

2.5.4   Time-Course Studies

Time-course studies require that transcript samples from the same source are taken at different points in time and are then hybridized with the probes on the array – one array per time point. The different snapshots can then be used to reveal temporal changes in gene expression of the investigated sample. This type of experiment could, for example, be used to study cell cycle phenomena. Time-course expression experiments are also useful in gene network identification studies (see below).

2.5.5 Dose-Response Studies

Dose-response experiments are designed to reveal changes in gene expression patters as response to exposing a sample, tissue, or patient to different dosages of a chemical compound, for example, a drug inhibiting cell growth. Drug-dose response studies may also involve a time-course study element.

2.5.6 Identification of Pathways and Gene Regulatory Networks

A pathway-identification study aims at revealing the routes and processes by which genes and their products (i.e. proteins) function in cells, tissues, and organisms. It involves the perturbation of a pathway and the monitoring of changes in gene expression of the investigated genes as a response to this intervention. Gene regulatory networks control gene expression. Identifying these networks requires that one finds the genes that are turned on and off at various time points after stimulation of a cell. These studies require time-course data to be generated.

2.5.7 Predictive Toxicology Studies

This type of study relies on a reference database, which stores the results of a large number of microarray screening experiments of organs and their responses to toxic agents. These studies are particularly interesting for the pharmaceutical industry, where the aim is to identify toxic effects of unknown compounds as early as possible. When investigating a new compound, its influence on the gene expression of key genes is compared with the expression profiles of known toxins in the reference database. Based on the degree of similarity of the compound's effects to the known profiles, an inference-by-analogy step is then employed to predict the toxicity of the new compound. Basically, this approach compares *array profiles* (see below). In addition, the identification of novel marker genes or pathways involved in the toxic effect of a compound can be tackled by this method.

2.5.8 Clinical Diagnosis

Gene expression experiments are also a powerful tool for clinical diagnostics, as they can discover expression patterns that are characteristic for a particular disease. Another analysis within the jurisdiction of this type of study is concerned with inferring unknown subtypes of known diseases. This is achieved by revealing characteristically different expression profiles that correlate with clinically distinct subtypes of a disease. Here, the clinical course of the disease is known to show differences in a small fraction of cases but conventional analysis of the disease could not reveal any distinct subtypes. The hypothetical four-gene experiment in Section 2.3 is an illustration of this kind of investigation.

2.5.9 Sequence-Variation Studies

Uncovering DNA sequence variations that correlate with phenotypic changes, e.g., diseases, is the aim of this type of study. Common types of sequence variations are *single nucleotide polymorphism* (SNP: pronounced "snip"), insertions and deletions of a few nucleotides, and variation in the repeat number of a *motif*. As a way of illustration, the two "complementary" letter sequences in the heading of Section 2.4.1 demonstrate a kind of "double single-letter polymorphism". This example also illustrates how drastic these minuscule changes in sequence may affect the function (here the meaning of words) of a gene. Generally, a motif refers to any sequence pattern that is predictive of a molecule's function, structural feature, or family membership. Motif-based analyses are often used to detect sequence patterns (motifs) that correspond to structural and functional features of proteins. The most common type of sequence variation in the human genome are SNPs, which occur with a frequency of roughly 1 per 1,000 nucleotides. In order to analyze SNP variation, at least three categories of microarray experiment designs are meaningful: (a) arrays including all known SNPs of a human genome sequence, (b) microarrays containing a sample of SNPs located across the entire human genome, and (c) devices for re-sequencing a sample of the human genomic sequence. However, for sequence analysis, microarrays containing only oligonucleotides are used. Complex sequence variations responsible for phenotypic changes could be uncovered by SNP microarray studies. Such studies promise a deeper understanding of these biological enigmas, which are hard to tackle by any other means so far.

2.6   A Pint of Genes and two Packets of Chips, Please

Array design is concerned with decisions on which genes or DNA sequences to put on the chip and how to represent each gene (how many probes and what kind of sequence(s) per probe).

For cDNA arrays, this exercise boils down to selecting a set of clones from relevant DNA libraries. From the clones the actual probe sequences are derived by appropriate laboratory procedures.

The oligonucleotide approach requires the definition of the probe sequences that are to be synthesized. While providing more flexibility and promising better results, this way of doing things is more complex. One of the complexities involved in designing good probe sequences lies in the variability of binding affinity of the underlying sequences. Binding affinity of probes to their counterpart on the array is determined largely by the extent to which target and probe sequence match, and by the adenine-thymine and cytosine-guanine content of the match. Other factors like secondary structure (the overall shapes of the molecules) are also involved in the process, but are poorly understood. Therefore, variation in binding affinity can vary considerably from one probe to another. In addition to this, the designer of oligonucleotide probe sequences has to take into account factors that affect the ability of a probe to accurately measure transcripts of a given gene. These factors include alternative splicing, presence of repetitive sequences that appear in otherwise unrelated genes, and the possibility of highly similar sequences in multiple members of gene families. This implies that the chosen sequences should be unique in the genome. Therefore, access to databases containing the genomic sequences of organisms is important for designing oligonucleotide arrays. To accomplish this task, appropriate sequence analysis software and retrieval of information from databases with gene and genomic sequence data is mandatory.

Using microarrays from an industrial source has some advantages over producing your own. Industry standards ensure high quality of arrays, making the experiments more reliable. However, for unsequenced organisms no sequence is available from existing repositories. Thus, producing your own microarray has the advantage of spotting the sequences you selected from available sources in addition to using sequences not known by others.

3. BASIC CONCEPTS OF MICROARRAY DATA ANALYSIS

Microarray data analysis is a truly complex process. Ideally, it starts long before the actual numerical data matrix (similar to the one depicted in Table 1.1) is at hand. Sound knowledge of the available data analysis methods and tools could help the investigator in selecting a good problem and in formulating clear and specific hypotheses. Of course, microarray experiments are often used as high-throughput exploratory tools, and in this case highly focused hypotheses are perhaps not desired, or possible. Furthermore, a number of statistical issues that impact on data analysis creep in at the experimental design stage. The investigator must decide (a) which factors (independent variables) will be varied in the experiment, (b) which factor combinations (experimental conditions) are to be tested, and (c) how many experiment replicas should be done for each tested condition. The specter of measurement variation looms large over microarray experiments, not least because of considerable variation inherent in the biological specimens themselves. Thus, microarrayers are becoming increasingly aware of the need for replications (Lee et al., 2000). Replicas may, where possible, involve taking multiple samples from the investigated subject. The purpose of this procedure is to increase precision of measurement estimates and to provide a sound basis for error estimation models. There are yet more statistical issues that deserve mentioning. However, the aim of this article is to look at the big picture of microarray data analysis; therefore we will not further elaborate on this topic. For a more in-depth discussion on this topic see, for example, Branca and Goodman (2001), Kerr and Churchill (2001), or Sabatti (2002).

Let us now step back – figuratively speaking, this time – and behold the "big picture" of the microarray analysis process in Figure 1.4. This should help us to regain orientation and to look ahead on things to come.

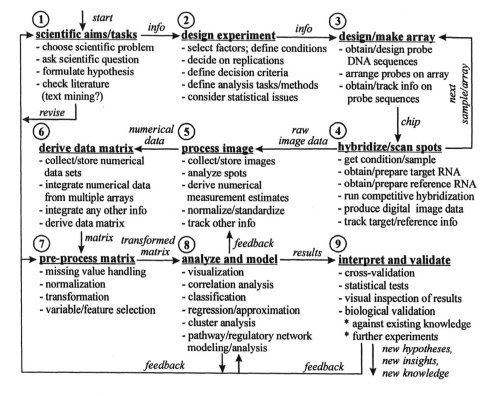

Figure 1.4. Microarray data analysis process – the big picture.

Consider the microarray analysis process diagram in Figure 1.4. Step 1 to Step 4 of the process are discussed in previous sections. Step 5 is mainly concerned with the analysis of the digitized image arising out or Step 4. The result of the image-processing step is a collection of numerical estimates representing the measured expression levels. Many computational and statistical issues need to be considered at the image processing and analysis steps. For reasons of limited space, we will not further dwell on these.

Instead, we will focus our discussion on the integrated data matrix produced in Step 6 and the subsequent analytical steps (Step 7 to Step 9).

4. DATA ANALYSIS = DATA + PROCESS

Microarray data analysis involves methodologies and techniques from life science fields and biotechnology on one hand, and from computer science and statistics on the other. With these broad disciplines comes a lot of heavy baggage in the form of terminology. Not only does this terminology describe a bewilderingly large number of complex concepts and mechanisms, it can also be redundant, incomplete, inconsistent, and sometimes downright incomprehensible. By the time you have figured out how exactly to approach your microarray data analysis problem, the genes you want to study may have drifted out of the gene pool into oblivion. As always, the basic guideline to follow is to keep things simple and to stick to general principles.

4.1 Da Da Data

The data synthesized by Step 6 of the overall microarray analysis process (Figure 1.4) must somehow be structured, physically and logically. Here we are concerned with the description of the general logical organization of gene expression data and some of the underlying notions.

4.1.1 The Matrix

The term gene *expression profile* is commonly used to describe the expression values for a single gene across many samples or experimental conditions, *and* for many genes under a single condition or sample (Branca and Goodman, 2001). Adopting the terminology of Branca and Goodman, and Quackenbush (Branca and Goodman, 2001; Quackenbush, 2001), we suggest the following terms to distinguish these types of gene expression profiles (see Figure 1.5):

- *One gene over multiple samples.* A *gene profile* is a gene expression profile that describes the expression values for a single gene across many samples or conditions.

- *Many genes over one sample.* An *array profile* is a gene expression profile that describes the expression values for many genes under a single (condition or) sample. Wu calls this expression *signature* (Wu, 2002).

Examining the co-regulation of genes, for example, requires the comparison of gene profiles, whereas differential expression studies typically compare array profiles.

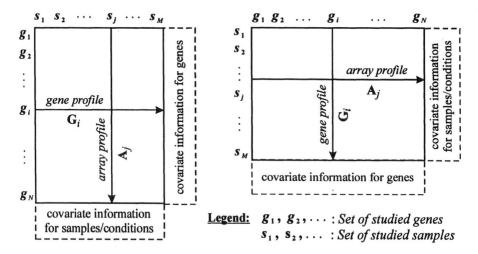

Figure 1.5. Typical gene expression data matrix formats. The solid-line boxes contain the actual numerical values representing the measured expression levels.

The diagram in Figure 1.5 depicts two commonly employed data formats for the integrated gene-expression data matrix, generated by Step 6 of the overall process. The matrix format shown in the left part of the diagram is perhaps more widely recognized; however, many analytical tools and people prefer to "think" in the transposed format (right part of diagram). Indeed, as scientists often ask questions that require the simultaneous comparison of array profiles *and* gene profiles, the distinction between variables and observations becomes blurred. Conceptually, this is a tricky business, since such a simultaneous view imposes different sets of randomness assumptions on the various dimensions. Many tools require the covariate information (dashed-line boxes) to be present in the data matrix. Typical sample/condition covariate information includes tissue and condition type disease-versus-normal labels, and clinical information like survival times, treatment response, tumor stage, and so on. For specific types of analyses it may be necessary to focus on a small set of the available covariate information. See Table 1.1 for an example.

To avoid confusion, this article will focus on the format where the horizontal axis represents samples and conditions, and the vertical axis represents genes (left part in Figure 1.5). Concentrating on the gene expression part only, this format can be described by an $N \times M$ *expression matrix* **E**, as defined in Equation 1.1.

$$\mathbf{E} = (x_{ij}) = \begin{pmatrix} x_{11} & x_{12} & \cdots & x_{1M} \\ x_{21} & x_{22} & \cdots & x_{2M} \\ \vdots & \vdots & \ddots & \vdots \\ x_{N1} & x_{N2} & \cdots & x_{NM} \end{pmatrix} \tag{1.1}$$

where x_{ij} denotes the expression level of sample j for gene i, such that $j = 1, \ldots M$, and $i = 1, \ldots N$ (Dudoit et al., 2000).

The expression matrix $\mathbf{E}$ is a convenient format to represent the expression profiles of N genes over M samples. Whether column vectors or row vectors in this matrix are interpreted as variables (or observations) is not pre-determined by the matrix format. However, the matrix does nail down the interpretation in terms of gene and array profile.

With Equation 1.1 we can define the ith *gene profile* of expression matrix $\mathbf{E}$ by the row vector, $\mathbf{G}_i$, and the jth *array profile* of $\mathbf{E}$ by the column vector, $\mathbf{A}_j$, as follows:

$$\mathbf{G}_i = (x_{i1}, x_{i2}, \ldots, x_{iM}) \tag{1.1a}$$

$$\mathbf{A}_j = (x_{1j}, x_{2j}, \ldots, x_{Nj}) \tag{1.1b}$$

4.1.2 On Variables and other Things

To establish a common terminology, we will briefly review some important aspects of variables.

Variables are used in statistics, machine learning, data mining, and other fields to describe or represent *observations*, objects, records, data points, samples, subjects, or entities. Sometimes variables are also referred to as attributes, features, fields, dimensions, descriptors, measures, or properties. *Variables* are often categorized with regard to their mathematical properties, that is, in terms of the intrinsic organization or structure of the associated values (or value range or scale). Generally speaking, there are *continuous* or *numeric* variables and *discrete* or *symbolic* variables. Numeric variables are coded as real and integral numbers and interpreted quantitatively. Symbolic variables, on the other hand, use some alphanumeric coding scheme and their interpretation emphasizes a qualitative view. To avoid confusion, this article concentrates on the terms *variable* and *observation*.

Commonly, four types of variable *types* or *scales* are distinguished: (a) *nominal* or *categorical* scale, (b) *ordinal* or *rank* scale, (c) *interval* scale, and (d) *real* or *ratio* scale (also called *true measures*).

Nominal variables tell us which of several unordered categories a thing belongs to. For example, we can say a tumor is of type or category *A*, *B*, or *C*. Such variables exhibit the lowest degree of organization, since the set of values such a variable may assume possesses no systematic intrinsic organization or order. The only relation between the values of nominal variables is the *identity relation*. Because of the lack of an order relation, it is not possible to tell if one attribute value is greater than another or that one value is closer to a certain value than another. However, we can tell if two values are equal or not equal. Given relevant background knowledge (human-based or computerized), it is possible to define more complex relations on nominal variables.

Ordinal variables or scales allow us to put things in order, because the set of values associated with an ordinal variable exhibits an intrinsic organization, which is defined by a *total order relation*. Therefore, we can tell if one value is bigger or smaller than another, but we can normally not quantify or measure the degree of difference or distance between two values. For example, if the observations x, y, and z are ranked, 5, 6, and 7, respectively, we can tell that $x < y < z$, but *not* if $(z - y) < (y - x)$.

Interval-scaled variables exhibit an intrinsic organization, which not only allows us to establish if a value is smaller or greater than another (*total order relation*) but also to determine a meaningful difference or distance between two values or to add and subtract values. This property of a meaningful difference is particularly important for distance-based or similarity-based approaches like clustering.

The values of real-scaled variables show a higher level of intrinsic organization than ordinal-scaled and interval-scaled variables. Besides order or rank information and the possibility to meaningfully add and subtract values, real variables allow us also to multiply and divide, as they are measured against a meaningful zero point. Typical examples for real variables include weight, age, length, volumes, and so on. In microarray experiments involving controls or references we can define a natural zero point, against which overexpression and underexpression can be defined. Therefore, it is perfectly legal to say that gene *a* is twice as highly expressed as gene *b*.

Although not a completely new category, *binary* or *dichotomous* variables could be considered as a special case. Binary variables can assume two possible values. For microarrays, this could be two categories, one for overexpression and one for underexpression, or one for overexpression and one for the absence of overexpression (lumping together balanced and underexpressed values). Often binary variables are coded with the numeric values 1 and 0. The advantage of this coding scheme is that binary variables can be treated as interval variables or attributes. However, treating binary variables as if they were interval-scaled implies that they are assumed to be symmetric (i.e. each value is equally important). If they are symmetric, then the interchanging of the codes will still result in the same score, for example, a degree of dissimilarity measured by the Euclidean distance measure. If the two entities, e.g., expression profiles, match on either code they are perceived to have something in common, and in case of a mismatch they are considered different (to the same extent independent on the coding). However, this could be a problem if the two values are not equally important. If one, for instance, uses the code 1 for overexpression and 0 for the absence of overexpression, it may be safe to say that two expression profiles have something in common if they match on a variable on code 1. This is not so clear if they match on the same variable on code 0. In case of asymmetric binary variables it is recommended to use non-invariant coefficients, such as the *Jaccard coefficient* to calculate similarities or dissimilarities (Kaufman and Rousseeuw, 1990).

4.2   Process

Once the raw digitized graphics images have been established by Step 1 to Step 4 of the overall analysis process (Figure 1.4), a highly iterative *inner loop* of processing kicks in. The analysis steps and information flow of this process are depicted in Figure 1.6. What characterizes this process is its computerized nature, meaning that most of the involved data and analytical steps are realized on a computer. This makes it possible to explore and refine the different processing alternatives until a satisfactory conclusion is reached.

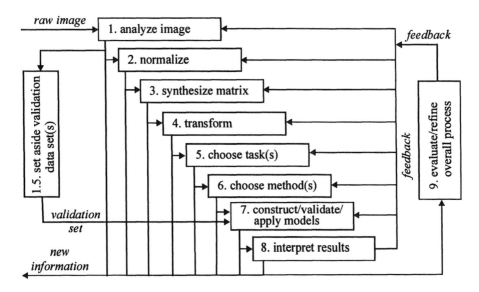

Figure 1.6. Microarray data analysis – the *inner loop* process (Step 5 to 9 of overall microarray data analysis process).

4.2.1   Pre-Processing – on the Rawness of Raw Data

People involved in microarray analysis often speak of "raw" data. The rawness characterizes the indigestible state of some data before a processing step, normally one that reduces the data volume. However, depending on the person you are dealing with, the precise meaning of "raw" may differ. For an image analyst, raw data probably refers to the analog signal produced by a laser scanner. A statistician might consider the digitized image or the integrated numerical data matrix as raw data. For a biologist raw data might be manifest in the descriptions of a discovered pattern or a constructed model (e.g., cluster definitions or regression coefficients). In the jet-set-paced world of a senior investigator, summary statistics and diagrams may fall into to raw-data category. And finally, the editor-in-chief of a scientific journal may refer to the initial submission of a paper reporting on the findings of a microarray experiment as raw data.

Here, we use the term *raw data* to denote the collection of digitized images – one image per hybridization experiment – arising out of Step 5 of the overall analysis process (Figure 1.4). These data must be computationally gathered, processed, and integrated with other relevant information.

First, each probe spot on the images must be analyzed to establish quantitative estimates of the red and green content. Here, *normalization* methods may be applied to compensate for systematic measurement errors due to equipment imperfection. This procedure summarizes the raw image data and stores them in an intermediate, more compact representation where each spot is described by a set of numeric values.

Second, if multiple spots on the image represent multiple expression measurements for the same gene, these measurements must be combined to obtain a single expression level estimate for the gene. This is where normalization procedures come in, to address issues of measurement variation due to instrumentation imperfection and biological variation. At this point one may also decide to represent each measurement by a set of absolute values, a red-green ratio, a fold change, or some other format.

Third, the sets of measurements from each hybridization experiment must be integrated into a single data matrix, similar to the one presented in Figure 1.5. To compensate for array-to-array measurement variation and to facilitate comparison between different hybridization experiments, a normalization procedure called *standardization* is employed at this point. An example of the resulting integrated data matrix is depicted in Table 1.2.

Now the data is almost "well-done", ready to be devoured by the numerical analysis methods further down the processing stream. However, to facilitate a more effective and efficient performance of the subsequent processing steps of the inner loop (Figure 1.6), a further data manipulation step called *transformation* may be necessary. The objective of data transformation is to reduce the complexity of the data matrix and to represent the information in a different, more useful format.

The first four steps of the inner loop shown in Figure 1.6 depict these data manipulation operations – image analysis, normalization and standardization, data matrix synthesis, and transformation. Collectively, we refer to these as *pre-processing*. The following subsections discuss some of the issues and concepts of pre-processing in more detail. However, for predictive modeling, an important conceptual issue has to be considered. In order to closely "simulate" a realistic prediction scenario, the first thing to do after image analysis is to set aside randomly selected cases. These data are called validation data or *validation set*, with which the final prediction model is to be validated. Generally speaking, 10-20% of the whole data set is sufficient as validation data. All pre-processing steps considered hereafter have to be performed on the validation data separately, which would be the case for new cases (*application set*) that have to be predicted in a real-life prediction scenario.

4.2.1.1 Missing Values

There are many reasons why the measurement of a gene's expression level for an individual sample-gene combination may have failed. In this case, the best that the data matrix can offer is to simply flag the failed measurement, that is, to report a *missing value*. Table 1.1 and Table 1.2 illustrate missing values in the context of our four-gene experiment. There are a number of strategies for dealing with missing values. In a way, the missing value handling approaches could be considered as a hybrid between normalization and transformation (discussed below).

The first obvious, albeit drastic, choice for dealing with such errors is to *remove* the affected expression profile (gene or array profile) from the data matrix altogether. The drawback of this radical measure is that it also removes other valuable data. In the worst case, this approach can lead to the removal of $N \times M - min(N,M)$ valid expression values in the presence of only $min(N,M)$ missing values, leaving little left to be analyzed.

The second approach is to *ignore* the problem and leave the data matrix as it is. Perhaps one wants to use a special code to indicate the missing value, so that it cannot be confused with a valid measurement. There are many analytical methods that can inherently deal with missing values. Decision trees, for example, do have a "natural" mechanism for handling missing values by focusing on the existing measurements. Intuitively, this approach could be likened to computing an arithmetic mean of N values, M of which being flagged as missing. We compute the sum of the $N - M$ valid values and divide the sum by $N - M$. Clearly, the "ignore" approach works only if the proportion of missing values is within acceptable limits.

A clever third way of correcting errors due to missing values is to "replace" the offending missing item with a reasonable or plausible substitute value. The methods for this kind of error correction are collectively referred to as *missing value imputation*. A simplistic approach to missing value imputation is to assign some average value in place of the missing value. For example, one could replace the missing value of gene b for patient 5 (Table 1.1) by the average of the expression levels of gene b across condition *tumor A* of patient 5. Another straightforward approach is to use the level representing *balanced* expression (red-green ratio equals 1.00) in place of the missing values. This method is what we use to replace the two missing values of our four-gene experiment. Notice that missing value imputation may be carried out after some other transformations have been completed.

4.2.1.2 Normalization and Standardization

Taken together, the technical gear, laboratory protocol, and human element employed to measure transcript abundance can be viewed as a complex scientific "instrument". In such a system errors creep in from all directions at the same time, and in all shapes and sizes. These errors are due to imperfections in the instrument and the processes and materials involved in using it. This abstract view of a gene expression instrument considers sample selection, sample preparation, hybridization, and so on, as integral part of the instrument. Given this view, we are able to distinguish *measurement errors* due to instrumentation imperfection from *measurement variations* due to biological variations within the studied specimens. Although measurement variation due to biological fluctuations has implications for microarray analysis, such variations are not measurement "errors" in the sense of a *deviation* from a *true* expression level. They are simply a part of reality. If nature has decided that the height of 12-year-olds varies in some fashion, one cannot get rid of this variation by measuring height more precisely or by somehow trying to correct it.

Ideally, a numerical value in the expression matrix reflects the true level of transcript abundance (or some abundance ratio) in the measured gene-sample combination. However, as our instrument is inherently imperfect, the measured value, *measurement*, that we obtain from the instrument usually deviates from the true expression level, *truth*, by some amount called *error*.

The relationship between the true and measured values and the error is described by Equation 1.2.

$$measurement = truth + error \tag{1.2}$$

Generally, a measurement *error* can be attributed to two distinct causes or error components – *bias* and *variance*; the relationship between bias and variance and error as described in Equation 1.3.

$$error = bias + variance \tag{1.3}$$

In this error model, bias describes a systematic tendency of the expression measurement instrument to detect either too low or too high values. The amount of this consistent deviation is more or less constant. We know, for example, that the emitted light intensities from the dyes Cy3 and Cy5 are not the same for the same reporter molecule quantities. This introduces a systematic measurement error that, if understood properly, can be compensated for at a later stage, for example, at the image-processing step.

The measurement error due to variance is often normally distributed, meaning that wrong measurements in either direction are equally frequent, and that small deviations are more frequent than large ones. Examples of this kind of error include a whole range of lab processes and conditions and manufacturing variations. A standard way of addressing this class of error is experiment replication.

With these considerations and the concepts presented in Equation 1.2 and 1.3, we summarize our general measurement error model as follows:

$$measurement = truth + bias + variance \tag{1.4}$$

Notice, the variance term in Equation 1.3 refers to the variations affecting our readout based on the fluctuations in the instrument (and the involved processes). There is no need to incorporate variations occurring in the underlying specimen into this error model, as the term *truth* reflects the real quantity, no matter how much this quantity varies from sample to sample. Clearly, as replication is often employed to address measurement variation, the task of quantitatively disentangling the two elements of variation (instrumentation imperfections and biological fluctuations) may be difficult.

Normalization and standardization are numerical methods designed to deal with measurement errors (both bias and variance) and with measurement variations due to biological variation. In contrast, transformation refers to a class of numerical methods that aim to represent the expression levels calculated by normalization and standardization, in a format facilitating more effective and efficient analysis downstream.

We differentiate between two broad classes of measurement variations caused by instrumentation imperfection and biological variation: (a) those that affect individual measurements, and (b) those that affect an entire array or parts of an array.

There are numerous sources for measurement variation that affect single measurements, including biological variation, probe imperfection, the tendency of low expression levels to vary more than high levels, and so forth. The principal approach to dealing with single-measurement variations is replication. But beware! Instead of carving up the same mouse over and over again, you should go *all* the way back when you replicate an experiment. This may be time-consuming, expensive, and ethically questionable. However, with multiple measurements it is possible to apply statistical methods to estimate the true quantity more accurately and to judge the error of this estimate. Simple t-statistics, for example, can be applied to accomplish this task. With this approach one could, for example, state a measured expression level as: "1.35 ± 0.03, *confidence* = 99%". Besides replication, improvement in experimental design is also an effective approach to addressing the variability issues of single measurements.

Each array or chip measures the gene expression levels of many genes for a *single* sample (under a certain experimental condition). In a way, the hybridization of a single microarray element could be considered as an experiment in its own right. A typical microarray study carries out many such experiments, several tens to several hundreds. Due to laboratory problems, manufacturing imperfections, multiple investigators carrying out the individual experiments, biological variation, and other sources, array-to-array measurement variations are bound to creep in, more than one would ideally like. For studies (like predictive toxicology) that require the comparison of array profiles, this *global* measurement variation is a problem. Clearly, replication of the entire array is an approach to consider here. For instance, by hybridizing two arrays per sample, you could arrange the probe spots differently, so as to compensate for certain structural biases. In any case, even if you can afford to do multiple arrays per sample, you still need to address the issue of inter-array variability. Numerical procedures known as *global normalization* or *standardization* are designed to address this problem. As these methods attempt to place each array on a comparable scale, this process is also called *scaling*.

The simplest approach to scaling involves the multiplication of each value on an array with an array-specific factor, *m*, such that the resulting mean is the same for all arrays. Another procedure along these lines multiplies each value on an array so that the mean for each array equals 0 and the standard deviation equals 1. More sophisticated approaches fit a straight line to the data points of an array, and then use the line's parameters to transform the data points so that they spread about a "normal" line with a slope of 45 degrees. This procedure is applied to all the arrays involved, so that the 45-degree line is used as a common scale for all arrays. Other, yet more sophisticated approaches exist. All these methods make certain assumptions about the data. As the measurement of gene expression is a highly complex process, it is sometimes difficult to say if these assumptions are justified or not (Branca and Goodman, 2001; Quackenbush, 2001; Wu, 2002).

The data matrix generated by our hypothetical four-gene experiment (Table 1.1) contains the "semi-raw" measurements obtained from the image analysis stage. This data has not been normalized. To illustrate normalization, we will transform the data in a three-step process.

First, we replace the two missing values by 1s, indicating a balanced expression.

Second, we log-transform each value using the base-2 logarithm. Strictly speaking, this is a form of data transformation (as opposed to normalization). The reason for performing this operation is the *positive skewness* of the ratio data. This means that a large proportion of the measured values is confined to the lower end of the observed scale. As we use a ratio representation, values representing underexpression are all crammed into the interval between 0 and 1, whereas values denoting overexpression are able to roam free in the range of 1 to 6. This situation is shown in the left part of Figure 1.7. There are statistical reasons, why a more evenly or normally spread of the data are to be preferred by analytical methods. This desired property is called *normality* and the underlying concept for describing it is the normal distribution. The base-2 log-transformation re-distributes the skewed data in the desired manner (right part of Figure 1.7). For more realistic microarray data sets, the visual difference between shapes of the skewed and log-transformed data is much more pronounced.

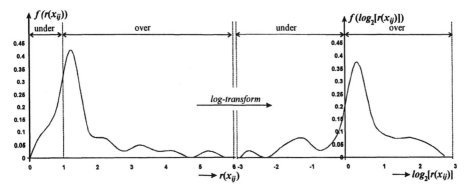

Figure 1.7. Log-transformation. Left: relative frequency, *f*, of all 40 expression levels shown in Table 1.1 (missing values replaced by 1); 22 of 40 values are within the interval [0,1], and

18 within (1,6]. Right: relative frequency of log-2-transformed values; 22 of 40 in [-3,0] and 18 in (0,3].

Third, we rescale the log-transformed data by applying a mean centering method to each of the 10 array profiles (in this case, sample or patient data sets). This method proceeds as follows. For each log-transformed array profile, $A_j = (x_{aj}, x_{bj}, x_{cj}, x_{dj})$, we (a) Compute the *mean, me*(A_j) and the *standard deviation, sd*(A_j), (b) Subtract the mean from each expression value, x_{ij}. This centers the already more or less symmetrically distributed values around 0. (c) Standardize the zero-centered values in terms of standard-deviation units relative to the mean by dividing each shifted value by the standard deviation. So a new expression value, $*x_{ij}$, is derived using: $*x_{ij} = (x_{ij} - me(A_j)) / sd(A_j)$.

Applying this three-step procedure to the four-gene expression matrix shown in Table 1.1, produces the normalized matrix depicted in Table 1.2.

Table 1.2. Numerical expression data matrix after missing value imputation, log-transformation, and standardization (mean-centering). Legend: *me* = mean of array profile, *sd* = standard deviation.

Patient#	1	2	3	4	5	6	7	8	9	10
Tumor	A	A	A	A	A	A	B	B	B	B
*r(a)	1.35	0.76	1.31	1.37	1.30	0.79	−1.40	−1.35	−1.27	−1.38
*r(b)	−0.10	−0.93	−0.07	−0.05	0.08	0.81	0.17	1.07	−0.01	0.14
*r(c)	−1.07	0.96	−1.12	−1.02	−1.12	−1.26	0.98	0.12	1.17	1.01
*r(d)	−0.18	−0.79	−0.12	−0.30	−0.26	−0.34	0.25	0.16	0.11	0.24
me	0.00	0.00	0.00	0.00	0.00	0.00	0.00	0.00	0.00	0.00
sd	1.00	1.00	1.00	1.00	1.00	1.00	1.00	1.00	1.00	1.00

After having performed the data transformation steps on our four-gene experiment data, we are now ready to have a closer look at the data. Figure 1.8 visualizes the expression levels for the gene profiles of gene *a* and *b*. By means of this visualization we can confirm our previous hunch that gene *a* is presumed to be differentially expressed across the two experimental conditions, whereas gene *b* is not.

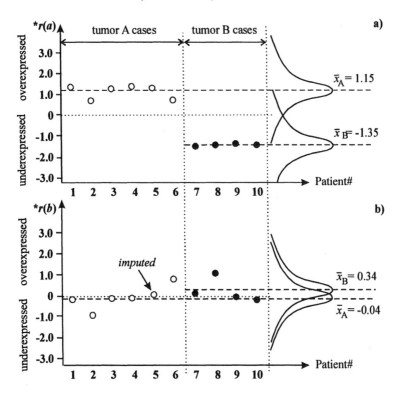

Figure 1.8. Visualization of two gene profiles, gene *a* (top) and gene *b* (bottom), from our four-gene experiment *after* normalization (i.e. the expression values are taken from Table 1.2). Data points from left to right correspond to the patient numbers. Horizontal dashed lines represent tumor-type-specific *mean* expression levels.

One way of approaching this decision more formally is to hypothesize that the expression levels, $G_g = (x_{g1}, x_{g2}, ..., x_{gM})$, of a gene, g, observed across two conditions, A and B, are in fact coming from the *same* population, and that the observed differences and variations are indeed within the limits of what we would expect in this case. Casting this into a testable hypothesis, we *assume* two *different*, normally distributed, equal-variance populations, one for condition A and one for B, with the corresponding population means μ_A and μ_B respectively. In Figure 1.8 these populations are depicted by the bell-shaped curves at the right-hand side of the diagram. Then, we formulate the *null hypothesis* H_0: $\mu_A = \mu_B$, and proceed testing it with, for example, a standard *t-test*.

Carrying out the two-tailed t-test calculation for the two gene profiles of gene *a* and *b* shown in Figure 1.8, yields the following results (*confidence level* = 95%; $\alpha = 0.05$; $t_{\alpha/2} = 2.306$):

- Gene *a*: $t = 16.72 \notin [-2.306, +2.306]$, therefore, we reject the null hypothesis H_0 for gene *a*. In other words, the expression levels of gene *a* for condition A and B are *not* likely to come from the same population.

- Gene *b*: $t = 1.125 \in [-2.306, +2.306]$, therefore, we accept the null hypothesis H_0 for gene *b*. That is, the expression levels measured for gene *b* for tumor A and B *are* likely to have come from the same population.

4.2.1.3 Transformation, Data Reduction, Data Enrichment

Once image analysis and normalization are performed, we are left with a data matrix similar to the one shown in Table 1.2. Recall, in our specific case we are dealing with ratio figures, reflecting the relative abundance of target versus reference mRNA. Depending on what the subsequent analytical steps are, we may still not be entirely happy with what we have.

First, we may want to bring in other covariate data that should be analyzed together with the expression profiles. This is sometimes called *data enrichment*. Typical covariate data added to gene expression profiles include clinical and similar data (e.g., tumor type/stage, tissue type, survival times, treatment response/dosage), information on the analyzed genes or gene products (e.g., sequence information, gene function, protein localization, protein interaction information), and results from other types of molecular experiments (e.g., comparative genomic hybridization, SNPs). Data enrichment methods will not be further pursued in this article.

Second, there may be a need to *reduce* the data of the given matrix (a) to focus the analysis on a particular subset of the data, and (b) to improve performance of subsequent analysis step. Four broad data reduction strategies can be distinguished:

- *Variable selection.* The objective of this approach is to select a subset of the available variables and subject only them to further analysis.

- *Observation selection.* Similar to variable selection, except that observation are in play here. Clearly, the issue of whether gene profiles or array profiles are considered as observations or variables must be addressed.

- *Variable combination.* This approach seeks to combine existing variables into a kind of "super" or composite variable. Subsequent analysis proceeds with the composite variables and the variables used to create the composite variables are excluded from further analysis. In principle, this approach is also possible for observations.

- *Value transformation.* The aim is to transform actual (variable) values into another format or representation. We have already met different types of transformation methods used for normalization. Another common value transformation strategy is to discretize continuous or real-numbered values.

Variable or feature selection is an important issue in microarray analysis, as the number of measured variables (i.e. genes) is usually much larger than the number of observations (i.e. single hybridization experiments or samples). Given N genes, there are $2^N - 1$ distinct gene combinations or unique subsets of genes (the minus 1 is accounted for by the *empty set*). Even a moderate number of genes will render any method attempting to evaluate each possible subset impossible, at least until the long-awaited time when we have quantum computers. If you have a modest number of 30 (yes: only thirty) genes, and you let a conventional computer evaluate one subset per second, chances are that you will not live to behold the final results. Unless, that is, you do exceptionally successful work in the area of aging. So plowing through all subsets is generally not a good idea.

Perhaps one of the best strategies to select good variables is by bringing to bear relevant background knowledge or heuristics. For example, certain families of genes may initially be chosen, rather than all genes measured by the experiment. The next best thing you can do is to analyze the N variables one by one and throw out those that fail to fulfill a predefined test or criterion. The most obvious choice is to filter out genes whose gene profile does not show much (differential) variation across the samples. Depending on the pursued analytical task, this could be achieved by simple threshold-based techniques, statistical tests, or other techniques such as interdependence analyses, distance-based methods, information gain and entropy-based approaches, or the *general separability method* (Grabczewski & Duch, 1999). The problem with selection methods focusing on one feature at a time is that (a) they may end up selecting highly redundant features, that is, variables that are strongly correlated, and (b) they may filter out variables that are meaningful only in conjunction with other features.

Multivariate feature selection methods attempt to take in more than one variable at once. Computing a *correlation matrix* or *covariance matrix* is often a good start towards detecting redundant and correlated variables. These matrices are generated by computing a coefficient or score for each variable-pair in the original data matrix. The elements in the covariance matrix retain information concerning the scales in which the variables are measured. Thus, variables that tend to have large values tend to have large covariance scores. Elements of the correlation matrix are computed in a very similar way, except that they are subject to a normalization step, which forces the values to fall into the interval $[-1, +1]$; $+1$ and -1 indicating perfect positive and negative linear correlation, respectively, and 0 indicating no linear correlation at all. All elements on the diagonal of the correlation matrix reflect a perfect positive correlation of the corresponding variable with itself. The values on the diagonal of the covariance matrix reflect the actual *sample variance* of the corresponding single variable.

	*r(a)	*r(b)	*r(c)	*r(d)
*r(a)	1.00			
*r(b)	-0.36	1.00		
*r(c)	-0.81	-0.23	1.00	
*r(d)	-0.75	0.54	0.34	1.00

	*r(a)	*r(b)	*r(c)	*r(d)
*r(a)	1.71	-0.25	-1.14	-0.32
*r(b)		0.29	-0.13	0.10
*r(c)			1.15	0.12
*r(d)				0.11

Figure 1.9. Correlation matrix (left) and covariance (right) for the 4 gene profiles of our four-gene experiment, computed from normalized data shown in Table 1.2.

Figure 1.9 depicts the correlation and covariance matrices computed from the four normalized gene profiles of our four-gene experiment. The correlation matrix (left) reveals a relatively strong *negative* correlation between the gene profiles of gene *a* and *c* (– 0.81), and *a* and *d* (– 0.75), respectively. Thus, as far as their variation patterns are concerned, the gene profiles of gene *a* and *c*, and that of *a* and *d* are highly redundant. Therefore, in order to reduce the burden on subsequent analytical steps, we could consider to remove the expression profile of gene *a* from the matrix, or, alternatively, that of the genes *c* and *d*. But hold on! Do not throw out the baby with the bath water. Before you proceed with reducing the burden, you should take note of the almost self-evident *hypothesis* emerging from the results in the correlation matrix: *The genes a and c, and a and d are co-regulated.*

Other multivariate feature selection strategies exist, including cluster analysis techniques, neural networks, and multivariate decision trees. Also, visualization techniques are powerful allies in the crusade against superfluous variables. Michael Eisen's gene expression map software is perhaps one of the most impressive demonstrations how analysis and visualization techniques can be combined to explore and select gene expression profiles (Eisen et al., 1998; Brown and Botstein, 1999).

Merging variables into a composite variable or "components" is an alternative approach to reducing the dimensionality of microarray data. The basic idea is to examine the original set of variables collectively, and to combine and transform them into a new, smaller set of mutually largely uncorrelated variables each of which retaining most of the original information content. For example, if a set of gene or array profiles turns out to be highly correlated, one could consider to drop some profiles or to replace the correlated set by some average profile that conveys most of the profiles' information.

A feature-merging method commonly employed to reduce microarray data is called *principal component analysis* (PCA). This technique discovers variables that are correlated with one another, but are independent from other subsets of variables. The correlated subsets are combined into factors or components, which are thought to reflect the underlying process that led to the correlations among variables. Generally, the major goals of PCA are to summarize patterns of correlation, and to reduce the number of observed variables, and to provide the basis for predictive models (Tabachnick and Fidell, 1996). The mathematical operations involved in PCA are intricate. Intuitively, one could imagine a principal component as a rotation of the discerned data points so that the variance among the data points, when viewed from a particular "perspective", is maximized. Once the highest-variance perspective (i.e. the first principal component) is found, other components are determined in the similar fashion, with the additional constraint that each new component is uncorrelated with the previous ones.

Microarrays measure expression levels on a continuous scale. However, many analytical methods benefit from or require discrete-scaled input variables. Examples include Bayesian networks and techniques and *logic methods* such as *association analysis, decision trees,* and *rule-based* approaches. Discretization methods are designed to perform the necessary data transformation.

Let U be the discerned *universe of expression levels*. Further, let L denote a predetermined set of *labels* describing discrete expression levels, such that $L = \{L_1, L_2, ..., L_n\}$. A discretization method is defined as a function, $f(x)$, that assigns each $x \in U$ to a single label in L as follows:

$$f(x) : U \to L$$

An obvious, albeit extremely simple, three-label discretization scheme for ratio-based expression data would use the labels *under* for expression ratios less than 1, *balanced* for 1, and *over* for values greater than 1. A more sophisticated version would express discrete expression levels in terms of standard deviations from the mean of either the array or gene profile. For example, *under* to represent levels below 1 standard deviation from the mean, *over* for levels above 1 standard deviation from the mean, and *balanced* elsewhere. The second scheme is more flexible, as it reflects the actual underlying distribution of expression levels and can be adapted to represent more than three discrete expression levels. It would also benefit from a global normalization in terms of units of standard deviations from the mean. The expression map in Table 1.3 illustrates a three-way discretization of the data from our four-gene experiment. The previously employed color mapping, *f*: {*black, gray, white*} → {*red, orange, green*}, such that *f*(*black*) = *red*, *f*(*gray*) = *orange*, *f*(*white*) = *green*, should enable you to trick your visual perception into seeing the gene expression color map in the accustomed fashion.

Table 1.3. Data matrix after discretization of normalized matrix (Table 1.2) using decision threshold of ±0.75 standard deviations from mean (*mean* = 0). <u>Legend</u>: *ovr* = over-expression (black), *bal* = balanced (gray), and *udr* = under-expression (white).

Patient#	1	2	3	4	5	6	7	8	9	10
Tumor	A	A	A	A	A	A	B	B	B	B
r(a)	ovr	ovr	ovr	ovr	ovr	ovr	udr	udr	udr	udr
r(b)	bal	udr	bal	bal	bal	ovr	bal	ovr	bal	bal
r(c)	udr	ovr	udr	udr	udr	udr	ovr	bal	ovr	ovr
r(d)	bal	udr	bal	bal	bal	bal	bal	bal	bal	bal

More advanced discretization strategies exist, one such strategy is the previously mentioned *general separability method* (Grabczewski & Duch, 1999; Berrar et al., 2001).

4.2.2   Selection of Data Analysis Task

Once data pre-processing (Steps 1 to 4 in Figure 6) is done, we are ready for numerical analysis. Ideally, when the microarray experiment is conceived, hypotheses are formulated, and specific scientific tasks are defined, one should also specify how the scientific tasks map onto specific (*numerical*) *analysis tasks*. So task selection should have happened well before preprocessing was started. However, pre-processing itself and the initial, explorative application of some analytical methods may well have improved our understanding of the data. This newly gained knowledge may alter the course of action we take in terms of task and method selection.

The term data analysis task refers to a generic type of data analysis process or approach for which many different specific methods exist. Typically, a data analysis task is characterized by the type of information it analyzes or by the information it extracts from a given data set. Data analysis tasks can be grouped into two broad categories namely, *hypothesis testing* and *knowledge discovery* or *hypothesis generation*. Which of these two approaches is appropriate depends on the nature of the scientific questions and the characteristics of the available microarray data.

Hypothesis testing is a top-down process. It starts with preconceived notions, ideas, and hypotheses about the studied processes and entities, and it then seeks to verify or disprove these mental models via experiment and analysis. In a microarray scenario, this entails the (a) conception, design, and execution of suitable microarray hybridization experiments, (b) the design and construction of an analytical computer model based on the generated data, and (c) thorough evaluation of the model to confirm or reject the hypothesis. A more theoretical and less data-driven variant of top-down hypothesis testing attempts to formalize the mental models so that they can be cast into executable analytical models (paper-based, computer-based, or other). Both the natural processes as well as the executable models are then probed with analogous input data (experimental conditions) and their response data are compared. This approach is perhaps more in line with what systems biologists have in mind. If realized on a computer, such executable models are also known as *computational theories. Numerical simulation systems* based on differential equations have been used extensively in computational biology to embody computational theories. Recently, there has been increased interest in *symbolic simulation systems* to embody computational theories about biological processes (Karp, 2001). Symbolic computation derives from work in the fields of artificial intelligence and machine learning.

In knowledge discovery (hypothesis generation), no prior assumptions about the data are made. The data are probed with the objective of revealing previously unknown, non-trivial relationships or patterns in the data. Knowledge discovery can be thought of as a bottom-up process, which starts with the data and tries to let the data suggest new hypotheses. This type of generic task reflects the exploratory, high-throughput character of many microarray studies – throw thousands of genes onto the chip, plow up the field of generated expression data, and sift through the unearthed debris in the hope of finding nuggets of gold. Knowledge discovery comes in two forms, *directed* and *undirected knowledge discovery.*

Directed knowledge discovery attempts to explain or categorize some particular aspect of the underlying process or entity. For instance, the analysis of gene expression profiles may lead to an explanation of two types of tumors in terms of differences in the underlying gene profiles.

Undirected knowledge discovery, on the other hand, attempts to find patterns or similarities among groups (e.g. gene or array expression profiles) without the use of a particular target variable or a collection of predefined classes.

Both hypothesis testing and knowledge discovery are high-level, generic tasks. They are usually accomplished by decomposing them into a set of more elementary numerical or analytical tasks. These elementary tasks can be grouped into two categories – *predictive* and *pattern-detection* tasks. Some of these tasks are shown in Table 1.4.

Table 1.4. Overview of some elementary analytical tasks important for microarray analysis.

predictive tasks	pattern-detection tasks
classification	clustering
regression or estimation	correlation analysis
time-series prediction	association analysis
	deviation detection
	visualization

The following discussion will focus on classification, clustering, and association analysis.

4.2.2.1 Classification Task

Classification is also known as prediction, class prediction, discriminant analysis, or supervised classification or learning. Generally, classification is a process of learning-from-examples, in which the objective is to induce a function, f, when provided with examples of the form $(\mathbf{X}_i, f(\mathbf{X}_i))$. In this pair the term $\mathbf{X}_i = (x_{i1}, x_{i2}, ..., x_{iK})$ denotes the observed variables of the ith example and $f(\mathbf{X}_i) \in \{1, 2, ..., Q\}$ denotes the class label associated with the observed variables. For microarrays, the observed variables refer to gene expression profiles, that is, gene or array profiles (see Equation 1.1a and 1.1b). Class labels correspond to array and gene profile covariant information, that is, experimental conditions for array profiles and gene or gene family descriptors for gene profiles (see Figure 1.5).

Classification task. A *classifier* or predictor for Q classes partitions the underlying K-dimensional gene expression profile space into Q disjoint subsets, $\{S_1, S_2, ..., S_Q\} = S$, such that for an observation with expression profile $\mathbf{X}_i \in S_q \in S$ the predicted class is q. (adapted from Dudoit et al., 2000)

Clearly, it is sometimes desirable to include covariate information to the K expression profile dimensions.

Classification is probably the most popular elementary data analysis task. The range of available classification methods is huge. Besides classical statistical techniques, methods from data mining, machine learning, and artificial intelligence are starting to enter the microarray field.

4.2.2.2 Clustering Task

Clustering is an analytical task which is also known as cluster analysis, automatic class prediction or classification, data segmentation or grouping, partitioning, or unsupervised classification or learning. Clustering techniques are concerned with finding meaningful groups in data. Clustering seeks a convenient and valid organization (and description) of data. In contrast to classification, clustering is not concerned with establishing rules for separating future data into predefined categories. Clustering can be thought of as learning from *observation*, as opposed to learning from pre-classified *examples.* A clustering algorithm is provided with some observations, and the goal is to look for similarities in the observed data and group them such that the patterns in one group are similar to one another and dissimilar from the patterns in any other group. Thus, clustering techniques provide a useful approach to exploring data in order to identify and understand relationships that may exist in the data. A common type of relationship discovered by cluster analysis is hierarchical topology. Clustering is a key component in model fitting, hypothesis generation and testing, data exploration and data reduction.

Clustering is a process or task that is concerned with assigning class membership to observations such as gene expression profiles, but also with the definition or description of the classes that are discovered. Because of this added requirement and complexity, clustering is considered a higher-level process than classification. In general, clustering methods attempt to produce classes that maximize similarity within classes but minimize similarity between classes. A typical clustering method is the *k-means algorithm*. Current statistical and syntactic clustering methods have trouble expressing structural information, while neural clustering approaches are limited in representing semantic information.

Clustering task. Given a set, $Y = \{\mathbf{X}_1, \mathbf{X}_2, ..., \mathbf{X}_P\}$, of expression profiles, each profile, $\mathbf{X}_i \in Y$, described by K measured expression levels, that is $\mathbf{X}_i = (x_{i1}, x_{i2}, ..., x_{iK})$, determine a classification (grouping, partitioning) that is most likely to have generated the observed objects (adapted from Upal and Neufeld, 1996).

Clearly, it is often desirable to include covariate information with the gene or array profiles that are to be clustered.

Most early microarray analysis studies were concerned with the clustering of gene expression data. Hierarchical and self-organizing map clustering methods have been employed in many studies. The clustering tool from Eisen and colleagues is perhaps one of the most widely used analysis method in the microarray arena (Eisen et al., 1998).

4.2.2.3 Correlation and Association Analysis Tasks

Correlation analysis attempts to detect correlated variables, that is, variables that change (across the underlying observations or samples) in a coordinated manner. This type of analytical task is relevant to many scientific tasks, including dosage-response co-regulation studies. We have already discussed some methods that are designed to detect linearly correlated variables (correlation matrix, principal component analysis).

Another similar analytical task is association analysis. This discovers the co-occurrence of expression patterns over a set of observations. The patterns are accompanied by two probabilistic measures – *support* and *confidence*.

For example, given the discretized data of our four-gene experiment (Table 1.3), association analysis might have uncovered the following association pattern

if **r(a)=udr* **and** **r(b)=bal* **and** **r(c)=ovr* **and** **r(d)=bal*
then *Tumor=*B
    [*support* = 0.30, *confidence* = 1.00]

Support (sometimes referred to as *coverage*) is a measure that reflects the probability of the entire pattern, that is, *support* = *p*(*then* ∩ *if*), where *then* refers to the *then*-part of the pattern and *if* to the *if*-part.

Referencing the discretized expression data depicted in Table 1.3 and the association pattern shown above, we obtain the following probabilities: *support* = *p*(**r(a)* = *udr*,   **r(b)* = *bal*,   **r(c)* = *ovr*,   **r(d)* = *bal*, *Tumor* =B) = 3/10 = 0.30. The *confidence* (or *accuracy*) of the pattern is determined by the probability of occurrence of the *then*-part under the condition   of   occurrence   of   the   *if*-part,   thus: *confidence* = *p*(*then* | *if*) = *p*(*then* ∩ *if*) / *p*(*if*).   For   our   example: *p*(*then* ∩ *if*) = 0.30 and *p*(*if*) = 0.30, so *confidence* = *p*(*then* | *if*) = 1.00.

Typically, one seeks to find association patterns that satisfy the constraint that the *support* and *confidence* exceed some predefined threshold values (for example, thresholds of 0.05 for *support* and 0.80 for *confidence*). Association patterns summarize co-occurrence patterns in the data. As such they are a useful technique to explore expression data.

4.2.3   Selection of a Specific Method

The armament of techniques we have at our disposal to implement the necessary analytical tasks is enormous. It ranges from classical statistical methods, to data mining, and to things like *support vector machines, relevance machines,* and *lattice machines*. No consensus has yet emerged as to which methods are the best for which tasks.

Once you have selected the specific analytical tasks required for solving your microarray analysis problem, there are a few criteria you could usefully apply to address your method-selection problem.

Visualization. For pre-processing and pattern discovery tasks, it is probably a good idea to get hold of tools that are able to visualize the data and intermediate analysis results in different formats. Good contenders are tree-based clustering methods like Eisen's tool. Visualization may also turn out to be a good ally when it comes to classification. Some decision tree tools provide excellent visualization of the induced trees.

Symbolic versus subsymbolic methods. For some problems, an artificial neural network may be the best choice in terms of obtaining acceptable results. However, due to their highly distributed encoding of learned patterns as numeric weight matrices, it is difficult to understand the models involved. Genetic-algorithm-inspired methods, support vector machines, lattice machines, and other algorithms do not lend themselves to easy and intuitive understanding. On the other hand, logic-based methods like decision trees and association patterns, and regression techniques, produce easy-to-understand models.

Computational complexity. Some methods make more demands on computational resources (processor, memory, and disk space) than others. Artificial neural networks, genetic algorithms, association algorithms, and other methods may require a lot of resources under certain circumstances. This situation is worsened when multiple analysis runs are needed, for example, for cross-validation or for trying out different parameter settings to explore the performance of a method.

Reproduceability. Some methods (e.g., some neural learning algorithms or genetic algorithms) employ non-deterministic elements, as it might be for selection initialization settings for some parameters. Because of this, the results for two different runs on the same data with the same parameter settings may not be identical.

4.2.4 Model Construction and Application

Once the pre-processing is done, and the analytical tasks and the methods of choice have been established, it is time to actually put the methods to work and let them analyze the data. Generally, we differentiate three different phases (a) *model construction* and *model verification*, (b) *model validation*, and (c) *model application*.

Models are constructed from a set of *training data* specifically selected for this purpose (this selection may involve a statistical sampling technique). Initially parameterized models frequently fail to accurately fit *test data* or do not meet other pre-defined criteria. To address this problem, some model parameters are adjusted or the data is manipulated in some way. Then a new model is constructed and its output is evaluated. This process is called model verification.

Once one is satisfied with the model's performance, the model is validated. For classification models, this is achieved by applying the classification model to independent *validation data*, that is, data that were not used in the model construction and verification phase. Model validation may also be performed by "manually" inspecting and evaluating the learned parameters and properties. Ideally, an independent analyst is summoned to carry out this evaluation.

For many scientific investigations, model construction, verification, and validation are all that is required. However, there are situations where a validated model may be usefully applied to new data (called *application data*) that has not been used for verification or validation purposes. This is what we call model application. For example, a reliable classification model may assist medical experts in diagnostic and prognostic tasks.

The literature uses different terms for the various data sets involved in these processes. Especially with classification tasks in mind, we use the term *learning set* to refer to the union of both *training set* and *test set*. Normally, the training set is used to construct a model and the test set to verify it. Depending on the strategy and method, training and test set may partially or fully overlap. We use the term *validation set* to denote a data set that is a part of the actual study but is not involved in model construction or verification. Once a model has been built and tuned, the validation set is employed to "simulate" application of the model to some independent data. Only if the model passes this validation step it can be considered useful. The problem with this is that if the validation does not produce the expected results, you are not allowed to go back and modify the model and then validate it again on the *same* validation set. If the validation fails, you should throw the model out of the window.

The *application set* (or *prediction set*) is normally not used in the study. It refers to data that may be available to you in the future, and for which no classification exists. Applying your classification model to completely new cases for which no classification exists is the real "acid test", in particular if real-world decisions are based on the model's prediction. This is where real prediction takes place, and you had better have evaluated and validated your model thoroughly.

Developing and verifying a model based on a single learning set, and only that, runs the risks that the model is too biased towards the data distribution within the two underlying training and test sets. This is also known as the *overfitting* problem. Using *cross-validation* is a more reliable method, especially when the amount of the available data is limited. Given that a validation data set has been set aside (see Step 1½ in Figure 1.6), cross-validation divides the available data set into *n folds* or subsets. Based on these *n* data sets, *n* different models are developed, each using $n - 1$ data sets for model construction and verification, and the remaining data set for quasi-validation. As the remaining data set does not participate in the construction and verification process, the performance on this data set provides an estimate for the subsequent real validation. The best-performing model is then selected and validated using the validation set that was set aside. This method is also referred to as *leave-n-out cross-validation*, as for the development of each of the *n* models, *n* data patterns are "left out" for quasi-validation. In the extreme situation, where *n* is equal to the number of available data patterns (for learning and quasi-validation), one has *leave-one-out cross-validation.*

4.2.5 Result Interpretation and Process Evaluation

The remaining steps of the inner loop (Step 8 and 9) are concerned with the interpretation of the analytical results and constructed models, and the revision of the entire inner loop.

The interpretation of microarray analysis results is relying upon the two disciplines: computer science or statistics and biology, more specifically, molecular biology. Statistics takes into account specific measures for cluster validation, specificity, sensitivity, positive/negative predictive values, ROC-analyses for clustering and classification tasks. The interpretation of the analysis results from the point of view of molecular biology investigates the functions of selected variables, possibly placing the genes found in a greater frame, e.g., known pathways, enzymatic reactions, or even provide evidence for a more systemic view on the organism under investigation.

4.2.6 Biological Validation

Microarrays measuring the expression of thousands of genes are applied as screening methods. Thus, biological validation of microarray data analysis results includes specific examination of identified genes by other highly biotechnological techniques like the quantitative reverse transcriptase polymerase chain reaction. In many studies, this additional yet important aspect has been neglected, but the necessity for validating results by other techniques has been recognized.

REFERENCES

Alberts B., Bray D., Lewis J., Raff M., Roberts K., Watson J.D. (1989). *Molecular biology of the cell.* New York: Garland Publishing.

Berrar D., Dubitzky W., Granzow M., Eils R. (2001). Analysis of gene expression and drug activity data by knowledge-based association mining. Proceedings of Critical Assessment of Microarray Data Analysis (CAMDA 2001), pp. 23-28.

Branca M.A. and Goodman N. (2001). DNA microarray informatics: Key technological trends and commercial opportunities. Cambridge Healthtech Institute, CHI Genomic Reports.

Brown P.O. and Botstein D. (1999). Exploring the new world of the genome with DNA microarrays. *Nature Genet* 21(1):33-37.

Eisen M.B., Spellman P.T., Brown P.O., Botstein D. (1998). Cluster Analysis and display of genome-wide expression patterns. Proc. Natl. Acad. Sci. USA 95:14863-14886.

Grabczewski K. and Duch W.A (1999). General purpose separability criterion for classification systems. 4th Conference on Neural Networks and Their Applications, Zakopane; pp. 203-208.

Dudoit S., Fridland J., Speed T.P (2000). Comparison of discriminant methods for classification of tumors using gene expression data. Technical Report No. 576, University of California, Berkeley.

Duggan D.J., Bittner M., Chen Y., Meltzer P., Trent J.M. (1999). Expression profiling using cDNA microarrays. *Nature Genet* 21(1):10-14.

Granjeaud S., Bertucci F., Jordan B.R. (1999). Expression profiling: DNA arrays in many guises. BioEssays 21(9):781-790.

Karp P.D. (2001) Pathway Databases: A Case Study in Computational Symbolic Theories, Science 293:2040-4.

Kaufman L., Rousseeuw P.J. (1990). *Finding groups in data. An introduction to cluster analysis.* John Wiley & Sons, Inc.

Kerr M.K. and Churchill G.A. (2001). Statistical design and the analysis of gene expression microarray data. Genetic Research 77:123-128.

Lee M.L., Kuo F.C., Whitmore G.A., Sklar J. (2000). Importance of replication in microarray gene expression studies: statistical methods and evidence from repetitive cDNA hybridizations. Proc. Natl. Acad. Sci. USA 97(18):9834-9839.

Quackenbush J. (2001). Computational analysis of microarray data. Nature Genet 2:418-427.

Raskó I. and Downes C.S. (1995). *Genes in medicine.* Chapman and Hall, London.

Sabatti C. (2002). Statistical Issues in Microarray Analysis. Current Genomics, to appear; available at http://www.bentham.org/cg3-1/sabatti/sabatti-ms.htm.

Schena M., Shalon D., Davis R.W., Brown P.O. (1995). Quantitative monitoring of gene expression patterns with a complementary DNA microarray. Science 270:467-70.

Tabachnick B.G., Fidel L.S. (1996). Using Multivariate Statistics 3rd Edition. Harper Collins College Publisher, pp. 635-708.

Upal M.A., Neufeld, E. (1996). Comparison of unsupervised classifiers. Proc. First International Conference on Information, Statistics and Induction in Science, World Scientific, Singapore, pp. 342-353.

Wu T.D. (2002). Large-scale analysis of gene expression profiles. Briefings in Bioinformatics 2(1):7-17.

Chapter 2

DATA PRE-PROCESSING ISSUES IN MICROARRAY ANALYSIS

Nicholas A. Tinker, Laurian S. Robert, Gail Butler, Linda J. Harris.
ECORC, Agriculture and Agri-Food Canada, Bldg. 20, 960 Carling Ave., Ottawa, ON, K1A 0C6, Canada
e-mail: {tinkerna,robertls,butlergm,harrislj}@agr.gc.ca

1. INTRODUCTION

Microarray experimentation is a young but rapidly maturing field, and the potential complexities of microarray data are spawning a rich statistical literature. Some of the concepts presented in this literature may be new and unfamiliar to the molecular biologist, who typically gleans information in a stepwise manner from many small, carefully controlled experiments. Microarrays yield enormous quantities of data and can address many simultaneous hypotheses, but results are variable and require careful preparation and statistical analysis. Gathering, organizing, and preparing data for statistical analysis is a large and important component of microarray experimentation. These steps are referred to collectively as *pre-processing*.

This chapter is written for the newcomer to microarray technology. It is intended as an introduction to some of the pre-processing steps that are detailed in further chapters. Broadly defined, pre-processing includes the planning and design of experiments[1], the acquisition and processing of images, data transformation, data inspection, and data filtering. We cannot possibly represent every variation in terminology and procedure that will be encountered in other literature, but we have attempted to introduce a variety

---

[1] The design and construction of microarrays themselves, and the extensive laboratory management that accompanies this task, are not discussed in this chapter. These steps vary substantially with the type of microarray technology being used; those who intend to prepare their own microarrays should consult literature on this subject (e.g., Allain et al., 2001; Dolan et al., 2001; Rockett et al., 2001; http://www.microarrays.org).

of terminology, and to explain the basis of some major procedural differences. A flow diagram for the topics covered in this chapter is presented in Figure 2.1. Because of the strong requirement for data management at all stages of pre-processing, we begin with a general discussion of this topic.

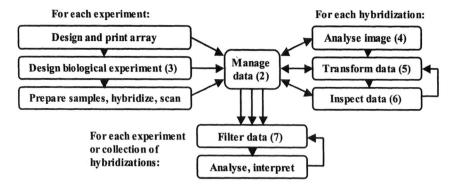

Figure 2.1. A diagram of preprocessing steps in microarray analysis. Numbers in parentheses indicate sections where topics are discussed.

2.   DATA MANAGEMENT

For most laboratories, microarray analysis will not be a one-time experience. Even if an experiment addresses one specific question, it will likely be followed by experiments to test additional factors. Properly done, a microarray experiment has the potential to become part of a larger study. This concept has inspired some authors to address the need for common data descriptions, formats, and repositories (Becker, 2001; Fellenberg et al., 2002; Kellam, 2001). Given the success and importance of large central databases for DNA and protein, there are strong reasons to nudge microarray data in the same direction. However, it is clear that microarray data is far more complex than sequence data, and that it can be meaningless or misleading unless: (1) the context of experimental conditions is fully described, and (2) measurements are standardized such that comparisons are valid. For this reason, standards such as the "Minimal Information About Microarray Experiments (MIAME)" (Brazma et al., 2001) are being developed. Meeting these standards is a responsibility that demands good data management – even if reduced to a basic need to publish reproducible results. At a minimum, forms that provide templates for the description of an experiment and its components should be used. These forms can be based on standards such as MIAME, with adaptations that are appropriate for unique characteristics of the laboratory, the experimental organism, or the methodology being employed. Separate forms can be made to describe experiments, arrays, samples, and hybridization conditions. Completion of these forms prior to conducting an experiment will aid in the planning of experiments and in standardizing the relevant information.

Following image analysis, management of a much larger quantity of data will be required. This will include raw images and large matrices of derived image parameters. Different versions of these data will need to be managed throughout the various pre-processing and filtering steps. Finally, a variety of results from data analysis (often performed by numerous collaborators) will need to be stored and accessed over time. There will inevitably be requirements to re-do, re-analyze, or re-interpret some aspect of an experiment. For many researchers, the electronic spreadsheet is the most familiar data management tool and, undoubtedly, many good microarray experiments have been managed entirely with spreadsheets. However, many other experiments have likely succumbed to the errors and difficulties that spreadsheets can cause, such as: forgetting which of many copies holds the original data, mutations caused by accidental keystrokes, lack of security, a propensity to sort separately two columns that belong together, and the relative difficulty of joining information from two separate tables. All of these issues can be solved by relational databases; hence some thought should be given toward developing or adopting a relational database management system that is appropriate to the skills and resources of the laboratory, and compatible with required procedures for data pre-processing and analysis.

Publicly available database systems have been described in the literature (e.g., Comander et al., 2001; Sherlock et al., 2001), and others are available through acquisition of commercial database-enabled microarray software. The user should consider the following when adopting data management systems: 1) the need to design arrays and track the intermediate steps of array construction; (2) ability to store experimental parameters and sample preparation steps; (3) ability to store images, and to link individual spots to stored data; (4) integration of procedures for data pre-processing; (5) options for data filtering; (6) integration with data analysis software and other databases; (7) cost, support, and ease of maintenance; (8) ability to support multiple users; (9) ability to share and protect data across a local network or the Internet.

3.   EXPERIMENTAL DESIGN

Statisticians insist that experimental planning and design is the most critical part of every scientific experiment. Experimental planning begins with careful consideration of the biological questions to be answered, the conditions or materials to be compared, the expected sources of variation, and the methodology to be used. It is important to establish the type and amount of replication needed while allocating limited resources to an appropriate balance between treatments and replication. Attention should be given to randomization to minimize the effects of extraneous variation, so as to avoid bias and possibly confounding comparisons of interest with other factors such as changes in arrays or in sample labelling. Most statisticians also insist that methods for data analysis should be planned in advance. This serves to identify potential shortcomings of the design, but is also intended to prevent the temptation to "torture" data into providing the desired conclusions.

Given the above, it might seem that microarray experiments are some of the worst experiments in existence. Experiments are often carried out without specific hypotheses, thousands of hypotheses can be tested with little regard to statistical error rate, planned treatments may be altered mid-experiment, and new methods for analysis are encountered after experiments are completed. Much may be forgiven due to the novelty of microarray technology. It may even be argued that microarray experimentation need not be hypothesis-driven (traditional scientific method has been challenged in large sequencing projects, with great reward). Nevertheless, some attention to basic principles of experimental design can only improve the chances of success. The discussion below is intended to introduce some general concepts of statistical design and their application to microarray experimentation. A more extensive discussion of this topic is provided by Nadon and Shoemaker (2002).

3.1   General Concepts of Statistical Design

Statistical analysis involves measuring the effect of *treatments* (often referred to somewhat synonymously as *experimental conditions, factors, varieties,* or *environments*). We then wish to compare these effects, draw conclusions, and guard these conclusions by a probability that they are wrong. In doing so, we speak of *type I error rate* (the probability of saying something is different when it is not), and *power* (the probability of discovering a true difference). Power and type I error are opposing forces – the experimenter must choose an acceptable type I error rate (e.g., 5%) with the knowledge that larger type I error rates will give increased power. Errors are estimated by measuring *residual effects* (the amount of variability that cannot be accounted for by treatments or controlled experimental factors). Estimation of error requires that a treatment be *replicated,* such that each *replicate* is a randomly chosen independent *experimental unit.* Conclusions will then be valid with regard to the *population* of similar experimental units. For example, if two chemical treatments are applied to eight yeast cultures, each culture must be treated equally except for the random assignment of a chemical treatment. When treatments are not (or cannot be) assigned randomly, results may be *confounded*, such that measurements might be related in part to some factor other than the treatments under consideration. Multiple observations made on material from the same experimental unit are called *sub-samples*, and should not be confused with replications. For example, if a tissue sample is pooled from five plants, then analyzed twice, we have a single experimental unit with two sub-samples. The sub-samples can tell us about the variability of measurements, but not about the variability among plants.

3.2 Practical Application to Microarrays

The most unique aspect of a microarray experiment is that it provides simultaneous information about transcript abundance for a large number of genes based on measurements of hybridization between a complex cDNA *sample* and a large set of *substances*[2] on an array. Since many aspects of the substances are variable (e.g., sequence length, concentration, and spot size), it is necessary to account for this variability by using experimental design. This accounting is often done by conducting every hybridization with two samples simultaneously; one from the treatment of interest, and the other from a common control. By expressing all measurements in relation to a common control, some biases associated with the target substances can be removed. Some alternative designs do not require a common control (e.g., Kerr and Churchill, 2001a,b) but these designs are not in common use, and they may require additional statistical consultation during planning and analysis (see Section 5.6). Even when hybridizations are conducted using a common control sample, it is essential to perform replications in a manner that avoids confounding errors. This should include *reverse-labelling* (replicates performed by reversing the dyes assigned to the control and treatment samples) and randomization over potential sources of bias such as different printings of an array.

Caution should be exercised with regard to what is considered a replicated observation. Multiple copies of a substance on the same array are subjected to some of the same biases as all other spots on the array, so they are considered as sub-samples rather than replicates. These sub-samples can provide insurance against missing data, and will provide greater precision of measurement when they are averaged. They can even remove some bias

---

[2] Here, we will refer to the spots on an array as substances, reflecting the fact that they may not all represent genes. However, terminology varies. Some authors refer to the spots on a microarray as probes, and the labeled samples as targets. Others reverse this terminology.

within an array (if their locations within an array are randomized), but they cannot be used to estimate experimental error.

While the experimenter has limited control over some sources of bias associated with the array, a higher degree of control may be achieved over the labelled cDNA samples. The treatments giving rise to these samples should be replicated and randomized using basic concepts described in Section 3.1. There may be pragmatic requirements to pool common elements of an experiment (e.g., grouping similar treatments in space or time, or pooling replicates for RNA isolation). These practices can confound treatments with important sources of bias, and should be avoided if possible. However, when compromise is necessary, the potential consequences should be evaluated. It is helpful to list all perceived sources of error and to speculate on which sources may have an impact on conclusions.

3.3   To Replicate or Explore?

The microarray experimenter is often pulled in two competing directions: to perform highly replicated experiments that allow valid and robust statistical conclusions, or to explore a large set of experimental conditions with the hope of developing hypotheses about groups of related genes or hidden mechanisms of gene expression. The experimenter must decide which of these objectives is of higher value, and compromise as necessary. Specific hypotheses (e.g., diseased *vs.* normal) lend themselves to the first approach while others (e.g., onset of development) invite exploration. The strategy chosen will influence the experimental design, data processing, and statistical analysis. For example, a replicated experiment comparing two conditions might allow more complex and efficient experimental designs, giving better resolution of specific treatment effects and differences among genes (e.g., Lee et al., 2000). Conversely, many methods of multivariate analysis that are applied to experiments with many treatment levels (e.g., hierarchical clustering) may not utilize replicated observations, since observations from identical conditions are averaged before analysis. Nevertheless, it is important to remember that replicates still add precision, remove bias, and allow the identification of aberrant observations. The optimum number of replications is affected by many factors, including the desired power for detecting differences. The optimum number of replications can be estimated if assumptions are made (e.g., Wolfinger 2001); alternatively, general recommendations can be followed (e.g., Lee et al., 2000, suggest at least 3 replicates). If replication is not present, extra effort is required to avoid confounding, and results should be considered tentative and subject to further validation.

4. IMAGE PROCESSING

For the pioneers of microarray analysis, image processing must have presented major technical challenges. These challenges have been largely supplanted by the development of algorithms and software, but it is worth understanding some of the potential pitfalls in image processing in order to select software, and to avoid introducing artefacts at this stage of analysis.

Image analysis begins with the scanning of hybridized arrays at a wavelength suitable to each labelled sample. Some optimization may be required at this stage to obtain images with approximately equal intensity and good signal-to-noise ratio. For two-sample hybridizations, separate images may be combined into a single false-colour image for further analysis. A template (or overlay) that identifies the grid positions of the microarray substances is then superimposed on the image. Typically, positions of substances are identified by block, row, and column. Needless to say, it is important to have an accurate template, and later, to check the accuracy of the template through the use of control substances. Errors in the spotting of microarrays can occur, but these will likely be gross errors (e.g., switching two entire sets of samples) that can be identified through changes in the expected position of blanks or known controls.

Various algorithms are used to orient the template with an image. Those who work with spotted microarrays will discover that the alignment and geometry of these arrays is rarely perfect, so it is useful to have software that allows interactive stretching of the template grid. If possible, it is useful to incorporate *"landing lights"* (substances expected to show strong hybridization) near the corners or edges of each block of spots. Otherwise, it may be difficult to align a template with an image resulting from a weak hybridization.

Once the template is approximately aligned, an algorithm will find and delineate regions of pixels that represent the spotted substances. The edges of those regions will be highlighted to allow manual inspection and editing. There are a variety of methods for detecting spot edges and for compensating for background (Dudoit et al., 2002; Kim et al., 2001; Tran et al., 2002; Jain et al., 2002; Bozinov and Rahnenfuhrer, 2002). Since different methods may be optimized for different types of arrays, it is probably best to test several methods with your "worst" images, prior to committing to one particular method or software.

A number of problems or issues may be apparent upon inspecting an image, and should be noted for consideration. One example is the presence of "black holes". These are spots that are actually darker (less fluorescent) than the surrounding background, highlighting the possibility that spot intensity and background noise are not necessarily additive. Image analysis software may automatically subtract background from spot foreground. In some cases, this background subtraction will result in negative values for spot intensity, such that these data points must be discarded. Options involve setting the software parameters such that background is not subtracted or adopting an alternate method for compensating for background (e.g., Tran et al., 2002).

Most software will derive many variables from each spot on an image. Many of these variables may have no particular use – they are just easy to compute. The variables most commonly used include mean and median pixel intensity, and the ratios of these values. Most authors chose to use median pixel intensities rather than mean pixel intensities because medians are less affected by outlying pixel values. Outlying pixel values are often caused by failure to delineate the exact edge of a spot, such that background pixels are included.

Although it is advisable to return to the original image for validation whenever possible, image analysis should be conducted as though it were the last time that the image will be seen. Image analysis software is seldom perfect, and artefacts will arise that require human judgement. Spots that are suspect can be flagged so that they are excluded from further analyses. Some software will concomitantly indicate the substance contained on each spot, which will assist in the identification of potential problems (e.g., when blanks or controls show inappropriate values). It is preferable that a single person conducts all image analyses within a given experiment, since this avoids the introduction of bias due to differences in human judgement.

The temptation to draw conclusions during image analysis should be avoided. The colour coding used to identify spot intensity and intensity ratios can be misleading, and it is susceptible to biases that need to be removed by appropriate methods of transformation.

5.   DATA TRANSFORMATION

Much of the confusion in microarray analysis arises from differences in opinions regarding what must be done to prepare these data for analysis. We have tried to reduce this confusion by describing a generalization of a commonly used approach to transform data from arrays that contain paired samples (Sections 5.1 to 5.6). Alternative approaches are discussed in Section 5.6.

5.1   A Useful Vocabulary

The term *normalization* is generally used synonymously with the term *transformation*, and a large body of literature has addressed this topic. It is easier to understand the need for transformation if it is divided into logical steps or components. Three important components are: *transformation to normality*, *centralization*, and *re-scaling*. *Transformation to normality* refers to the adjustment of data such that it approximates a normal distribution (a prerequisite for many types of statistical analysis). The purpose of *centralization* is to remove biases from the data. *Re-scaling* is a final step that may be applied to ensure that data from different hybridizations have equal variances. When considered as three independent steps, the experimenter can apply judgment in selecting and applying (or not applying) an appropriate method in each category. Some variations of this terminology should be expected. In particular, the term scaling has been used by some to refer to what we (and others) call centralization.

5.2 Transformation to Normality

There is general agreement that a log transformation of most microarray data provides a good approximation of the normal distribution (e.g., Figure 2.2) with minor exceptions (Hoyle et al., 2002). Different authors have preference for different log bases (usually $\log_{10}$, $\log_2$, or $\log_e$ [=ln]). Although all of these work equally well, the base that is used needs to be explicitly recorded for future reference. The log transformation can be applied to original observations (e.g., median pixel intensities) or to ratios – remembering that $\log(x) - \log(y)$ is equivalent to $\log(x/y)$. Some authors will plot original values on log axes, while others prefer to work directly with log-transformed variables. For simplicity, we now assume that the log transformation is applied to all variables.

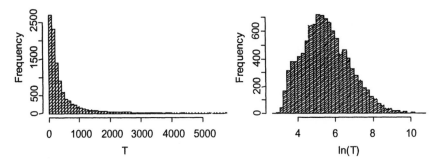

Figure 2.2. Distributions of median values of pixel intensity (*T*) and their natural logarithms, ln(*T*), derived from the image analysis of 10,476 spots on a microarray. The microarray contained 4,628 unique substances (maize [*Zea mays*] cDNA clones and controls) and was hybridized with cDNA derived from two RNA samples labeled with different dyes. The treatment sample (shown here as "*T*") was isolated from developing maize kernels that were manually inoculated 48-hours previously with the pathogen *Fusarium graminearum*, and the control sample (shown elsewhere as "*C*") was isolated from mock-inoculated kernels at the same developmental stage.

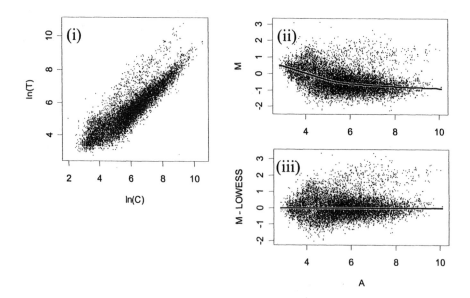

Figure 2.3. Median pixel intensities from paired treatment (*T*) and control (*C*) samples hybridized to a 10,476-spot maize microarray (see Figure 2.2) presented as scatter plots of (i) ln(*T*) *vs.* ln(*C*), (ii) M = ln(*T*/*C*) *vs.* A = ln(*T***C*)/2, and (iii) an adjusted value of *M vs. A*. The adjustment in (iii) is made by subtracting a fitted local regression called LOWESS from *M* shown in (ii).

5.3   Data Centralization

Most published references to normalization deal primarily with what we are calling centralization: the removal of biases in the data. Centralization is particularly important when using ratios to monitor changes in gene expression. Bias arises from a number of sources, including variation within and among arrays, differences in mRNA concentration or quality, unequal dye incorporation, and wavelength-related differences in scanner strength. Without correcting these biases, it may appear as though too many genes are up- (or down-) regulated. This can be especially misleading when one is inspecting the data using a colour scheme[3], which is why it is necessary to identify this problem using a graphical aid such as the *M vs. A* plot.

Most microarray software packages provide an option for global centralization, and this should be considered as a bare minimum. For log ratios (*M*), the expected mean of a data set is zero, so global centralization is

---

[3] Most microarray software packages provide a feature whereby one variable (e.g., the ratio of medians) is shown in spreadsheet format, with hybridizations in columns, and substances in rows. Each cell is artificially coloured based on the value in the cell (e.g., such that red indicates up-regulation and green indicates down-regulation).

performed by subtracting the overall mean from each data point. One problem with this approach is that the true expected mean may differ from zero due to an abundance of genes that are actually up- or down-regulated. This is sometimes addressed by performing global centralization based on the mean of some control substances, whose expression is not expected to change. Technically, this amounts to subtracting the mean of the control substances from each data point. Two types of control substances have been considered: *housekeeping genes*, and *alien sequences* (or *spiking controls*). *Housekeeping genes* are those required for basic cellular activity. Unfortunately, the definition of a housekeeping gene is subjective, and the assumption that its expression is not affected by the treatment is not certain. A *spiking control* is a sequence that is not expected to be present in the sample. It is applied to specific spots on the array, and it is introduced into the sample in a known concentration. This allows for correction of bias related to dyes or wavelengths, but it does not remove bias that may arise due to the quality or concentration of mRNA samples. Furthermore, unless a large number of controls are used, the mean of the controls may actually give a poorer estimate than the mean of the entire data set. One possible alternative is based on systematic identification of classes of genes that do not change (Tseng et al., 2001). Another alternative, called *self-normalization,* is based on the assumption that most bias that is related to dye intensity can be removed by averaging the ratios from two reverse-labelled hybridizations. Regardless of whether controls are used for centralization, the inclusion of controls on an array should be considered essential. They provide an important point of reference for use in future comparisons, and they may help to identify unusual or erroneous results.

Global centralization cannot correct for biases that are present within specific parts of the data. Two general categories of this are: bias that is due to intensity (e.g., stronger bias at lower intensities) and bias that is spatial (i.e. dependent on the position within the printed array). The former is illustrated in Figure 2.3ii, where bias in measured expression (M) seems to be related to intensity (A) as estimated by the fitted local regression (LOWESS). The use of the LOWESS function to correct this bias has been suggested (Yang et al., 2002), and the result of this correction is shown in Figure 2.3iii. Bias may also relate to position on the microarray. This will not be apparent unless observations from different parts of the array are selectively plotted. When this type of bias is present, it may require centralization that is weighted based on array coordinates or block position. This should not be attempted unless this source of bias is clearly identifiable. Depending on how the arrays were printed, and on other factors that affect the composition of the array (e.g., DNA concentration, source, or purity), spatial bias may take many different forms. Furthermore, if spatial bias is manifested as changes in spot intensity, then the two types of bias may be entirely related, and spatial bias would be corrected based on the LOWESS function.

5.4   Data Re-scaling

Due to technical variability, hybridizations may show differences in the range or variance of response variables. This phenomenon is not corrected by centralization (e.g., after centralization, M values from one hybridization may range from -2 to +2, whereas another may have values ranging from –3 to +3). Ideally, the variance from different hybridizations should be equal to facilitate comparisons among substances. Furthermore, if replicated treatments are to be averaged, re-scaling ensures that each replicate contributes equal information.

A simple method of re-scaling involves dividing a variable by its variance, thus giving the data a standard variance of unity. Algorithms for multivariate analysis (e.g., hierarchical clustering) may provide an option to perform this transformation automatically at the time of analysis. Unfortunately (or fortunately, depending on your viewpoint), the variance within a hybridization is a result of two components: substances and error, and it is only the error variance that should be standardized. Variance due to substances may differ due to treatment effects, which should not be altered. Conceivably, a variance adjustment could be made by dividing a variable by the variance of control substances, similar to the way in which control samples can be used in centralization (see Section 5.4). However, this might be subject to the same uncertainties about whether controls are appropriate. Furthermore, estimates of variance can be unreliable unless they are based on a large set of observations.

Re-scaling is potentially the least necessary component of normalization, and many researchers may choose to avoid the possibility that re-scaling may have unintentional consequences. Consideration can be given to testing whether variances differ prior to re-scaling (Snedecor and Cochran, 1989).

5.5   Alternative Approaches

The use of paired samples with data transformation is probably the most common approach to the preparation of microarray data at the time of writing. It is based on the assumption that biases can be removed separately from each hybridization prior to assembling data from multiple hybridizations for analysis. Notable alternatives to this approach involve removing bias from the data after it has been assembled for analysis by using an analysis of variance (ANOVA). The ANOVA is a technique whereby the total variability (variance) in an experiment is partitioned into components that can be attributed to treatments and to other recognized sources of variability (e.g., developmental stages, sample dyes, or different batches of arrays). After attributing variance to these factors, residual variance (that which remains) is considered to be a measurement of error that can be used to formulate statistical tests of significance among treatment means. Good discussions of this approach are provided by Kerr and Churchill (2001a,b) and Wolfinger et al. (2001). The models that these authors describe would provide more direct comparisons of expression levels among different treatments and would eliminate the need for including a control sample with every hybridization. This would also reduce the need for further data transformation because biases would be removed by fitting their sources as terms in an ANOVA.

In choosing an appropriate approach, a general recommendation is to consider different options and to store data in such a way that options are not excluded. We believe that the ANOVA-based approaches being developed have distinct advantages, particularly for testing specific hypotheses in a planned experiment. Conversely, the approach of normalizing each hybridization may be more conducive to building databases for retrospective exploration. While an ANOVA or mixed model can account for virtually any source of bias, some may find the approach of normalizing each hybridization to be more intuitive, and more readily adaptable to complex sources of bias. In general, an ANOVA can be applied to an experiment that is designed with paired samples, but normalized ratios cannot be constructed without the presence of a common control. Most existing software and databases are oriented toward the paired-sample approach, but data can easily be exported for alternate analysis using statistical analysis packages. In situations where different methods can be applied, it may be useful to determine if different methods give similar results[4].

6. DATA INSPECTION

Data inspection is an important step that should be applied at many stages of microarray analysis, but especially during and after transformation. Several tests can be applied to ensure that transformation has been successful, and that additional steps are not required. These inspection steps are aided by software that provides interactive sorting, highlighting groups of substances, and plotting data points in a variety of configurations. As an alternative to commercial software, which some people find inflexible or expensive,

---

[4] Generally, it will be possible to develop an ANOVA that will give similar results to methods where bias is removed through normalization. One technical difference is that pre-normalization fails to account for degrees of freedom that are lost through this process.

various free alternatives are available (e.g., Liao et al., 2000; Breitkreutz et al., 2001). A further alternative is the use of a graphical statistics environment such as "R" (Ihaka and Gentleman, 1996; http://www.r-project.org), which allows infinite variation in methods for data analysis and inspection. These solutions may require a greater amount of training or practice.

One of the data inspection steps that should be considered is the use of an *M vs. A* plot (Figure 2.3) in which individual substances can be interactively highlighted within the complete distribution. The basic *M vs. A* plot allows one to visualize whether the entire distribution of "*M*" values are centred with a mean of zero, and that the mean is not influenced by intensity ("*A*"). By highlighting control substances, one can determine if centralization has caused their mean to differ from zero. It will also be clear whether controls represent a range of intensities, thus, whether they can be used to provide intensity-dependent corrections. By highlighting blocks of data from various spatial regions of an array, one can determine whether spatial-dependent centralization is required. For spotted microarrays, it is useful to know the method and sequence by which the array was printed, as this can have more influence than mere proximity on the slide. Some software applications provide direct interaction between graphing tools and individual spots on a microarray image. For example, one could highlight all spots within a certain region of a microarray to see their positions on the *M vs. A* plot.

Another step that is useful when multiple replicates have been performed is the use of scatter plots and correlation analysis. A scatter plot of variables (e.g., *M* or *A*) from one replicate *vs.* those from another is an informative visual aid that can quickly show whether replications are consistent. This is also an easy verification of whether the ratio of treatment to control has been constructed in the same direction for both arrays (a negative correlation is a good indication that one ratio is inverted). Direct interaction between the scatter plot and the microarray image can be useful to determine whether specific substances or specific parts of the array are contributing to greater or lesser agreement between the arrays. The Pearson correlation coefficient (*r*) between two hybridizations provides a simple quantification of this agreement. This is also a useful statistic to gauge the success of different transformation methods. The correlation coefficient is not affected by global centralization or global re-scaling, but it might be useful to validate whether improvements have been made based on local centralization or transformation to normality. A matrix of correlation coefficients can be used to compare more than two replicated arrays, and is useful to gauge the results of different types of replication.

7. DATA FILTERING

Data filtering is a term that can take many different meanings. Indeed, like data inspection, it is a concept that can be applied at any stage of pre-processing. Our reason for discussing this topic at the end of this chapter is that there is a general danger of "losing data" before one is certain that it is not useful. Hence, while suspect observations are flagged during image analysis, and annotations about the substances and experimental conditions are stored, all (or most) data are kept intact within a primary database. Thus, filtering can refer to any step that is required to prepare a generic data set for a specific type of analysis.

The concept of filtering can be visualized as taking a large matrix of data (possibly an entire database) and making a smaller matrix. The large matrix contains hybridizations (usually arranged as columns) and substances (usually arranged as rows), and possibly many different types of observations from each hybridization (also arranged as columns). The smaller, filtered matrix probably consists of only one type of observation (e.g., *M*, a normalized log ratio of medians), a subset of substances, and a subset of the hybridizations.

Filtering involves three general concepts: selection, averaging, and estimation. Examples of each are given in the following paragraphs. It is important to consider that filtering may be a recursive process, and that different types of filtering are appropriate for different types of analysis. Minimal filtering would be required to prepare data for a statistical analysis designed to estimate error and determine significance thresholds. This might be followed by more intense filtering designed to produce reduced data sets that are appropriate for some of the more advanced functional or classifying analyses discussed in later chapters.

Selection is functionally synonymous to elimination, and can be based on substances, hybridizations, or individual observations. Selection of hybridizations would involve choosing those that provide information on specific conditions (e.g., those based on samples collected less than 50 hours after exposure to a pathogen). One might also eliminate entire hybridizations that failed to meet some aspect of quality control. Substances might be selected on the basis of pre-defined subsets or prior annotations (e.g., selecting only the controls, or only those that are known to be transcription factors). Substances can also be selected or eliminated based on observed data. For example, one might eliminate all substances that did not show a change in expression within any hybridization. This might be done based on a rule of thumb (e.g., ratio must be greater than 2 or less than 0.5) or after one had conducted an ANOVA on the complete data set to establish a significance threshold. This type of filtering might be used to prepare data for hierarchical clustering, where unchanging substances would be considered uninteresting. Finally the elimination of specific observations based on some measurement of quality (such as those flagged at the image analysis stage) would be considered.

Averaging is used to combine observations from replicated hybridizations (columns) or replicated substances (rows). Some software packages will automatically average all replicated substances, so this step may not be required. Replicated hybridizations should only be averaged if required by the analysis; otherwise, replications should be left un-averaged so that they can be used in the estimation of error.

Estimation of missing values can be considered as a final filtering step that may be required when individual observations have been eliminated, and when empty cells remain after averaging. Depending on the extent of missing values, one may consider another round of selection to remove those rows or columns that contain large proportions of missing values. Or, one may chose to replace missing values with "neutral" averages that are constructed by averaging entire rows and/or columns. Some software packages do this automatically, but this can lead to artificial results if missing values are numerous. Appropriate methods for dealing with missing values are discussed in the next chapter.

Since more than one type of filtering may be required for several types of analysis, and since data analysis may be done at various stages of data collection, it is useful to implement filtering as a set of "rules" that can be saved, modified, and re-applied to the data as needed. Ease of data filtering is one of the advantages of storing data in a relational database, where such rules can be saved as "queries" or "stored procedures". A further advantage of using a database is that filtering may depend on several types of information that may be stored in different tables (e.g., gene categories in one table, experimental conditions in another). Most relational databases, or microarray software packages that are associated with a database, will provide some type of "query wizard" that can assist the user in defining a query to filter the data. In the absence of a database, filtering will probably be done in spreadsheets, using various manipulations such as copying formulas, sorting, cutting, pasting, inserting, and deleting.

8.   CONCLUSIONS

Defined broadly, pre-processing involves many potential steps that are essential for successful microarray experimentation. The need for some steps (e.g., experimental design, image analysis) is unquestionable. Other steps are less dogmatic. Data transformation, inspection, and filtering should occur based on individual analytical goals and data management systems. These steps may take on new meaning as different techniques for analysis become widely accepted. A complete and perfect recipe for pre-processing and analyzing microarray experiments does not exist. Therefore, each experimenter must develop systems and procedures that are both appropriate and correct. We hope that the concepts introduced in this chapter will help the reader to better understand the detailed presentations found in later chapters.

9. REFERENCES

Allain L.R., Askari M., Stokes D.L., Vo-Dinh T. (2001). Microarray sampling-platform fabrication using bubble-jet technology for a biochip system. Fresenius J Anal Chem 371:146-50.

Becker K.G. (2001). The sharing of cDNA microarray data. Nat Rev Neurosci 2:438-40.

Bozinov D., Rahnenfuhrer J. (2002). Unsupervised technique for robust target separation and analysis of DNA microarray spots through adaptive pixel clustering. Bioinformatics 18:747-56.

Brazma A., Hingamp P., Quackenbush J., Sherlock G., Spellman P., Stoeckert C., Aach J., Ansorge W., Ball C.A., Causton H.C., Gaasterland T., Glenisson P., Holstege F.C., Kim I.F., Markowitz V., Matese J.C., Parkinson H., Robinson A., Sarkans U., Schulze-Kremer S., Stewart J., Taylor R., Vilo J., Vingron M. (2001). Minimum information about a microarray experiment (MIAME)-toward standards for microarray data. Nat Genet 29:365-71.

Breitkreutz B.J. (2001). Jorgensen P., Breitkreutz A., Tyers M. AFM 4.0: a toolbox for DNA microarray analysis. Genome Biol 2:Software0001.1-0001.3.

Comander J., Weber G.M., Gimbrone M.A. Jr, Garcia-Cardena G. (2001). Argus – a new database system for Web-based analysis of multiple microarray data sets. Genome Res 11:1603-10.

Dolan P.L., Wu Y., Ista L.K., Metzenberg R.L., Nelson M.A., Lopez G.P. (2001). Robust and efficient synthetic method for forming DNA microarrays. Nucleic Acids Res 29:E107-7.

Dudoit S., Yang Y.H., Callow M.J., Speed T.P. (2002). Statistical methods for identifying genes with differential expression in replicated cDNA microarray experiments. Statistica Sinica 12:111-139.

Fellenberg K., Hauser N.C., Brors B., Hoheisel J.D., Vingron M. (2002). Microarray data warehouse allowing for inclusion of experiment annotations in statistical analysis. Bioinformatics 18:423-33.

Hoyle D.C., Rattray M., Jupp R., Brass A. (2002). Making sense of microarray data distributions. Bioinformatics 18:576-84.

Ihaka R., Gentleman R. (1996). R: A language for data analysis and graphics. Journal of Computational and Graphical Statistics 5:299-314.

Jain A.N., Tokuyasu T.A., Snijders A.M., Segraves R., Albertson D.G., Pinkel D. (2002). Fully automatic quantification of microarray image data. Genome Res 12:325-32.

Kellam P. (2001). Microarray gene expression database: progress towards an international repository of gene expression data. Genome Biol 2:Reports 4011.

Kerr M.K., Churchill G.A. (2001a). Statistical design and the analysis of gene expression microarray data. Genet Res 77:123-8.

Kerr M.K., Churchill G.A. (2001b). Experimental design for gene expression microarrays. Biostatistics 2:183-201.

Kim J.H., Kim H.Y., Lee Y.S. (2001). A novel method using edge detection for signal extraction from cDNA microarray image analysis Exp Mol Med 33:83-8.

Lee M.L., Kuo F.C., Whitmore G.A., Sklar J. (2000). Importance of replication in microarray gene expression studies: statistical methods and evidence from repetitive cDNA hybridizations. Proc Natl Acad Sci 97:9834-9.

Liao B., Hale W., Epstein C.B., Butow R.A., Garner H.R. (2000). MAD: a suite of tools for microarray data management and processing. Bioinformatics 16:946-7.

Nadon R. (2002). Shoemaker J. Statistical issues with microarrays: processing and analysis. Trends Genet 18:265-71.

Rockett J.C., Christopher Luft J., Brian Garges J., Krawetz S.A., Hughes M.R., Hee Kirn K., Oudes A.J., Dix D.J.(2001). Development of a 950-gene DNA array for examining gene expression patterns in mouse testis. Genome Biol 2:Research0014.1-0014.9.

Sherlock G., Hernandez-Boussard T., Kasarskis A., Binkley G., Matese J.C., Dwight S.S., Kaloper M., Weng S., Jin H., Ball C.A., Eisen M.B., Spellman P.T., Brown P.O., Botstein D., Cherry J.M. (2001). The Stanford Microarray Database. Nucleic Acids Res 29:152-5.

Snedecor G.W., Cochran W.G. (1989). *Statistical Methods*. 8th edition. Iowa State University Press, Ames. 503 pp.

Tran P.H., Peiffer D.A., Shin Y., Meek L.M., Brody J.P., Cho K.W. (2002). Microarray optimizations: increasing spot accuracy and automated identification of true microarray signals. Nucleic Acids Res 30:e54.

Tseng G.C., Oh M.K., Rohlin L., Liao J.C., Wong W.H. (2001). Issues in cDNA microarray analysis: quality filtering, channel normalization, models of variations and assessment of gene effects. Nucleic Acids Res 29:2549-57.

Wolfinger R.D., Gibson G., Wolfinger E.D., Bennett L., Hamadeh H., Bushel P., Afshari C., Paules R.S. (2001). Assessing gene significance from cDNA microarray expression data via mixed models. J Comput Biol 8:625-37.

Yang Y.H., Dudoit S., Luu P., Lin D.M., Peng V., Ngai J., Speed T.P. (2002). Normalization for cDNA microarray data: a robust composite method addressing single and multiple slide systematic variation. Nucleic Acids Res 30:e15.

Chapter 3

MISSING VALUE ESTIMATION[1]

Olga G. Troyanskaya, David Botstein, Russ B. Altman

Department of Genetics, Stanford University School of Medicine, Stanford, CA, 94305 USA,
e-mail: olgat@smi.stanford.edu, botstein@genome.stanford.edu, russ.altman@stanford.edu

1.   INTRODUCTION

Gene expression microarray data are usually in the form of large matrices of expression levels of genes (rows) under different experimental conditions (columns), frequently with some values missing. Missing values have a variety of causes, including image corruption, insufficient resolution, or simply dust or scratches on the slide. Missing data may also occur systematically as a result of the robotic methods used to create the arrays. Our informal analysis of the distribution of missing data in real data sets shows a combination of all of these reasons, with none dominating. Experimentalists usually manually flag such suspicious data points and exclude them from subsequent analysis. However, many analysis methods, such as *principal components analysis* or *singular value decomposition*, require complete matrices to function (Alter et al., 2000, Raychaudhuri et al., 2000).

Of course, one solution to missing data problem is to repeat the experiment, but this strategy can be expensive. Such approach has been used in the validation of microarray analysis algorithms (Butte et al., 2001). Alternatively, as most often done in day-to-day practice, missing $\log_2$ transformed data are replaced by zeros or, less often, by an average expression over the row, or "row average". This approach is not optimal, since these methods do not take into consideration the correlation structure of the data. Many analysis techniques that require complete data matrices, as well as other analysis methods such as *hierarchical clustering, k-means*

---

[1] Parts of the work presented in this chapter were originally published in *Bioinformatics* (Troyanskaya et al., 2001).

clustering, and *self-organizing maps*, may benefit from using more accurately estimated missing values.

The missing value problem is well addressed in statistics in the contexts of non-response issues in sample surveys and missing data in experiments (Little and Rubin 1987). Common methods include iterative analysis of variance methods, filling in least squares estimates (Yates 1933), randomized inference methods, and likelihood-based approaches (Wilkinson 1958). An algorithm similar to *nearest neighbors* was used to deal with missing values in CART-like algorithms (Loh and Vanichsetakul 1988). The most commonly applied statistical techniques for dealing with missing data are model-based approaches, and there is not a large published literature concerning missing value estimation for microarray data.

To address the problem of missing value estimation in the context of microarray data, we introduce *KNNimpute*, a missing value estimation method we developed to minimize data modeling assumptions and take advantage of the correlation structure of the gene expression data. The work that introduced KNNimpute as well as another method *SVDimpute* (Troyanskaya et al., 2001) was the first published comprehensive evaluation of the issue of missing value estimation specifically as applied to microarray data. Since then, a few other works have at least in part addressed this issue (e.g., Bar-Joseph et al., 2002). Many statistical packages for microarray analysis also have routines for missing value estimation. An important issue in choosing the method to use for estimation is evaluation of the method's performance on microarray data. In this context, issues such as data sets used in evaluation (data type, data size, diversity of data sets, etc.) and quality of the evaluation metric are important.

This chapter will focus on KNNimpute. We suggest specific guidelines for the use of KNNimpute software and recommend specific parameter choices. This chapter also presents results of comparative evaluation of KNNimpute with a *singular value decomposition* (SVD) based method (SVDimpute), the method of replacing missing values with zeros, and a row average method.

2.   ALGORITHMS

2.1   KNNimpute Algorithm

The KNN-based method takes advantage of the correlation structure in microarray data by using genes with expression profiles similar to the gene of interest to impute missing values. Let us consider gene i that has one missing value in experiment j. Then this method would estimate the value (i, j) with a weighted average of values in experiment j of K other genes that have a value present in experiment i and expression most similar to j in all experiments other than i. In the weighted average, the contribution of each gene is weighted by similarity of its expression to that of gene A. The exact neighborhood of K neighbors used for estimation in KNNimpute is recalculated for each missing value for each gene to minimize error and memory usage.

We examined a number of metrics for gene similarity (*Pearson correlation, Euclidean distance, variance minimization*) and Euclidean distance appeared to be a sufficiently accurate norm (data not shown). This finding is somewhat surprising, given that the Euclidean distance measure is often sensitive to outliers, which could be present in microarray data. However, we found that *log-transforming* the data seems to sufficiently reduce the effect of outliers on gene similarity determination.

2.2   SVDimpute Algorithm[2]

In this algorithm, we use singular value decomposition (3.1) to obtain a set of mutually orthogonal expression patterns that can be linearly combined to approximate the expression of all genes in the data set. Following previous work (Alter et al., 2000), we refer to these patterns, which in this case are identical to the principal components of the gene expression matrix, as eigengenes (Alter et al., 2000, Anderson 1984, Golub and Van Loan 1996).

$$A_{m \times n} = U_{m \times m} \Sigma_{m \times n} V_{n \times n}^{T} \qquad (3.1)$$

The matrix V^T contains eigengenes, and their contribution to the expression in the eigenspace is quantified by the corresponding eigenvalues on the diagonal of matrix Σ. We can then identify the most significant eigengenes by sorting the eigengenes based on their corresponding eigenvalue. It has been shown by (Alter et al., 2000) that several significant eigengenes are sufficient to describe most of the expression data, but the exact number (k) of most significant eigengenes best for estimation needs to be determined empirically by evaluating performance of SVDimpute algorithm while varying k[3]. We can then estimate a missing value j in gene i by first regressing this gene against the k most significant eigengenes and then use the coefficients of the regression to reconstruct j from a linear combination of the k eigengenes. The jth value of gene i and the jth values of the k eigengenes are not used in determining these regression coefficients. SVD can only be performed on complete matrices; thus we originally substitute row average for all missing values in matrix A, obtaining $A^{'}$.

---

[2] For more on singular value decomposition, see Chapter 5.
[3] Details of these experiments are presented in (Troyanskaya et al., 2001)

We then utilize an *expectation maximization* (EM) method to arrive at the final estimate. In the EM process, each missing value in $A^{'}$ is estimated using the above-described algorithm, and then the procedure is repeated on the newly obtained matrix, until the total change (sum of differences between individual values of A and $A^{'}$) in the matrix falls below the empirically determined threshold of 0.01.

3.   EVALUATION

3.1   Evaluation Method

Evaluation was performed on three microarray data sets (DeRisi et al., 1997; Gasch et al., 2000; Spellman et al., 1998). Two of the data sets were time-series data (DeRisi et al., 1997, Spellman et al., 1998), and one contained a non-time series subset of experiments from Gasch et al. (2000). One of the time-series data sets contained less apparent noise (Botstein, personal communication) than the other. We further refer to those data sets by their characteristics: time series, noisy time series, and non-time series.

We removed any rows and columns containing missing expression values from each of the data sets, yielding "complete" matrices. Then, between 1% and 20% of the data were deleted at random to create test data sets, and each method was used to recover the introduced missing values for each data set. To assess the accuracy of estimation, the estimated values were compared to those in the original data set using the root mean squared (RMS) difference metric. This metric is consistent across data sets because the average data value in data sets is the same after centering. This experimental design allowed us to assess the accuracy of each method under different conditions (type of data, fraction of data missing) and determine the optimal parameters (number of nearest neighbors or eigengenes used for estimation) for KNNimpute and SVDimpute.

3.2   Evaluation Results

3.2.1   KNNimpute

KNNimpute is very accurate, with the estimated values showing only 6-26% average deviation from the true values, depending on the type of data and fraction of values missing (Figure 3.1). In terms of errors for individual values, approximately 88% of the data points are estimated with RMS error under 0.25 with KNN-based estimation for a noisy time series data set with 10% entries missing (Figure 3.2). Under low apparent noise levels in time series data, as many as 94% of the values are estimated within 0.25 of the original value. Notably, this method is successful in accurate estimation of missing values for genes that are expressed in small clusters (groups of co-expressed genes), while other methods, such as row average and SVD, are likely to be more inaccurate on such clusters because the clusters themselves do not contribute significantly to the global parameters (such as top eigenvalues) upon which these methods rely.

The algorithm is robust to increasing the percentage of values missing, with a maximum of 10% decrease in accuracy with 20% of the data missing (Figure 3.1). However, a smaller percentage of missing data does make imputation more precise (Figure 3.1). We assessed the variance in RMS error over repeated estimations for the same file with the same percent of missing values removed. We performed 60 additional runs of missing value removal and subsequent estimation using one of the time series data sets. At 5% values missing and $K = 123$, the average RMS error was 0.203, with variance of 0.001. Thus, our results appear reproducible.

KNNimpute is relatively insensitive to the exact value of K within the range of 10 to 20 neighbors (Figure 3.1). When a lower number of neighbors is used for estimation, performance declines primarily due to overemphasis of a few dominant expression patterns. When the same gene is present more than once on the arrays, the method appropriately gives a very strong weight to other clones for that gene in the estimation (assuming that the multiple clones exhibit similar expression patterns). The deterioration in performance at larger values of K (above 20) may be due to noise present in microarray data outweighing the signal in a "neighborhood" that has become too large and not sufficiently relevant to the estimation problem.

Thus, the optimal value for K likely depends on the average cluster size for the given data set.

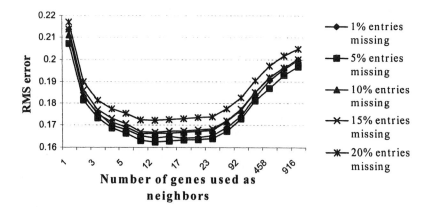

Figure 3.1. Effect of number of nearest neighbors (*K*) parameter on KNN-based estimation (on noisy time series data). Different curves correspond to experiments performed for data sets with different percent of entries missing.

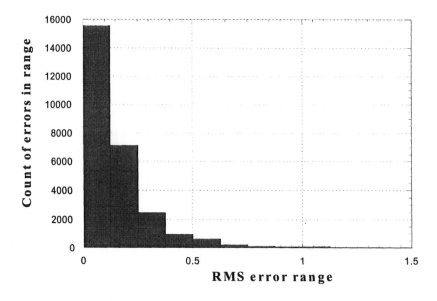

Figure 3.2. Distribution of individual errors for KNN-based estimation (on a noisy time-series data set). The histogram displays individual errors from estimation with *K*=15 at 10% of data missing. Most of the RMS errors are under 0.25.

Microarray data sets typically involve a large number of experiments, but sometimes researchers need to analyze data sets with small numbers of arrays. KNNimpute is robust to the number of experiments in the data set and can accurately estimate data for matrices with as low as six columns (Figure 3.3). However, we do not recommend using this method on matrices with less than four columns.

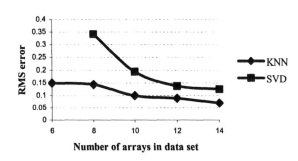

Figure 3.3. Effect of reduction of number of arrays in the data set on KNN- and SVD-based estimation (on a time series data set). Estimation was performed on matrices with successively lower number of columns. The SVD algorithm could not be applied to matrices with less than 8 columns due to mathematical constraints of the algorithm.

3.2.2 SVDimpute and Row Average

SVD-based estimation provides considerably higher accuracy than row average on all data sets (Figure 3.4), but SVDimpute yields best results on time-series data with a low noise level, most likely reflecting the signal-processing nature of the SVD-based method[1]. In addition, the performance of SVDimpute is very sensitive to the exact choice of parameters (number of top eigengenes) used for estimation and deteriorates sharply as the parameter is varied from the optimal value (which is unknown for a non-synthetic data set)[4].

Estimation by row (gene) average performs better than replacing missing values with zeros, but yields drastically lower accuracy than either KNN- or SVD- based estimation (Figure 3.4). As expected, the method performs most poorly on non-time series data (RMS error of .40 and more), but error on other data sets is also drastically higher than both of the other methods. This is not surprising because row averaging assumes that the expression of a gene in one of the experiments is similar to its expression in a different experiment, which is often not true in microarray experiments. In contrast to SVD and KNN, row average does not take advantage of the rich information provided by the expression patterns of other genes (or even duplicate runs of the same gene) in the data set. An in-depth study was not performed on column average, but some experiments were performed with this method, and it does not yield satisfactory performance (results not shown).

---

[4] Detailed results presented in (Troyanskaya et al., 2001)

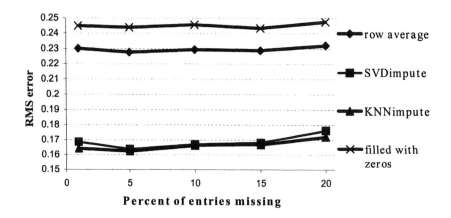

Figure 3.4. Performance of KNN, SVD, and row average based estimations on a noisy time series data set. The same data set with identical entries missing was used to assess the accuracy of each method, and RMS error was plotted as a function of fraction of values missing in the data. Most accurate parameter choices (as determined in the above-described experiments) were used for both KNNimpute and SVDimpute.

4.   PRACTICAL GUIDELINES FOR USE OF KNNimpute SOFTWARE

4.1   Software Description

KNNimpute[5] software is a C++ program that can be used on any UNIX- or LINUX-based platform. The program requires a C++ compiler (preferably gcc), and can be used on a Windows platform with a UNIX emulator, for example Cygwin[6]. Detailed instructions for installation and use of the KNNimpute software are supplied with the download.

4.2   Recommendations for Software Use

KNNimpute software requires data input in a tab-delimited text file formatted in preclustering data format (pcl), described at http://genome-www5.stanford.edu/MicroArray/help/formats.shtml#pcl. It is essential to follow the format specifications exactly. To minimize the effect of outliers, it is important to log-transform the data set prior to imputation (a different transformation or other data processing steps may be equally appropriate).

---

[5] KNNimpute can be downloaded at http://smi-web.stanford.edu/projects/helix/pubs/impute/
[6] http://www.cygwin.com/

The only parameter the user can vary is K, the number of nearest neighbors used in estimation. We recommend setting K in the range of 10-20, with lower K preferable for very large data sets to minimize the amount of time required for estimation. For smaller data size, any K between 10 and 20 is appropriate. We do not recommend the use of KNNimpute with very small (as compared to the number of genes in the genome) data sets, where the method may not be able to identify groups of similarly expressed genes for accurate estimation. However, small data sets that consist of tight clusters of functionally similar genes may be appropriate for estimation. The method should not be used on data sets containing fewer than four experiments.

While KNNimpute provides robust and accurate estimation of missing expression values, the method requires a sufficient percent of data to be present for each gene to identify similarly expressed genes for imputation. Therefore, we recommend exclusion from the data set any genes or arrays with a very large fraction of missing values (for example, genes or arrays with more than 25% of values missing).

4.3 Performance of KNNimpute

For a data set with m genes and n experiments and number of nearest neighbors set to k, the computational complexity of the KNNimpute method is approximately $O(m^2 n)$, assuming $m \gg k$ and fewer than 20% of the values missing. Thus, if the size of the dataset doubles in the number of genes, the algorithm will take approximately four times as long to run, while the same increase in the number of experiments would lead to doubling of the original running time. The KNNimpute software (implemented in C++) takes 3.23 minutes on a Pentium III 500 MHz computer to estimate missing values for a data set with 6,153 genes and 14 experiments, with 10% of the entries missing. However, for very large data sets (more than 20,000 genes) the program may need to run overnight or even longer, depending on the performance characteristics of the specific computer system used.

5. CONCLUSIONS

KNNimpute is a fast, robust, and accurate method of estimating missing values for microarray data. Both KNNimpute and SVDimpute methods far surpass the currently accepted solutions (filling missing values with zeros or row average) by taking advantage of the structure of microarray data to estimate missing expression values.

We recommend KNNimpute over SVDimpute method for several reasons. First, the KNNimpute method is more robust than SVD to the type of data for which estimation is performed, performing better on non-time series or noisy data. Second, while both KNN and SVD methods are robust to increasing the fraction of missing data, KNN-based imputation shows less deterioration in performance with increasing percent of missing entries. And third, KNNimpute is less sensitive to the exact parameters used (number of nearest neighbors), whereas the SVD-based method shows sharp deterioration in performance when a non-optimal fraction of missing values is used. In addition, KNNimpute has the advantage of providing accurate estimation for missing values in genes that belong to small tight expression clusters. Such genes may not be similar to any of the eigengenes used for regression in SVDimpute, and their missing values could thus be estimated poorly by SVD-based estimation.

KNNimpute is a robust and sensitive approach to estimating missing data for gene expression microarrays. However, scientists should exercise caution when drawing critical biological conclusions from partially imputed data. The goal of this and other estimation methods is to provide an accurate way of estimating missing data points in order to minimize the bias introduced in the performance of microarray analysis methods. Estimated data should be flagged where possible, and their impact on the discovery of biological results should be assessed in order to avoid drawing unwarranted conclusions.

REFERENCES

Alizadeh A.A., Eisen M.B., Davis R.E., Ma C., Lossos I.S., Rosenwald A., Boldrick J.C., Sabet H., Tran T., Yu X., Powell J.I., Yang L., Marti G.E., Moore T., Hudson J., Jr., Lu L. Lewis D.B., Tibshirani R., Sherlock G., Chan W.C., Greiner T.C., Weisenburger D.D., Armitage J.O., Warnke R., Staudt L.M. et al. (2000). Distinct types of diffuse large B-cell lymphoma identified by gene expression profiling. Nature 403: 503-11.

Alter O., Brown P.O., Botstein D. (2000). Singular value decomposition for genome-wide expression data processing and modeling. Proc Natl Acad Sci USA 97: 10101-6.

Anderson T.W. (1984). An introduction to multivariate statistical analysis. Wiley, New York.

Bar-Joseph Z., Gerber G., Gifford D.K., Jaakkola T.S., Simon I. (2002). A new approach to analyzing gene expression time series data. Proceedings of the Sixth Annual International Conference on Computational Biology (RECOMB), Washingon DC, USA, ACM Press.

Brown M.P., Grundy W.N., Lin D., Cristianini N., Sugnet C.W., Furey T.S., Ares M., Jr., Haussler D. (2000). Knowledge-based analysis of microarray gene expression data by using support vector machines. Proc Natl Acad Sci USA 97: 262-7.

Butte A.J., Ye J. et al. (2001). "Determining Significant Fold Differences in Gene Expression Analysis." Pacific Symposium on Biocomputing 6: 6-17.

Chu S., DeRisi J., Eisen M., Mulholland J., Botstein D., Brown P.O., Herskowitz I. (1998) The transcriptional program of sporulation in budding yeast. Science 282: 699-705.

DeRisi J.L., Iyer V.R., Brown P.O. (1997). Exploring the metabolic and genetic control of gene expression on a genomic scale. Science 278: 680-6.

Eisen M.B., Spellman P.T., Brown P.O., Botstein D. (1998). Cluster analysis and display of genome-wide expression patterns. Proc Natl Acad of Sci USA 95: 14863-8.

Gasch, A. P., P. T. Spellman, C. M. Kao, O. Carmel-Harel, M. B. Eisen, G. Storz, D. Botstein, and P. O. Brown. 2000. Genomic expression programs in the response of yeast cells to environmental changes. Mol. Biol. Cell. in press.

Golub G.H., Van Loan C.F. (1996). Matrix computations. Johns Hopkins University Press, Baltimore.

Golub T.R., Slonim D.K., Tamayo P., Huard C., Gaasenbeek M., Mesirov J.P., Coller H., Loh M.L., Downing J.R., Caligiuri M.A., Bloomfield C.D., Lander E.S. (1999). Molecular classification of cancer: class discovery and class prediction by gene expression monitoring. Science 286: 531-7.

Hastie T., Tibshirani R., Eisen M., Alizadeh A., Levy R., Staudt L., Chan W.C., Botstein D., Brown P. (2000). "Gene shaving" as a method for identifying distinct sets of genes with similar expression patterns. Genome Biol 1: research0003.1-research0003.21.

Heyer L.J., Kruglyak S., Yooseph S. (1999). Exploring expression data: identification and analysis of coexpressed genes. Genome Res 9: 1106-15.

Little R.J.A., Rubin D.B. (1987). Statistical analysis with missing data. Wiley, New York.

Loh W., Vanichsetakul N. (1988). Tree-Structured Classification via generalized discriminant analysis. Journal of the American Statistical Association 83: 715-725.

Perou C.M., Sorlie T., Eisen M.B., van de Rijn M., Jeffrey S.S, Rees C.A., Pollack J.R., Ross D.T., Johnsen H., Akslen L.A., Fluge O., Pergamenschikov A., Williams C., Zhu S.X., Lonning P.E., Borresen-Dale A.L., Brown P.O., Botstein D. (2000). Molecular portraits of human breast tumours. Nature 406: 747-52.

Raychaudhuri S., Stuart J.M., Altman R.B. (2000). Principal components analysis to summarize microarray experiments: application to sporulation time series. Pacific Symposium on Biocomputing : 455-66.

Spellman P.T., Sherlock G., Zhang M.Q., Iyer V.R., Anders K., Eisen M.B., Brown P.O., Botstein D., Futcher B. (1998). Comprehensive identification of cell cycle-regulated genes of the yeast Saccharomyces cerevisiae by microarray hybridization. Molecular Biology of the Cell 9: 3273-97.

Tamayo P., Slonim D., Mesirov J., Zhu Q., Kitareewan S., Dmitrovsky E., Lander E.S., Golub T.R. (1999). Interpreting patterns of gene expression with self-organizing maps: methods and application to hematopoietic differentiation. Proc Natl Acad Sci USA 96: 2907-12.

Troyanskaya O., Cantor M., Sherlock G., Brown P., Hastie T., Tibshirani R., Botstein D., Altman R.B. (2001). Missing Value Estimation methods for DNA microarrays. Bioinformatics 17(6):520-5.

Wilkinson G.N. (1958). Estimation of missing values for the analysis of incomplete data. Biometrics 14: 257-286.

Yates Y. (1933). The analysis of replicated experiments when the field results are incomplete. Emp. J. Exp. Agric. 1: 129-142.

Chapter 4

NORMALIZATION

Concepts and Methods for Normalizing Microarray Data

Norman Morrison and David C. Hoyle.

University of Manchester, Department of Computer Science, Kilburn Building, Oxford Road, Manchester M13 9PL. U.K.
e-mail: morrison@cs.man.ac.uk , david.c.hoyle@man.ac.uk

1. INTRODUCTION

Microarray technology offers the ability to capture transcriptional information for thousands of genes at once. Furthermore, for an increasing number of organisms it is possible to monitor the activity of their entire transcriptome.

Due to the natural biological variability and the numerous procedures involved in a typical microarray experiment, microarray data is inherently noisy and high dimensional. Therefore, it is desirable to carry out the analysis of such data within a statistical framework. This chapter is concerned with statistical techniques for normalization of microarray data – techniques which aim to allow the user to extract as much of the biological signal from a microarray experiment as possible. The normalized data often acts as input into further stages of analysis – so as clean a signal as possible is obviously desirable. Alternatively, the normalized data may be used for comparison with normalized data from another microarray experiment – in which case removal of signal from possibly confounding non-biological sources is important.

Our treatment here will, given the limited space, not be exhaustive. Rather, we attempt to give a feel for some of the current major themes within normalization, to point the reader towards the relevant research literature and guide them to the appropriate normalization algorithms.

1.1 Why Normalization?

Microarray experiments are complicated tasks involving a number of stages, most of which have the potential to introduce error. These errors can mask the biological signal we are interested in studying. A component of this error may be systematic, i.e. bias is present, and it is this component that we attempt to remove in order to gain as much insight as possible into the underlying biology in a microarray experiment. This can be achieved by using a number of well-established statistical techniques and is the aim of normalization.

Considering some of the stages of the experimental process a few potential sources of systematic error include:

- *Sample preparation* – the processes of mRNA extraction, reverse transcription, cDNA amplification and labeling efficiencies can all dramatically affect the sample.

- *Variability in hybridization* – conditions such as temperature, hybridization buffers, batch effects, uneven hybridization and DNA quantity on the microarrays can all introduce systematic error.

- *Spatial effects* – pin geometry and print tip problems can play a significant role in spotted microarrays.

- *Scanner settings* – parameters are subject to change from hybridization to hybridization and can introduce bias.

- *Experimenter Bias* – hybridizations carried out by the same experimenter often cluster together. Indeed experimenter bias has been reported as one of the largest sources of systematic variation (Ball 2002).

Since systematic error can occur at numerous stages within the experimental process it is important to define more precisely where in the analysis process normalization lies, and what we consider the starting form of the data to be.

1.2 What Normalization Is and What It Isn't

The boundary between pre-processing and normalization of raw data is not always clear-cut. Indeed, the definition of what is raw data can be quite different between one practitioner and the next and is often a topic that sparks heated debate. Some consider the only true source of raw data to be the image file generated by the scanning software. To that end, there are an increasing number of methods available for image processing and data acquisition (Chen et al., 1997; Schadt et al., 2001; Yang et al., 2001a). These methods can produce large tables of data per array, containing a number of columns of information such as: Spot intensity, localized background intensities, standard deviations, error estimates, etc. For example the output from GenePix scanning software (Axon Instruments Inc 2002) can contain over 40 columns of data for a single array. Some regard the calculation of these quantities as forming part of the normalization procedure. However, for the purposes of this chapter we consider normalization to be the treatment of data that has gone through some form of feature extraction process from an array image in order to compute a background subtracted intensity value for each feature. We will present an overview of normalization methods that deal with the treatment of such 'pre-processed' data. Figure 4.1 shows a schematic representation of where we regard 'normalization' to fall within the microarray experiment process.

The aim of normalization is, as we have already said, to remove as much of the systematic error as possible. It is not a panacea for poor data. Errors in microarray data consist of both systematic errors, i.e. bias, and random errors that are the result of uncontrollable variation. Although we can attempt to account for the systematic error, no amount of post-processing of the data will be able to remove random errors, which may be large as a result of poor experimentation.

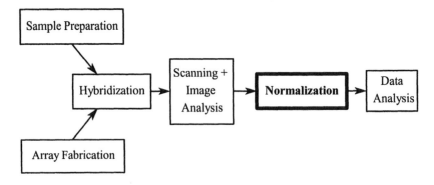

Figure 4.1. Typically, normalization takes place after image analysis and prior to further manipulation of the data.

2.   NORMALIZATION METHODS

In this section we will illustrate a selection of the more popular normalization methods. We will discuss methods that can be applied to both oligo- and spotted- array data. To that end we have chosen to use the human acute leukemia Affymetrix data set of Golub et al. (1999), a spotted array chosen from the *S. cerevisiae* data set of Gasch et al. (2000), and dye-swap hybridization data from S. *cerevisiae* subjected to heat-shock that has been supplied to us by Dr. Abdulla Bashein.

Early approaches to normalization of microarray data have used House-keeping genes (Heller et al., 1997), i.e. genes whose expression level are expected to be approximately constant between different physiologies. This approach follows that widely used in Northern blots (Goldsworthy et al., 1993), where measurements are normalized to the level of genes such as Actin or GAPDH. However, it is known that the expression level of so-called house-keeping genes, such as Actin, can in fact vary significantly between different physiologies (Delbruck and Ernst 1993; Goldsworthy et al., 1993). Normalization using only one or two genes performs poorly and using a larger set of genes for normalization is more appropriate. This illustrates that the choice of normalization method is intricately linked to the chosen set of genes used to perform the normalization. Whilst larger subsets of genes can be chosen for normalization that perform better than the use of a few house-keeping genes, e.g. rank invariant sets (Schadt et al., 2001), spiked in controls (Hill et al., 2001), we will concentrate here on normalization methods that use all or the vast majority of spot values to perform the normalization.

Most practitioners work with log-transformed data (Kerr et al., 2000; Quackenbush, 2001; Yang et al., 2002). One reason for this is that the spread of values from up- and down-regulated genes is more symmetrical (Quackenbush, 2001). We use natural logarithms although logarithms of any base are proportional and any base will give identical results. Throughout this section we will use x to denote the logarithm of un-normalized spot intensities or ratios, and y to denote to the logarithm of normalized spot intensities or ratios. For two-fluor spotted arrays we will also use R (red) and G (green) to denote the background corrected spot intensities from the two fluors.

2.1   Total Intensity Methods

The simplest normalization method corresponds to adding a constant to all the x values, i.e. a simple linear transformation,

$$y = x + b \qquad (4.1)$$

Such a transformation is often termed a *global* normalization method since it is applied globally, i.e. each x value is transformed in the same way, independent of log intensity. The constant b allows us to adjust the mean, $\bar{y}$, of the normalized data. We choose to set the mean of the normalized data from each chip to the same value. Commonly the constant b is set to $-\bar{x}$, so that $\bar{y} = 0$. The transformation in Equation 4.1 has a very intuitive origin. When dealing with comparable samples of extracted mRNA adjusting the mean of the log-transformed data allows us to correct for different image scanning settings and total amounts of hybridized mRNA, which have a multiplicative effect on the spot intensity values and therefore an additive effect on the log- transformed values (Chen et al., 1997).

Setting the mean of the normalized data to 0 is equivalent to centering the distribution of the normalized data over 0. Therefore, we tend to call this method mean centering. Sometimes the sample mean can provide an inaccurate estimate of the true center of a distribution due to the presence of large outliers in the sample data. In such cases a more robust estimate of the true distribution center is required. Such an estimate is provided by the median of the sample data (Huber, 1981). Thus a related normalization method is that of median centering. Again the normalization transformation takes the form of Equation 4.1 but with $b = -x_{50}$, the median value of the un-normalized data, i.e. the 50th centile.

After normalizing several arrays by mean or median centering we may wish to bring the normalized data sets into a common scale (spread of values) in order to prevent any single hybridization dominating when data from several arrays are combined, e.g. when averaging log ratio values over replicate hybridizations, or when performing clustering. Overall this two step process corresponds to a linear-affine transformation,

$$y = ax + b \qquad (4.2)$$

Obviously, if $a = 1$ this corresponds to mean or median centering of the data. If $a \neq 1$ this corresponds to changing the scale of the data. It is worth noting that some practitioners believe this to be unprincipled in certain biological contexts (Sherlock, 2001). The aim of the transformation in Equation 4.2 is to bring the data from each chip into a common location and scale. Again the parameter b is chosen so that the mean or median of the normalized data is 0. The choice of scale is unimportant so long as we are consistent and thus we choose a so that the normalized data has a variance of 1. The parameters a and b are easily estimated from the mean, $\bar{x}$, and standard deviation, σ_x, of the un-normalized data,

$$a = 1/\sigma_x, \quad b = -a\bar{x} \qquad (4.3)$$

In practice we use a trimmed mean and standard deviation of the un-normalized data, e.g., when calculating the trimmed mean and standard deviation we ignore data that is greater than ±3 standard deviations from the mean of the un-normalized data in order to avoid bias introduced by outliers.

The normalization methods, represented by the transformation in Equation 4.2, can be applied to both spotted and oligo-based data. They are – due to their mathematical simplicity – extremely quick and easy to implement. Speed and computational performance are not usually issues for the user with such normalization methods. They can even be easily included in simple spreadsheet analysis of the microarray data. Most microarray analysis software packages will offer some form of these simple transformations.

2.2 Regression Methods

If the majority of genes are not considered to be differentially expressed when comparing two microarray signals – either R and G intensity values from a two-fluor spotted array, or intensity values from two separate oligo-based hybridizations – then a scatter plot of one signal against the other is expected to produce a cluster along a straight line. If systematic errors are small, then we might expect the line to have slope 1 and intercept 0 since the majority of genes are assumed to have similar expression levels in the two signals. Regression based normalization methods then attempt to fit a curve through the scatter plot and adjust the intensities. Regression normalization techniques fall into two general categories 1) Linear fits, which then apply the same global normalization transformation to all spots irrespective of their log intensity value and hence are (log) intensity independent methods, 2) Non-linear fits which are usually applied to a rotated form of the scatter plot and are log intensity dependent, i.e. log spot intensities are adjusted according to their value.

2.2.1 Intensity Independent Methods (Linear)

Here a straight line is fitted through the scatter plot of one signal, x, against the other which is considered as a reference signal, x_{ref}. This can be done for both spotted and oligo-based data. The values x are adjusted so that a scatter plot of normalized data y against x_{ref} has the desired slope of 1 and zero intercept. The transformation takes the same form as Equation 4.2 but with the parameters a and b determined from standard linear regression formulae,

$$a = \frac{\sigma^2_{ref}}{Cov(x, x_{ref})}, \quad b = \overline{x}^{ref} - a\overline{x} \qquad (4.4)$$

Here $\overline{x}$ and $\overline{x}_{ref}$ denote means of the un-normalized and reference data respectively. σ^2_{ref} is the variance of x_{ref} and $Cov(x, x_{ref})$ its covariance with x. The validity of such an approach can be checked simply by doing a scatter plot of expression data x against the reference expression data x_{ref}. From the data set of Golub et al., (1999) we have chosen, at random, two hybridizations and plotted in Figure 4.2a the log of the PM-MM[1] value from one chip against the log of the PM-MM value from the other. Any negative or missing values of PM-MM are set to a threshold value of 20. The correlations between the log expression values is clear from Figure 4.2a, as are the two lines at $\ln(20) \approx 3.0$, reflecting the effect of the thresholding. Figure 4.2b shows a scatter plot of $\log(R)$ against $\log(G)$ for the yeast data set of Gasch et al., (2000).

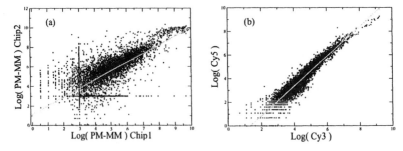

Figure 4.2. Figure a shows a scatter plot of Log PM-MM for 2 hybridizations from the data set of Golub et al. (1999). The solid grey line shows the linear fit. Figure b shows a scatter plot of Log Cy5 against Log Cy3. Here only positive values of intensity have been used. Again the solid grey line shows the linear fit. In both cases the linear fits have non-zero intercept and slope $\neq 1$.

Standard linear regression onto a reference chip is somewhat unprincipled since this assumes that only one experiment is noisy (the measurement) whilst the reference chip is not. In practice all experiments will have comparable noise levels. The use of standard linear regression therefore leads to an asymmetrical method in which the result of normalization is not equivalent for different choices of reference chip. Given the concerns with standard regression we have introduced a more principled least squares method Rattray et al. (2001) which is one of a class of *total least squares* methods (Golub and Van Loan, 1979), which are appropriate for problems in which there is noise in all measurement variables. This approach uses the data from multiple chips in a more robust fashion to estimate the parameters *a* and *b* for each chip.

---

[1] Early data from Affymetrix systems provided two signal intensities, Perfect Match (PM) and Mismatch (MM) values, that quantify probes which are complimentary to the sequence of interest and those that differ by a single homomeric base mismatch.

2.2.2   Intensity Dependent Methods (Non-Linear)

2.2.2.1 Single Slide Normalization

The global linear normalization methods discussed above apply the same transformation to the data, irrespective of the actual log intensity value. They assume there are no effects that are dependent on the measured log intensity or spatial location of the spot. For two-fluor spotted arrays the presence of any intensity dependent patterns in the log-ratio data can quickly be ascertained through the use of 'M-A' plots (Yang et al., 2002). An example M-A plot is given below in Figure 4.3a for the yeast data set of Gasch et al. (2000). For each spot M is the log-ratio value, i.e. $M = \log(R/G)$, and A is the average log intensity value across the two channels, i.e. $A = \frac{1}{2}(\log(R) + \log(G))$. For a given A value, any value of M can be obtained. If no systematic log intensity dependent error were present and the majority of genes were not differentially expressed, then we would expect equal scatter above and below the global (chip wide) average value of M. We would expect an M-A plot qualitatively similar to that in Figure 4.3b.

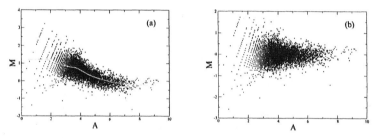

Figure 4.3. M-A plots for a hybridization from the dataset of Gasch et al. (2000). Figure a shows un-normalized data. Figure b shows Lowess (see below) normalized data.

For the yeast data set we can see from Figure 4.3a that this is clearly not the case. The solid grey line shown in Figure 4.3a represents the behaviour of a 'local average' of M. From the solid curve there is a systematic increase in M with decreasing intensity. A global transformation of the data is obviously not appropriate here – the linear transformations of the previous section merely shift and re-scale all the M values so that the M-A plot is still qualitatively the same as that in Figure 4.3a. How does one correct for such systematic variation? After correction of the log-ratios we would obviously like to see an M-A plot closer to that shown in Figure 4.3b. From Figure 4.3a one can see that if one subtracts from each log-ratio value the corresponding 'local average' value represented by the solid grey line, then the scatter of corrected log-ratios will be closer to that in Figure 4.3b Indeed that is precisely how the corrected log-ratios in Figure 4.3b were obtained. The question is, how does one determine the solid line representing the 'local average' of M? One of the most popular methods is *Lowess* (Locally Weighted Regression) also referred to as *Loess*, which is a generic statistical technique for smoothing of data (Cleveland and Devlin, 1988; Hastie et al., 2001) and has been applied to microarray data by Dudoit et al. (2002) and Yang et al. (2002). Typically, Lowess fits a separate straight line through the data at every value of A. The slope, $\alpha(A_0)$, and intercept, $\beta(A_0)$, for a particular value of interest, A_0 say, are determined predominantly by those data points with A values close to A_0. This is done using a window (or *kernel*) function, $K(A - A_0)$, whose width λ is set so that only a fixed percentage f of the total data points are used in determining the slope and intercept at A_0. A popular choice of window function is the tri-cube function,

$$K(A - A_0) = \left(1 - \left| \frac{A - A_0}{\lambda} \right|^3 \right)^3 , \quad |A - A_0| \le \lambda$$

$$K(A - A_0) = 0, |A - A_0| > \lambda$$

(4.5)

which we have used for the yeast data set. Typically the *span f* is set somewhere between 20% and 40% of the total data. The larger f, the more data is used to determine the straight line passing through A_0. If f is set too large, e.g., f close to 100%, the 'local average' produced by Lowess will be something close to a single straight line through all the data and consequently not follow the local variation or *curvature* of the M-A plot. If f is too small Lowess will follow the local variation of the M-A plot too closely and the 'local average' produced may be too wildly varying and affected by outliers in the M-A plot. For the example in Figure 4.3a we have used $f = 30\%$. The 'local average' at A_0 is then simply the value on the straight line at A_0, i.e. $\alpha(A_0)A_0 + \beta(A_0)$. For more specific details on the application of Lowess we refer the reader to Yang et al. (2002).

Lowess smoothing of the M-A plot is a very general technique and one can use the approach to correct for systematic variation introduced through sources other than dye bias, for example bias due to the particular print-tip with which the probe was printed onto the array, by making an M-A plot using those spots that have been printed using that particular print-tip. Such an approach has been investigated by Yang et al. (2002), to which we refer the reader for further details. After Lowess smoothing of M-A plots from several print-tips or several chips the user may still wish to bring the Lowess corrected ratios from the different data sets into a common scale. This can be done using the techniques outlined in the previous section, such as the total least-squares normalization method, or using the *median absolute deviation* (MAD) estimates of scale employed by Yang et al. (2002). For Lowess corrected ratios from different print-tips, but a single hybridization, this would seem a very natural step.

When should one use Lowess smoothing of an M-A plot? Simply plotting the M-A values will give a good indication of whether a log intensity dependent correction of the ratio is required. This may be impractical when normalizing many chips and the user may wish to implement Lowess smoothing of M-A plots by default. The concern would then be the computational cost in doing this. In theory a locally weighted regression is performed at every value of A for which a corrected log-ratio is required, i.e. for every spot on the array. In theory this makes normalization using Lowess smoothing of an M-A plot considerably more computationally intensive than say mean or median centering of the log-ratio values. In practical terms, on modern desktop machines, there is not a dramatic difference in performance when normalizing a single chip. Lowess smoothing of the M-A plot shown in Figure 4.3a took 35 seconds on a 700MHZ Pentium III laptop using code written in Java. When performing normalization of many chips and/or print-tips, or normalization across multiple chips, speed maybe more of an issue. In such cases a locally weighted regression need not be done for every spot. Instead the observed range of A can be divided into a grid containing a finite number of points and a locally weighted regression performed at each of these grid points. Values of the smooth curve for values of A which do not lie at a grid point can be calculated by simple interpolation. Smoothing of the M-A plot in Figure 4.3a using a grid of 100 points took less than a second. If the same grid is used when normalizing many arrays a significant proportion of the calculation, the *equivalent kernel*, (Hastie et al., 2001) need only be done once, greatly improving the overall computational cost.

2.2.2.2 Paired-Slide Normalization

If we expect a significant number of genes to be differentially expressed, then correcting for dye-bias by smoothing of the M-A plot may not be appropriate, since we may not expect symmetric scatter of M values. In these circumstances elimination of dye-bias can be done with the use of paired slides, i.e. a dye-swap experiment in which two hybridizations are performed, with the labelling of sample and reference mRNA populations in one hybridization being the reverse of that in the other.

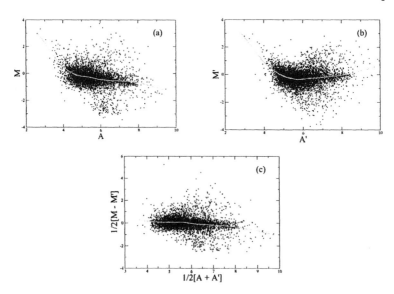

Figure 4.4. M-A plots for the yeast heat-shock dye-swap hybridizations. Figures a and b show un-normalized data from the two dye-swap hybridizations. Figure c shows the normalized ratios. In each case the solid grey line represents the 'local average' behaviour. In both Figures a and b the systematic bias can be seen to increase at low intensities, whilst in Figure c no significant curvature of the M-A plot is apparent, indicating that the systematic bias has been removed.

$M(A_0)$ is the log-ratio value, at log intensity A_0, from one hybridization, and $M'(A'_0)$ the log-ratio value, at A'_0, from the reverse labelled hybridization for the same probe. If the experiment were free from error, then we would expect $A_0 \approx A'_0$ and $M'(A'_0) \approx -M(A_0)$, due to the reverse labelling. Therefore $\frac{1}{2}[M(A_0) - M'(A'_0)]$ would provide us with a good estimate of the true log-ratio at A_0. Unfortunately, as we have seen from the previous M-A plots, bias will be present in the log ratios $M(A_0)$ and $M'(A'_0)$. However, from the M-A plots we have seen the bias would appear to depend only on log intensity A and therefore is approximately the same for A_0 and A'_0 (recall $A_0 \approx A'_0$), i.e. is irrespective of which way round we have labeled the mRNA populations. When calculating $\frac{1}{2}[M(A_0) - M'(A'_0)]$ the systematic intensity errors will approximately cancel out, even when we are considering genes that are differentially expressed, and so $\frac{1}{2}[M(A_0) - M'(A'_0)]$ still represents a good estimate of the true log ratio at $\frac{1}{2}[A_0 + A'_0]$ (Yang et al., 2001b). This is easily demonstrated using the yeast heat-shock dye-swap data. Figures 4.4a and 4.4b are M-A plots for the two individual dye-swap hybridizations. Plotted in Figure 4.4c is $\frac{1}{2}[M(A_0) - M'(A'_0)]$ against $\frac{1}{2}[A_0 + A'_0]$. From Figure 4.4c we can see that any log intensity dependent bias present in $M(A_0)$ and $M'(A'_0)$ is absent from $\frac{1}{2}[M(A_0) - M'(A'_0)]$.

2.3   Ratio Statistics Based Normalization

Often the purpose of normalization is to allow the robust identification of differentially expressed genes. We may approach this, say, by performing a Lowess normalization of our two-fluor data and examining the resultant normalized ratios to identify the largest fold changes. An alternative approach is to ask if the observed fold change in gene k, estimated (for two-fluor data) by the ratio $T_k = R_k/G_k$, is genuine or the result of random fluctuations in the signal intensities R_k and G_k for an otherwise non-differentially expressed gene. In other words perform a significance test under a hypothesis that on average gene k is not differentially expressed between two labelled populations of mRNA. To do this requires knowledge of the statistical properties of the quantity T_k, which is a ratio of two numbers – hence the name Ratio Statistics based normalization. This has been done by Chen et al. (1997), who assume a constant *coefficient of variation*[2], *c*. Chen et al. (1997) provide a formula to calculate the critical value of T_k – the value above which the observed ratio T_k is considered to genuinely represent differential expression – for a given level of confidence (*p*-value). The formula is dependent upon the model parameters, e.g. *c*, which are determined through an iterative process using the observed ratios from a set of house-keeping genes. We refer the reader to Chen et al. (1997) for more specific details.

Within the context of Ratio Statistics based normalization it is worth mentioning the variance stabilization method of Huber et al. (2002). The method is applied to un-logged spot intensities, *I*, from both oligo- and spotted- arrays. Huber et al. (2002) consider a non-constant coefficient of variation, from which they construct a transformation $h(I) = \operatorname{arsinh}(a + bI)$ so that the difference, Δh, between two signals has a constant variance over the whole intensity range – in contrast to log-ratio values even after Lowess normalization. The parameters a and b are estimated through an iterative *Maximum Likelihood* procedure, and so this method is more computationally intensive compared to simpler normalization methods. However, Huber et al. (2002) report that a statistical test, based on Δh, to identify differentially expressed genes performs better than most other normalization methods, including Lowess. The transformation $h(I)$ also has the advantage that it can be applied to negative values of *I*, as might arise after background correction.

---

[2] The ratio of the standard deviation to the mean for a random variable.

2.4   Summary

The simple global linear transformations we started this section with can be used to correct for systematic errors that affect all spot intensities equally, e.g. differences in total mRNA between samples. However, the use of M-A plots has revealed significant log intensity dependant biases and therefore normalization using <u>only</u> a global transformation is likely to be inadequate. The use of Lowess to correct log intensity dependent bias is fast becoming a *de facto* standard for two-fluor spotted array data. We would recommend that users of spotted arrays should implement Lowess or another statistical smoothing technique. After correcting bias in several hybridizations these can be brought into a common scale using the global linear transformations discussed at the beginning of this section.

For non-spotted arrays, such as Affymetrix chips, correction of log intensity dependant bias can in principle be done by multi-dimensional non-linear regression or non-linear regression to a reference chip.

So far we have not touched on the issue of spatial normalization. We have mentioned applying Lowess to spots just from the same print-tip, but this does not address the issue of bias that is dependent upon the actual physical location (planar coordinates) of the spot on the array. Such bias can arise due to uneven hybridization conditions across the array or contamination in a particular sector of the array. Plotting of log ratio values (from two-fluor spotted array data) against the 2D planar coordinates of the spots can reveal any biases. The local spatial behaviour of spot intensities, as a function of the spot coordinates, can be modelled using locally weighted regression in the 2D spot coordinates. Each regression calculation now involves two input variables – the spot coordinates – as opposed to one – the log intensity A – when performing standard Lowess normalization of an array. Consequently the computational cost is much higher, although for normalization of multiple arrays this can again be reduced if the regression calculations are always performed at the same spot locations – as might be appropriate when using arrays of similar design.

3.   CONCLUSIONS

Normalization of microarray data is a complex issue and is unlikely to ever be cut and dried. Common sense and the user's biological knowledge of the experiment can act as invaluable guides in the normalization process. As a minimum working practice we would recommend the following:

– **Always** keep the raw data – normalization methods change. They can be improved or new ones invented. New sources of systematic error may be identified and algorithms constructed to eliminate them.

– Maintain some sort of audit trail of what transformations (with what particular parameter settings) have been applied to your data.

Finally, as well as the research literature, one of the best places to keep abreast of developments in normalization techniques is by monitoring the efforts of the MGED normalization working group (http://www.mged.org).

ACKNOWLEDGEMENTS

We would like to thank Dr Abdulla Bashein for supplying the S. *cerevisiae* heat shock dye-swap data. We have also benefited from discussions with Prof. A. Brass, Dr. M. Rattray and Dr. Y. Fang. This work has been supported by the MRC (UK) and NERC (UK).

REFERENCES

Axon Instruments Inc. (2002). GenePix Software. Available at http://www.axon.com/GN_Genomics.html#software.

Ball C.(2002). Systematic Bias in Microarray Data, presentation given at MGED IV, 13th-16th Feb, Boston, USA. Available at http://www.dnachip.org/mged/normalization.html.

Chen Y., Dougherty E.R., Bittner M.L. (1997). Ratio-based decisions and the quantitative analysis of cDNA microarray images. J. Biomed. Optics 2: 364-374.

Cleveland W.S., Devlin S.J. (1988). Locally weighted regression: an approach to regression analysis by local fitting. J. Am. Stat. Assoc. 83: 596-610.

Delbruck S., Ernst J.F. (1993). Morphogenesis-independent regulation of actin transcript levels in the pathogenic yeast Candida albicans. Mol Microbiol 10: 859-866.

Dudoit S., Yang Y.H., Callow M.J., Speed T.P. (2002). Statistical methods for identifying genes with differential expression in replicated cDNA microarray experiments. *Statistica Sinica* 12: 111-139. See also Technical Report #578. Available at http://stat-www.berkeley.edu/users/sandrine/publications.html.

Gasch A.P., Spellman P.T., Kao C.M., Carmel-Harel O., Eisen M.B., Storz G.,. Botstein D, Brown P.O. (2000). Genomic expression programs in the response of yeast cells to environmental changes. Mol Biol Cell 11: 4241-4257.

Goldsworthy S.M., Goldsworthy T.L., Sprankle C.S., Butterworth B.E. (1993). Variation in expression of genes used for normalization of Northern blots after induction of cell proliferation. Cell Prolif 26: 511-518.

Golub G.H., Van Loan C. (1979). Total Least Squares. In *Smoothing Techniques for Curve Estimation*, pp. 69-76. Springer-Verlag, Heidelberg.

Golub T.R., Slonim D.K., Tamayo P., Huard C., Gaasenbeek M., Mesirov J.P., Coller H., Loh M.L., Downing J.R., Caligiuri M.A., Bloomfield C.D., Lander E.S. (1999). Molecular classification of cancer: class discovery and class prediction by gene expression monitoring. Science 286: 531-537.

Hastie T., Tibishirani R., Friedman J. (2001). *The Elements of Statistical Learning: Data Mining, Inference and Prediction.* Springer, New York.

Heller R.A., Schena M., Chai A., Shalon D., Bedilion T., Gilmore J., Woolley D.E., Davis R.W. (1997). Discovery and analysis of inflammatory disease-related genes using cDNA microarrays. Proc Natl Acad Sci USA 94: 2150-2155.

Hill A.A., Brown E.L., Whitley M.Z., Tucker-Kellogg G., Hunter C.P., Slonim D.K. (2001). Evaluation of normalization procedures for oligonucleotide array data based on spiked cRNA controls. Genome Biol 2: RESEARCH0055.0051-0055.0013.

Huber P.J. (1981). *Robust Statistics*. Wiley, New York.

Huber W., Von Heydebreck A., Sultmann H., Poustka A., Vingron M. (2002). Variance stabilization applied to microarray data calibration and to the quantification of differential expression. *Bioinformatics* 1(1):1-9.

Kerr M.K., Martin M., Churchill G.A. (2000). Analysis of variance for gene expression microarray data. J Comput Biol 7: 819-837.

Quackenbush J. (2001). Computational analysis of microarray data. Nat Rev Genet 2: 418-427.

Rattray, M., N. Morrison, D.C. Hoyle, and A. Brass. 2001. DNA microarray normalization, PCA and a related latent variable model.: Technical Report. Available from http://www.cs.man.ac.uk/~magnus/magnus.html.

Schadt E.E., Li C., Ellis B., Wong W.H. (2001). Feature Extraction and Normalization Algorithms for High-Density Oligonucleotide Gene Expression Array Data. Journal of Cellular Biochemistry 37: 120-125.

Sherlock, G. 2001. Analysis of large-scale gene expression data. Brief Bioinform 2: 350-362.

Yang Y.H., Buckley M.J., Speed T.P. (2001a). Analysis of cDNA microarray images. Brief Bioinform 2: 341-349.

Yang Y.H., Dudoit S., Luu P., Lin D.M., Peng V., Ngai J., Speed T.P. (2002). Normalization for cDNA microarray data: a robust composite method addressing single and multiple slide systematic variation. Nucleic Acids Research 30: e15.

Yang Y.H., Dudoit S., Luu P., Speed T.P. (2001b). Normalization for cDNA microarray data. In *Microarrays: Optical technologies and informatics*. (eds. Bittner M.L., Chen Y., Dorsel A.N., and Dougherty E.R.), pp. See also Technical Report available at http://www.stat.berkeley.edu/users/terry/zarray/Html/papersindex.html. SPIE, Society for Optical Engineering, San Jose, CA.

Chapter 5

SINGULAR VALUE DECOMPOSITION AND PRINCIPAL COMPONENT ANALYSIS

Michael E. Wall[1,2], Andreas Rechtsteiner[1,3], Luis M. Rocha[1]

[1]*Computer and Computational Sciences Division,* [2]*Bioscience Division, Los Alamos National Laboratory, Mail Stop B256, Los Alamos, New Mexico, 87545 USA,*
e-mail: {mewall, rocha}@lanl.gov

[3]*Systems Science Ph.D. Program, Portland State University, Post Office Box 751, Portland, Oregon 97207 USA,*
e-mail: andreas@sysc.pdx.edu

1. INTRODUCTION

One of the challenges of bioinformatics is to develop effective ways to analyze global gene expression data. A rigorous approach to gene expression analysis must involve an up-front characterization of the structure of the data. Singular value decomposition (SVD) and principal component analysis (PCA) can be valuable tools in obtaining such a characterization. SVD and PCA are common techniques for analysis of multivariate data. A single *microarray*[1] experiment can generate measurements for tens of thousands of genes. Present experiments typically consist of less than ten assays, but can consist of hundreds (Hughes et al., 2000). Gene expression data are currently rather noisy, and SVD can detect and extract small signals from noisy data.

The goal of this chapter is to provide precise explanations of the use of SVD and PCA for gene expression analysis, illustrating methods using simple examples. We describe SVD methods for visualization of gene expression data, representation of the data using a smaller number of variables, and detection of patterns in noisy gene expression data. In addition, we describe the mathematical relation between SVD analysis and Principal Component Analysis (PCA) when PCA is calculated using the

---

[1] For simplicity, we use the term *microarray* to refer to all varieties of global gene expression technologies.

covariance matrix, enabling our descriptions to apply equally well to either method. Our aims are 1) to provide descriptions and examples of the application of SVD methods and interpretation of their results; 2) to establish a foundation for understanding previous applications of SVD to gene expression analysis; and 3) to provide interpretations and references to related work that may inspire new advances.

In Section 1, the SVD is defined, with associations to other methods described. In Section 2, we discuss applications of SVD to gene expression analysis, including specific methods for SVD-based visualization of gene expression data, and use of SVD in detection of weak expression patterns. Our discussion in Section 3 gives some general advice on the use of SVD analysis on gene expression data, and includes references to specific published SVD-based methods for gene expression analysis. Finally, in Section 4, we provide information on some available resources and further reading.

1.1   Mathematical definition of the SVD[2]

Let X denote an $m \times n$ matrix of real-valued data and *rank*[3] r, where without loss of generality $m \geq n$, and therefore $r \leq n$. In the case of microarray data, x_{ij} is the expression level of the ith gene in the jth assay. The elements of the ith row of X form the n-dimensional vector $\mathbf{g}_i$, which we refer to as the *transcriptional response* of the ith gene. Alternatively, the elements of the jth column of X form the m-dimensional vector $\mathbf{a}_j$, which we refer to as the *expression profile* of the jth assay.

The equation for singular value decomposition of X is the following:

$$X = USV^{\mathrm{T}} \tag{5.1}$$

where U is an $m \times n$ matrix, S is an $n \times n$ diagonal matrix, and V^{T} is also an $n \times n$ matrix. The columns of U are called the *left singular vectors*, $\{\mathbf{u}_k\}$, and form an orthonormal basis for the assay expression profiles, so that $\mathbf{u}_i \cdot \mathbf{u}_j = 1$ for $i = j$, and $\mathbf{u}_i \cdot \mathbf{u}_j = 0$ otherwise. The rows of V^{T} contain the elements of the *right singular vectors*, $\{\mathbf{v}_k\}$, and form an orthonormal basis for the gene transcriptional responses. The elements of S are only nonzero on the diagonal, and are called the *singular values*. Thus, $S = \mathrm{diag}(s_1,...,s_n)$. Furthermore, $s_k > 0$ for $1 \leq k \leq r$, and $s_i = 0$ for $(r + 1) \leq k \leq n$.

---

[2] Complete understanding of the material in this chapter requires a basic understanding of linear algebra. We find mathematical definitions to be the only antidote to the many confusions that can arise in discussion of SVD and PCA.

[3] The *rank* of a matrix is the number of linearly independent rows or columns.

By convention, the ordering of the singular vectors is determined by high-to-low sorting of singular values, with the highest singular value in the upper left index of the S matrix. Note that for a square, symmetric matrix X, singular value decomposition is equivalent to diagonalization, or solution of the eigenvalue problem for X.

One important result of the SVD of X is that

$$X^{(l)} = \sum_{k=1}^{l} \mathbf{u}_k s_k \mathbf{v}_k^T \qquad (5.2)$$

is the closest rank-l matrix to X. The term "closest" means that $X^{(l)}$ minimizes the sum of the squares of the difference of the elements of X and $X^{(l)}$, $\sum_{ij}|x_{ij} - x^{(l)}_{ij}|^2$. This result can be used in image processing for compression and noise reduction, a very common application of SVD. By setting the small singular values to zero, we can obtain matrix approximations whose rank equals the number of remaining singular values. Each term $\mathbf{u}_k s_k \mathbf{v}_k^T$ is called a *principal image*. Very good approximations can often be obtained using only a small number of terms (Richards, 1993). SVD is applied in similar ways to signal processing problems (Deprettere, 1988) and information retrieval (Berry et al., 1995).

One way to calculate the SVD is to first calculate V^T and S by diagonalizing $X^T X = V S^2 V^T$, and then to calculate $U = XVS^{-1}$. The $(r + 1),...,n$ columns of V for which $s_k = 0$ are ignored in the matrix multiplications. Choices for the remaining $n - r$ singular vectors in V or U may be calculated using the Gram-Schmidt orthogonalization process or some other extension method. In practice there are several methods for calculating the SVD that are of higher accuracy and speed. Section 4 lists some references on the mathematics and computation of SVD.

Relation to principal component analysis. There is a direct relation between PCA and SVD in the case where principal components are calculated from the *covariance matrix*[4]. If one conditions the data matrix X by *centering*[5] each column, then $X^T X = \sum_i \mathbf{g}_i \mathbf{g}_i^T$ is proportional to the covariance matrix of the variables of $\mathbf{g}_i$ (i.e. the covariance matrix of the assays[6]). Diagonalization of $X^T X$ yields V^T (see above), which also yields the principal components of $\{\mathbf{g}_i\}$. So, the right singular vectors $\{\mathbf{v}_k\}$ are the same as the principal components of $\{\mathbf{g}_i\}$. The eigenvalues of $X^T X$ are equivalent to s_k^2, which are proportional to the variances of the principal components.

---

[4] $C(x,y) = (N-1)^{-1} \sum_i (x_i - \bar{x})(y_i - \bar{y})$ is the *covariance* between variables x and y, where N is the # of observations, and $i = 1,..., N$. Elements of the *covariance matrix* for a set of variables $\{z^{(k)}\}$ are given by $c_{ij} = C(z^{(i)}, z^{(j)})$.

[5] A *centered* vector is one with zero mean value for the elements.

[6] Note that $(X^T X)_{ij} = \mathbf{a}_i \cdot \mathbf{a}_j$

The matrix US then contains the *principal component scores*, which are the coordinates of the genes in the space of principal components.

If instead each row of X is centered, $XX^T = \Sigma_j \mathbf{a}_j \mathbf{a}_j^T$ is proportional to the covariance matrix of the variables of $\mathbf{a}_j$ (i.e. the covariance matrix of the genes[7]). In this case, the left singular vectors $\{\mathbf{u}_k\}$ are the same as the principal components of $\{\mathbf{a}_j\}$. The s_k^2 are again proportional to the variances of the principal components. The matrix SV^T again contains the principal component scores, which are the coordinates of the assays in the space of principal components.

2.   SVD ANALYSIS OF GENE EXPRESSION DATA

As we mention in the introduction, gene expression data are well suited to analysis using SVD/PCA. In this section we provide examples of SVD-based analysis methods as applied to gene expression analysis. Before illustrating specific techniques, we will discuss ways of interpreting the SVD in the context of gene expression data. This interpretation and the accompanying nomenclature will serve as a foundation for understanding the methods described later.

A natural question for a biologist to ask is: "What is the biological significance of the SVD?" There is, of course, no general answer to this question, as it depends on the specific application. We can, however, consider classes of experiments and provide them as a guide for individual cases. For this purpose we define two broad classes of applications under which most studies will fall: *systems biology applications*, and *diagnostic applications* (see below). In both cases, the n columns of the gene expression data matrix X correspond to assays, and the m rows correspond to the genes. The SVD of X produces two orthonormal bases, one defined by right singular vectors and the other by left singular vectors. Referring to the definitions in Section 1.1, the right singular vectors span the space of the gene transcriptional responses $\{\mathbf{g}_i\}$ and the left singular vectors span the space of the assay expression profiles $\{\mathbf{a}_j\}$. We refer to the left singular vectors $\{\mathbf{u}_k\}$ as *eigenassays* and to the right singular vectors $\{\mathbf{v}_k\}$ as *eigengenes*[8]. We sometimes refer to an eigengene or eigenassay generically as a singular vector, or, by analogy with PCA, as a *component*. Eigengenes, eigenassays and other definitions and nomenclature in this section are depicted in Figure 5.1.

---

[7] Note that $(XX^T)_{ij} = \mathbf{g}_i \cdot \mathbf{g}_j$

[8] This notation is similar to that used in (Alter et al., 2000), save that we use the term *eigenassay* instead of *eigenarray*.

In systems biology applications, we generally wish to understand relations among genes. The signal of interest in this case is the gene transcriptional response $\mathbf{g}_i$. By Equation 5.1, the SVD equation for $\mathbf{g}_i$ is

$$\mathbf{g}_i = \sum_{k=1}^{r} u_{ik} s_k \mathbf{v}_k, \quad i : 1, ..., m \tag{5.3}$$

which is a linear combination of the eigengenes $\{\mathbf{v}_k\}$. The ith row of U, $\mathbf{g}'_i$ (see Figure 5.1), contains the coordinates of the ith gene in the coordinate system (basis) of the scaled eigengenes, $s_k \mathbf{v}_k$. If $r < n$, the transcriptional responses of the genes may be captured with fewer variables using $\mathbf{g}'_i$ rather than $\mathbf{g}_i$. This property of the SVD is sometimes referred to as *dimensionality reduction*. In order to reconstruct the original data, however, we still need access to the eigengenes, which are n-dimensional vectors. Note that due to the presence of noise in the measurements, $r = n$ in any real gene expression analysis application, though the last singular values in S may be very close to zero and thus irrelevant.

$$X = USV^{\text{T}}$$

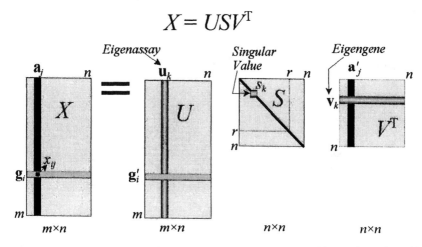

Figure 5.1. Graphical depiction of the SVD of a matrix X, annotated with notations adopted in this chapter.

In diagnostic applications, we may wish to classify tissue samples from individuals with and without a disease. Referring to the definitions in Section 1.1, the signal of interest in this case is the assay expression profile $\mathbf{a}_j$. By Equation 5.1, the SVD equation for $\mathbf{a}_j$ is

$$\mathbf{a}_j = \sum_{k=1}^{r} v_{jk} s_k \mathbf{u}_k, \quad j : 1, ..., n \tag{5.4}$$

which is a linear combination of the eigenassays $\{\mathbf{u}_k\}$. The jth column of V^T, $\mathbf{a}'_j$ (see Figure 5.1), contains the coordinates of the jth assay in the coordinate system (basis) of the scaled eigenassays, $s_k\mathbf{u}_k$. By using the vector $\mathbf{a}'_j$, the expression profiles of the assays may be captured by $r \leq n$ variables, which is always fewer than the m variables in the vector $\mathbf{a}_j$. So, in contrast to gene transcriptional responses, SVD can generally reduce the number of variables used to represent the assay expression profiles. Similar to the case for genes, however, in order to reconstruct the original data, we need access to the eigenassays, which are m-dimensional vectors.

Indeed, analysis of the spectrum formed by the singular values s_k can lead to the determination that fewer than n components capture the essential features in the data, a topic discussed below in Section 2.1.1. In the literature the number of components that results from such an analysis is sometimes associated with the number of underlying biological processes that give rise to the patterns in the data. It is then of interest to ascribe biological meaning to the significant eigenassays (in the case of diagnostic applications), or eigengenes (in the case of systems biology applications). Even though each component on its own may not necessarily be biologically meaningful, SVD can aid in the search for biologically meaningful signals. This topic is touched on in describing scatter plots in Section 2.1.2. Also in Section 2.1.2 we discuss the application of SVD to the problem of grouping genes by transcriptional response, and grouping assays by expression profile. When the data are noisy, it may not be possible to resolve gene groups, but it still may be of interest to detect underlying gene expression patterns; this is a case where the utility of the SVD distinguishes itself with respect to other gene expression analysis methods (Section 2.2). Finally we discuss some published examples of gene expression analysis using SVD, and a couple of SVD-based gene grouping methods (Section 2.3).

2.1   Visualization of the SVD

Visualization is central to understanding the results of application of SVD to gene expression data. For example, Figure 5.2 illustrates plots that are derived from applying SVD to Cho et al.'s budding yeast cell-cycle data set (Cho et al., 1998). In the experiment, roughly 6,200 yeast genes were monitored for 17 time points taken at ten-minute intervals. To perform the SVD, we have pre-processed the data by replacing each measurement with its logarithm, and normalizing each gene's transcriptional response to have zero mean and unit standard deviation. In addition, a serial correlation test (Kanji, 1993) was applied to filter out ~3,200 genes that showed primarily random fluctuations. The plots reveal interesting patterns in the data that we may wish to investigate further: a levelling off of the relative variance after the first five components (Figure 5.2a); a pattern in the first eigengene primarily resembling a steady decrease, or decay (Figure 5.2b); and patterns with cyclic structure in the second and third eigengenes (Figure 5.2c,d).

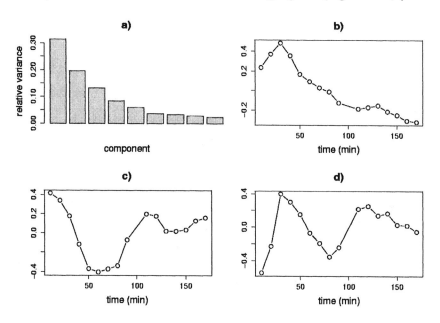

Figure 5.2. Visualization of the SVD of cell cycle data. Plots of relative variance (a); and the first (b), second (c) and third (d) eigengenes are shown. The methods of visualization employed in each panel are described in Section 2.1. These data inspired our choice of the sine and exponential patterns for the synthetic data of Section 2.1.

To aid our discussion of visualization, we use a synthetic time series data set with 14 sequential expression level assays (columns of X) of 2,000 genes (rows of X). Use of a synthetic data set enables us to provide simple illustrations that can serve as a foundation for understanding the more complex patterns that arise in real gene expression data. Genes in our data set have one of three kinds of transcriptional response, inspired by experimentally observed patterns in the Cho et al. cell-cycle data: 1) noise (1,600 genes); 2) noisy sine pattern (200 genes); or 3) noisy exponential pattern (200 genes). Noise for all three groups of genes was modelled by sampling from a normal distribution with zero mean and standard deviation 0.5. The sine pattern has the functional form $a\sin(2\pi t/140)$, and the exponential pattern the form $be^{-t/100}$, where a is sampled uniformly over the interval [1.5,3], b is sampled uniformly over [4,8], t is the time (in minutes) associated with each assay, and time points are sampled every ten minutes beginning at $t = 0$. Each gene's transcriptional response was centered to have a mean of zero. Figure 5.3 depicts genes of type 2) and 3).

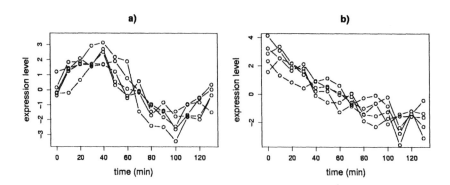

Figure 5.3. Gene transcriptional responses from the synthetic data set. Overlays of five noisy sine wave genes (a) and five noisy exponential genes (b).

2.1.1   Visualization of the matrices S, V^T and U

Singular value spectrum. The diagonal values of S (i.e. s_k) make up the singular value spectrum, which is easily visualized in a one-dimensional plot. The height of any one singular value is indicative of its importance in explaining the data. More specifically, the square of each singular value is proportional to the variance explained by each singular vector. The relative variances $s_k^2(\sum_i s_i^2)^{-1}$ are often plotted (Figure 5.4a; see also Figure 5.2). Cattell has referred to these kinds of plots as *scree plots* (Cattell, 1966) and proposed to use them as a graphical method to decide on the significant components. If the original variables are linear combinations of a smaller number of underlying variables, combined with some low-level noise, the plot will tend to drop sharply for the singular values associated with the underlying variables and then much more slowly for the remaining singular values. Singular vectors (in our case eigenassays and eigengenes) whose singular values plot to the right of such an "elbow" are ignored because they are assumed to be mainly due to noise. For our synthetic data set, the spectrum begins with a sharp decrease, and levels off after the second component, which is indicative of the two underlying signals in the data (Figure 5.4a).

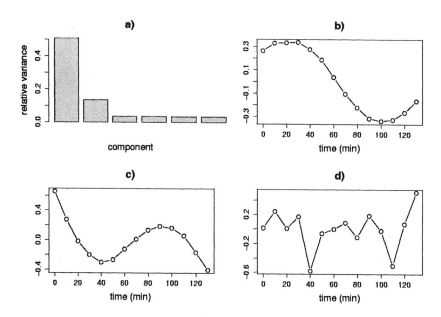

Figure 5.4. Visualization of the SVD of the synthetic data matrix. a) Singular value spectrum in a relative variance plot. The first two singular values account for 64% of the variance. The first (b), second (c), and third (d) eigengenes are plotted vs. time (assays) in the remaining panels. The third eigengene lacks the obvious cyclic structure of the first and second.

Other heuristic approaches for deciding on the significant components have been proposed. One approach is to ignore components beyond where the cumulative relative variance or singular value becomes larger than a certain threshold, usually defined upon the dimensionality of the data. For our example data set, the first two singular vectors explain about 64% of the total variance in the data (Figure 5.4a). Everitt and Dunn propose an alternate approach based on comparing the relative variance of each component to $0.7/n$ (Everitt and Dunn, 2001). For our example data set this threshold is $(0.7/14) = 0.05$, which selects the first two singular vectors as significant. Notice that if we re-construct the matrix X by using only the first two singular vectors, we would obtain $X^{(2)}$ (the best rank-2 approximation of X), which would account for 64% of the variance in the data.

Eigengenes. When assays correspond to samplings of an ordinal or continuous variable (e.g., time; radiation dose; toxin concentration), a plot of the elements of the eigengenes $\{v_k\}$ may reveal recognizable patterns. In our example, the first two eigengenes show an obvious cyclic structure (Figure 5.4b,c; see also Figure 5.2). Neither eigengene is exactly like the underlying sine or exponential pattern; each such pattern, however, is closely approximated by a linear combination of the eigengenes. Sine wave and exponential patterns cannot simultaneously be right singular vectors, as they are not orthogonal. This illustrates the point that, although the most significant eigengenes may not be biologically meaningful in and of themselves, they may be linearly combined to form biologically meaningful signals.

When assays correspond to discrete experimental conditions (e.g., mutational varieties; tissue types; distinct individuals), visualization schemes are similar to those described below for eigenassays. When the jth element of eigengene k is of large-magnitude, the jth assay is understood to contribute relatively strongly to the variance of eigenassay k, a property that may be used for associating a group of assays.

Eigenassays. Alter et al. have visualized eigenassays $\{\mathbf{u}_k\}$ resulting from SVD analysis of cell-cycle data (Alter et al., 2000) by adapting a previously developed color-coding scheme for visualization of gene expression data matrices (Eisen et al., 1998). Individual elements of U are displayed as rectangular pixels in an image, and color-coded using green for negative values, and red for positive values, the intensity being correlated with the magnitude. The rows of matrix U can be sorted using correlation to the eigengenes. In Alter et al.'s study, this scheme sorted the genes by the phase of their periodic pattern. The information communicated in such visualization bears some similarity to visualization using scatter plots, with the advantage that the table-like display enables gene labels to be displayed along with the eigenassays, and the disadvantage that differences among the genes can only be visualized in one dimension.

2.1.2 Scatter plots

Visualization of structure in high-dimensional data requires display of the data in a one-, two-, or three-dimensional subspace. SVD identifies subspaces that capture most of the variance in the data. Even though our discussion here is about visualization in subspaces obtained by SVD, the illustrated visualization techniques are general and can in most cases be applied for visualization in other subspaces (see Section 4 for techniques that use other criteria for subspace selection).

For gene expression analysis applications, we may want to classify samples in a diagnostic study, or classify genes in a systems biology study. Projection of data into SVD subspaces and visualization with scatter plots can reveal structures in the data that may be used for classification. Here we discuss the visualization of features that may help to distinguish gene groups by transcriptional response. Analogous methods are used to distinguish groups of assays by expression profile. We discuss two different sources of gene "coordinates" for scatter plots: projections of the transcriptional response onto eigengenes, and correlations of the transcriptional response with eigengenes.

Projection and correlation scatter plots. Projection scatter plot coordinates q_{ik} for transcriptional response $\mathbf{g}_i$ projected on eigengene $\mathbf{v}_k$ are calculated as $q_{ik} = \mathbf{g}_i \cdot \mathbf{v}_k$. The SVD of X readily allows computation of these coordinates using the equation $XV = US$, so that $q_{ik} = (US)_{ik}$. The projection of gene transcriptional responses from our example data onto the first two eigengenes reveals the *a priori* structure in the data (Figure 5.5a). The groups of the 200 sine wave genes (bottom right cluster), and the 200 exponential decay genes (top right cluster) are clearly separated from each other and from the 1,600 pure noise genes, which cluster about the origin.

Correlation scatter plots may be obtained by calculating the Pearson correlation coefficient of each gene's transcriptional response with the eigengenes:

$$r_{ik} = \delta\mathbf{g}_i \cdot \delta\mathbf{v}_k |\delta\mathbf{g}_i|^{-1} |\delta\mathbf{v}_k|^{-1} \qquad (5.5)$$

where r_{ik} denotes the correlation coefficient of the transcriptional response $\mathbf{g}_i$ with eigengene $\mathbf{v}_k$; $\delta\mathbf{g}_i$ is the mean-centered $\mathbf{g}_i$, the elements of which are $\{x_{ij} - <x_{ij}>_j\}_i$, where $<>_j$ indicates an average over index j, and $\delta\mathbf{v}_k$ is the mean-centered $\mathbf{v}_k$, the elements of which are $\{v_{jk} - <v_{jk}>_j\}_k$. The normalization leads to $-1 \leq r_{ik} \leq 1$. Note that if each $\mathbf{g}_i$ is pre-processed to have zero mean and unit variance, it follows that the correlation scatter plot is equivalent to the projection scatter plot ($\mathbf{g}_i = \delta\mathbf{g}_i$ implies $\mathbf{v}_k = \delta\mathbf{v}_k$; and $|\delta\mathbf{g}_i|^{-1} = |\delta\mathbf{v}_k|^{-1} = 1$).

In the projection scatter plot, genes with a relatively high-magnitude coordinate on the k-axis contribute relatively strongly to the variance of the kth eigengene in the data set. The farther a gene lies away from the origin, the stronger the contribution of that gene is to the variance accounted for by the subspace. In the correlation scatter plot, genes with a relatively high-magnitude coordinate on the k-axis have transcriptional responses that are relatively highly correlated with the kth eigengene.

Due to the normalization in correlation scatter plots, genes with similar patterns in their transcriptional responses, but with different amplitudes, can appear to cluster more tightly in a correlation scatter plot than in a projection scatter plot. Genes that correlate well with the eigengenes lie near the perimeter, a property that can be used in algorithms that seek to identify interesting genes. At the same time, low-amplitude noise genes can appear to be magnified in a correlation scatter plot. For our example data, the sine wave and exponential gene clusters are relatively tightened, the scatter of the noise genes appears to be increased, and the separation between signal and noise genes is decreased for the correlation *vs.* the projection scatter plot (Figure 5.5).

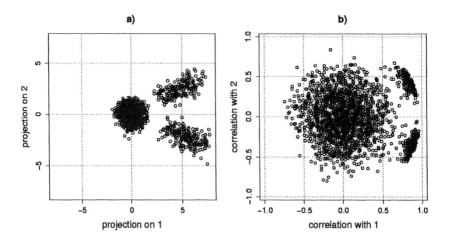

Figure 5.5. SVD scatter plots. Genes from our synthetic example data set are displayed in a projection scatter plot (a); and a correlation scatter plot (b). The bottom right cluster corresponds to sine wave genes, and the top right cluster corresponds to exponential decay genes. The cluster of genes around the origin corresponds to the noise-only genes.

The projection scatter plot (Figure 5.5a) illustrates how SVD may be used to aid in detection of biologically meaningful signals. In this case, the position (q_1, q_2) of any *cluster center*[9] may be used to construct the cluster's transcriptional response from the right singular vectors as $\mathbf{g} = q_1\mathbf{v}_1 + q_2\mathbf{v}_2$. If the first and second singular vectors are biologically meaningful in and of themselves, the cluster centers will lie directly on the axes of the plot. For our synthetic data, the first and second singular vectors are combined to approximately generate the sine wave and exponential patterns. SVD and related methods are particularly valuable analysis methods when the distribution of genes is more complicated than the simple distributions in our example data: for instance, SVD has been used to characterize ring-like distributions of genes such as are observed in scatter plots of cell-cycle gene expression data (Alter et al., 2000; Holter et al., 2000) (see Section 2.3).

Scatter plots of assays. Assays can be visualized in scatter plots using methods analogous to those used for genes. Coordinates for projection scatter plots are obtained by taking the dot products $\mathbf{a}_j \cdot \mathbf{u}_k$ of expression profiles on eigenassays, and coordinates for correlation scatter plots are obtained by calculating the Pearson correlation coefficient $\delta\mathbf{a}_j \cdot \delta\mathbf{u}_k |\delta\mathbf{a}_j|^{-1} |\delta\mathbf{u}_k|^{-1}$. Such plots are useful for visualizing diagnostic data, e.g., distinguishing groups of individuals according to expression profiles. Alter et al. used such

---

[9] A *cluster center* is the average position of the points in a cluster.

a technique to visualize cell-cycle assays (Alter et al., 2000), and were able to associate individual assays with different phases of the cell cycle.

2.2   Detection of weak expression patterns

As noise levels in the data increase, it is increasingly difficult to obtain separation of gene groups in scatter plots. In such cases SVD may still be able to detect weak patterns in the data that may be associated with biological effects. In this respect SVD and related methods provide information that is unique among commonly used analysis methods. To demonstrate this type of analysis, we generated a data matrix using two kinds of transcriptional response: 1,000 genes exhibiting a sine pattern, $\sin(2\pi t/140)$, with added noise sampled from a normal distribution of zero mean and standard deviation 1.5; and 1,000 genes with just noise sampled from the same distribution. Upon application of SVD, we find that the first eigengene shows a coherent sine pattern (Figure 5.6a). The second eigengene is dominated by high-frequency components that can only come from the noise (Figure 5.6b), and the singular value spectrum has an elbow after the first singular value (Figure 5.6c), suggesting (as we know *a priori*) that there is only one interesting signal in the data. Even though the SVD detected the cyclic pattern in the first eigengene (Figure 5.6a), the sine wave and noise-only genes are not clearly separated in the SVD eigengene projection scatter plot (Figure 5.6d).

2.3   Examples from the literature

Cell-cycle gene expression data display strikingly simple patterns when analyzed using SVD. Two different SVD studies have found cyclic patterns in cell-cycle data (Alter et al., 2000; Holter et al., 2000). In correlation scatter plots, previously identified cell cycle genes tended to plot towards the perimeter of a disc. Alter et al. used information in SVD correlation scatter plots to obtain a result that 641 of the 784 cell-cycle genes identified in (Spellman et al., 1998) are associated with the first two eigengenes. Holter et al. displayed previously identified cell-cycle gene clusters in scatter plots, revealing that cell-cycle genes were relatively uniformly distributed in a ring-like feature around the perimeter, leading Holter et al. to suggest that cell-cycle gene regulation may be a more continuous process than had been implied by the previous application of clustering algorithms.

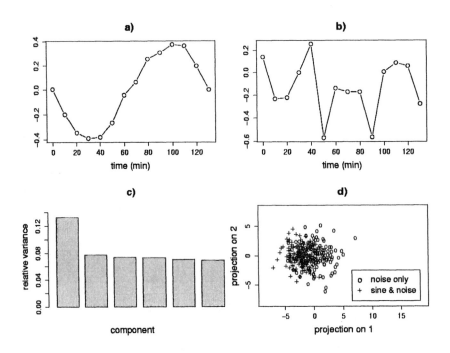

Figure 5.6. SVD-based detection of weak signals. a) A plot of the first eigengene shows the structure of the weak sine wave signal that contributes to the transcriptional response for half of the genes. b) The second eigengene resembles noise. c) A relative variance plot for the first six singular values shows an elbow after the first singular value. d) The signal and noise genes are not separated in an scatter plot of 150 of the signal genes, and 150 of the noise-only genes.

Raychaudhuri et al.'s study of yeast sporulation time series data (Raychaudhuri et al., 2000) is an early example of application of PCA to microarray analysis. In this study, over 90% of the variance in the data was explained by the first two components of the PCA. The first principal component contained a strong steady-state signal. Projection scatter plots were used in an attempt to visualize previously identified gene groups, and to look for structures in the data that would indicate separation of genes into groups. No clear structures were visible that indicated any separation of genes in scatter plots. Holter et al.'s more recent SVD analysis of yeast sporulation data (Holter et al., 2000) made use of a different pre-processing scheme from that of Raychaudhuri et al. The crucial difference is that the rows and columns of X in Holter et al.'s study were iteratively centered and normalized. In Holter et al.'s analysis, the first two eigengenes were found to account for over 60% of the variance for yeast sporulation data. The first two eigengenes were significantly different from those of Raychaudhuri et al., with no steady-state signal, and, most notably, structure indicating separation of gene groups was visible in the data. Below we discuss the discrepancy between these analyses of yeast sporulation data.

3.   DISCUSSION

Selection of an appropriate pre-processing method for gene expression analysis is critical. By inspecting the SVD, one can evaluate different pre-processing choices by gaining insight into, e.g., separability in scatter plots. The utility of SVD itself, however, depends on the choice of pre-processing, as the apparent discrepancy between the sporulation analyses described in Section 2.3 illustrates. While structure was revealed in yeast sporulation data using the SVD on centered, normalized data (Holter et al., 2000), structure was not visible using SVD on the original data (Raychaudhuri et al., 2000), where the first component accounted for the steady-state gene expression levels. The decision of how to pre-process the data should always be made based on the statistics of the data, what questions are being asked, and what analysis methods are being used. As an example, performing a centering of gene transcriptional responses for time series data is often sensible because we are typically more interested in how a gene's transcriptional response varies over time than we are in its steady-state expression level.

An important capability distinguishing SVD and related methods from other analysis methods is the ability to detect weak signals in the data. Even when the structure of the data does not allow clustering, it may be possible to detect biologically meaningful patterns in the data. In Section 2.2 we have given an example of this phenomenon using synthetic data. As a practical example, it may be possible to detect whether the expression profile of a tissue culture changes with radiation dose, even when it is not possible to detect which specific genes change expression in response to radiation dose.

SVD allows us to obtain the true dimensionality of our data, which is the rank r of matrix X. As the number of genes m is generally (at least presently) greater than the number of assays n, the matrix V^T generally yields a representation of the assay expression profiles using a reduced number of variables. When $r < n$, the matrix U yields a representation of the gene transcriptional responses using a reduced number of variables. Although this property of the SVD is commonly referred to as dimensionality reduction, we note that any reconstruction of the original data requires generation of an $m \times n$ matrix, and thus requires a mapping that involves all of the original dimensions. Given the noise present in real data, in practice the rank of matrix X will always be n, leading to no dimensionality reduction for the gene transcriptional responses. It may be possible to detect the "true" rank r by ignoring selected components, thereby reducing the number of variables required to represent the gene transcriptional responses. As discussed above, existing SVD-based methods for pre-processing based on this kind of feature selection must be used with caution.

Current thoughts about use of SVD/PCA for gene expression analysis often include application of SVD as pre-processing for clustering. Yeung and Ruzzo have characterized the effectiveness of gene clustering both with and without pre-processing using PCA (Yeung and Ruzzo, 2001). The pre-processing consisted of using PCA to select only the highest-variance principal components, thereby choosing a reduced number of variables for each gene's transcriptional response. The reduced variable sets were used as inputs to clustering algorithms. Better performance was observed without pre-processing for the tested algorithms and the data used, and the authors generally recommend against using PCA as a pre-processing step for clustering. The sole focus on gene clustering, however, in addition to the narrow scope of the tested algorithms and data, limit the implications of the results of this study. For example, when grouping assays is of interest, using $\{Sa'_j\}$ instead of $\{a_j\}$ (see Section 2; Figure 5.1) enables use of a significantly reduced number of variables (r vs. m) that account for *all* of the structure in the distribution of assays. Use of the reduced variable set for clustering must therefore result in not only decreased compute time, but also clusters of equal or higher quality. Thus the results in (Yeung and Ruzzo, 2001) for gene clustering do not apply to assay clustering.

In Section 2.3 we discuss how, rather than separating into well-defined groups, cell-cycle genes tend to be more continuously distributed in SVD projections. For instance, when plotting the correlations of genes with the first two right singular vectors, cell-cycle genes appear to be relatively uniformly distributed about a ring. This structure suggests that, rather than using a classification method that groups genes according to their co-location in the neighborhood of a point (e.g., k-means clustering), one should choose a classification method appropriate for dealing with ring-like distributions. Previous cell-cycle analyses therefore illustrate the fact that one important use of SVD is to aid in selection of appropriate classification methods by investigation of the dimensionality of the data.

In this chapter we have concentrated on conveying a general understanding of the application of SVD analysis to gene expression data. Here we briefly mention several specific SVD-based methods that have been published for use in gene expression analysis. For gene grouping, the *gene shaving* algorithm (Hastie et al., 2000) and SVDMAN (Wall et al., 2001) are available. An important feature to note about both gene shaving and SVDMAN is that each gene may be a member of more than one group. For evaluation of data, SVDMAN uses SVD-based interpolation of deleted data to detect sampling problems when the assays correspond to a sampling of an ordinal or continuous variable (e.g., time series data). A program called

SVDimpute (Troyanskaya et al., 2001) implements an SVD-based algorithm for imputing missing values in gene expression data. Holter et al. have developed an SVD-based method for analysis of time series expression data (Holter et al., 2001). The algorithm estimates a time translation matrix that describes evolution of the expression data in a linear model. Yeung et al. have also made use of SVD in a method for reverse engineering linearly coupled models of gene networks (Yeung et al., 2002).

It is important to note that application of SVD and PCA to gene expression analysis is relatively recent, and that methods are currently evolving. Presently, gene expression analysis in general tends to consist of iterative applications of interactively performed analysis methods. The detailed path of any given analysis depends on what specific scientific questions are being addressed. As new inventions emerge, and further techniques and insights are obtained from other disciplines, we mark progress towards the goal of an integrated, theoretically sound approach to gene expression analysis.

4.   FURTHER READING AND RESOURCES

The book (Jolliffe, 1986) is a fairly comprehensive reference on PCA (a new edition is meant to appear in summer of 2002); it gives interpretations of PCA and provides many example applications, with connections to and distinctions from other techniques such as correspondence analysis and factor analysis. For more details on the mathematics and computation of SVD, good references are (Golub and Van Loan, 1996), (Strang, 1998), (Berry, 1992), and (Jessup and Sorensen, 1994). SVDPACKC has been developed to compute the SVD algorithm (Berry et al., 1993). SVD is used in the solution of unconstrained linear least squares problems, matrix rank estimation, and canonical correlation analysis (Berry, 1992).

Applications of PCA and/or SVD to gene expression data have been published in (Alter et al., 2000; Hastie et al., 2000; Holter et al., 2000; Holter et al., 2001; Raychaudhuri et al., 2000; Troyanskaya et al., 2001; Wall et al., 2001; Yeung and Ruzzo, 2001; Yeung et al., 2002). SVDMAN is free software available at http://home.lanl.gov/svdman. Knudsen illustrates some of the uses of PCA for visualization of gene expression data (Knudsen, 2002).

Everitt, Landau and Leese (Everitt et al., 2001) present PCA as a special case of Projection Pursuit (Friedman and Tukey, 1974), which in general attempts to find an "interesting projection" for the data. A related method is Independent Component Analysis (ICA) (Hyvärinen, 1999), which attempts to find a linear transformation (non-linear generalizations are possible) of the data so that the derived components are as statistically independent from each other as possible. Hyvärinen discusses ICA and how it relates to PCA and Projection Pursuit (Hyvärinen, 1999). Liebermeister has applied ICA to gene expression data (Liebermeister, 2002).

Other related techniques are Multidimensional Scaling (Borg and Groenen, 1997) and Self-Organizing Maps (SOM) (Kohonen, 2001), both of which use non-linear mappings of the data to find lower-dimensional representations. SOM's have been applied to gene expression data in (Tamayo et al., 1999). There are also non-linear generalizations of PCA (Jolliffe, 1986; Scholkopf et al., 1996).

ACKNOWLEDGMENTS

We gratefully acknowledge Raphael Gottardo and Kevin Vixie for critically reading the manuscript. The writing of this chapter was performed within the auspices of the Department of Energy (DOE) under contract to the University of California, and was supported by Laboratory-Directed Research and Development at Los Alamos National Laboratory.

REFERENCES

Alter O., Brown P.O., Botstein D. (2000). Singular value decomposition for genome-wide expression data processing and modeling. Proc Natl Acad Sci 97:10101-06.

Berry M.W. (1992). Large-scale sparse singular value computations. International Journal of Supercomputer Applications 6:13-49.

Berry M.W., Do T., Obrien G.W., Krishna V., Varadhan S. (1993). SVDPACKC: Version 1.0 User's Guide. Knoxville: University of Tennessee.

Berry M.W., Dumais S.T., Obrien G.W. (1995). Using linear algebra for intelligent information-retrieval. Siam Review 37:573-95.

Borg I., Groenen P. (1997). Modern Multidimensional Scaling: Theory and Applications. New York: Springer Verlag.

Cattell R.B. (1966). The scree test for the number of factors. Multivariate Behavioral Research 1:245-76.

Cho R.J., Campbell M.J., Winzeler E.A., Steinmetz L., Conway A., Wodicka L. et al. (1998). A genome-wide transcriptional analysis of the mitotic cell cycle. Mol Cell 2:65-73.

Deprettere F. (1988). SVD and Signal Processing: Algorithms, Analysis and Applications. Amsterdam: Elsevier Science Publishers.

Eisen M.B., Spellman P.T., Brown P.O., Botstein D. (1998). Cluster analysis and display of genome-wide expression patterns. Proc Natl Acad Sci 95:14863-68.

Everitt B.S., Dunn G. (2001). Applied Multivariate Data Analysis. London: Arnold.

Everitt S.E., Landau S., Leese M. (2001). Cluster Analysis. London: Arnold.

Friedman J.H., Tukey J.W. (1974). A projection pursuit algorithm for exploratory data analysis. IEEE Transactions on Computers 23:881-89.

Golub G., Van Loan C. (1996). Matrix Computations. Baltimore: Johns Hopkins Univ Press.

Hastie T., Tibshirani R., Eisen M.B., Alizadeh A., Levy R., Staudt L. et al. (2000). "Gene shaving" as a method for identifying distinct sets of genes with similar expression patterns. Genome Biol 1:research0003.1-03.21.

Holter N.S., Mitra M., Maritan A., Cieplak M., Banavar J.R., Fedoroff N.V. (2000). Fundamental patterns underlying gene expression profiles: simplicity from complexity. Proc Natl Acad Sci 97:8409-14.

Holter N.S., Maritan A., Cieplak M., Fedoroff N.V., Banavar J.R. (2001). Dynamic modeling of gene expression data. Proc Natl Acad Sci 98:1693-98.

Hughes T.R., Marton M.J., Jones A.R., Roberts C.J., Stoughton R., Armour C.D. et al. (2000). Functional discovery via a compendium of expression profiles. Cell 102:109-26.

Hyvärinen (1999). A. Survey on Independent Component Analysis. Neural Computing Surveys 2:94-128.

Jessup E.R., Sorensen D.C. (1994). A parallel algorithm for computing the singular-value decomposition of a matrix. Siam Journal on Matrix Analysis and Applications 15:530-48.

Jolliffe I.T. (1986). Principal Component Analysis. New York: Springer.

Kanji G.K. (1993). 100 Statistical Tests. New Delhi: Sage.

Knudsen S. (2002). A Biologist's Guide to Analysis of DNA Microarray Data. New York: John Wiley & Sons.

Kohonen T. (2001). Self-Organizing Maps. Berlin: Springer-Verlag.

Liebermeister W. (2002). Linear modes of gene expression determined by independent component analysis. Bioinformatics 18:51-60.

Raychaudhuri S., Stuart J.M., Altman R.B. (2000). Principal components analysis to summarize microarray experiments: application to sporulation time series. Pac Symp Biocomput 2000:455-66.

Richards J.A. (1993). Remote Sensing Digital Image Analysis. New York: Springer-Verlag.

Scholkopf B., Smola A.J., Muller K.-R. (1996). "Nonlinear component analysis as a kernel eigenvalue problem" Technical Report. Tuebingen: Max-Planck-Institut fur biologische Kybernetik.

Spellman P.T., Sherlock G., Zhang M.Q., Iyer V.R., Anders K., Eisen M.B. et al. (1998). Comprehensive identification of cell cycle-regulated genes of the yeast Saccharomyces cerevisiae by microarray hybridization. Mol Biol Cell 9:3273-97.

Strang G. (1998). Introduction to Linear Algebra. Wellesley, MA: Wellesley Cambridge Press.

Tamayo P., Slonim D., Mesirov J., Zhu Q., Kitareewan S., Dmitrovsky E. et al. (1999). Interpreting patterns of gene expression with self-organizing maps: methods and application to hematopoietic differentiation. Proc Natl Acad Sci 96:2907-12.

Troyanskaya O., Cantor M., Sherlock G., Brown P., Hastie T., Tibshirani R. et al. (2001). Missing value estimation methods for DNA microarrays. Bioinformatics 17:520-25.

Wall M.E., Dyck P.A., Brettin T.S. (2001). SVDMAN – singular value decomposition analysis of microarray data. Bioinformatics 17:566-68.

Yeung K.Y., Ruzzo W.L. (2001). Principal component analysis for clustering gene expression data. Bioinformatics 17:763-74.

Yeung M.K., Tegner J., Collins J.J. (2002). Reverse engineering gene networks using singular value decomposition and robust regression. Proc Natl Acad Sci 99:6163-68.

Chapter 6

FEATURE SELECTION IN MICROARRAY ANALYSIS

Eric P. Xing

Computer Science Division, University of California, Berkeley, USA
e-mail: epxing@cs.berkeley.edu

1. INTRODUCTION

Microarray technology makes it possible to put the probes for the genes of an entire genome onto a chip, such that each data point provided by an experimenter lies in the high-dimensional space defined by the size of the genome under investigation. However, the sample size in these experiments is often severely limited. For example, in the popular leukemia microarray data set,[1] there are only 72 observations of the expression levels of each of 7,130 genes. This problem exemplifies a situation that will be increasingly common in the analysis of microarray data using machine learning techniques such as classification or clustering.

In high-dimensional problems such as these, feature selection methods are essential if the investigator is to make sense of his/her data, particularly if the goal of the study is to identify genes whose expression patterns have meaningful biological relationships to the classification or clustering problem. For example, for a microarray classification problem, it is of great clinical and mechanistic interest to identify those genes that directly contribute to the phenotype or symptom that we are trying to predict. Computational constraints can also impose important limitations. Many induction methods[2] suffer from the *curse of dimensionality*, that is, the time

---

[1] We use the leukemia microarray profile from the Whitehead Institute (Golub et al., 1999) as our running example in this chapter.

[2] *Induction* (or *inductive inference*, *inductive learning*) refers to the following learning task: given a collection of examples $(x, f(x))$, find a function h that approximates f. The function h is called a *hypothesis*.

required for an algorithm grows dramatically, sometimes exponentially with the number of features involved, rendering the algorithm intractable in extremely high-dimensional problems we are facing with microarray data. Furthermore, a large number of features inevitably lead to a complex hypothesis and a large number of parameters for model induction or density estimation, which can result in serious overfitting over small data sets and thus a poor bound on generalization error.(Indeed we may never be able to obtain a "sufficiently large" data set. For example, theoretical and experimental results suggest that the number of training examples needed for a classifier to reach a given accuracy, or *sample complexity*, grows exponentially with the number of irrelevant features.)

The goal of feature selection is to select relevant features and eliminate irrelevant ones. This can be achieved by either explicitly looking for a good subset of features, or by assigning all features appropriate weights. Explicit feature selection is generally most natural when the result is intended to be understood by humans or fed into different induction algorithms. Feature weighting, on the other hand, is more directly motivated by pure modeling or performance concerns. The weighting process is usually an integral part of the induction algorithm and the weights often come out as a byproduct of the learned hypothesis.

In this chapter, we survey several important feature selection techniques developed in the classic *supervised learning* paradigm. We will first introduce the classic *filter* and *wrapper* approaches and some recent variants for explicit feature selection. Then we discuss several feature weighting techniques including WINNOW and Bayesian feature selection. We also include a brief section describing recent works on feature selection in the *unsupervised learning* paradigm, which will be useful for clustering analysis in the high-dimensional gene space.

Before proceeding, we should clarify the scope of this survey. There has been substantial work on feature selection in machine learning, pattern recognition and statistics. Due to space limit and the practical nature of this book, we refrain from detailed formal discussions and focus more on algorithmic solutions for practical problems from a machine learning perspective. Readers can follow the references of this chapter for more theoretical details and examples of the techniques introduced in this chapter.

2. EXPLICIT FEATURE SELECTION

In explicit feature selection, we look for the subset of features that leads to optimal performance in our learning task, such as classifying biological samples according to their mRNA expression profiles.

Explicit feature selection can be formulated as a heuristic search problem, with each state in the search space specifying a specific subset of features (Blum and Langley, 1997). Any feature selection algorithm needs to deal with the following four issues which determine the nature of the heuristic search process: 1) How to start the search. One can either begin with an empty set and successively add features (*forward selection*) or start with all features and successively discard them (*backward elimination*) or other variations in between. 2) How to explore the search space. Popular strategies include a *hill-climbing* type of greedy scheme or a more exhaustive *best-first search*. 3) How to evaluate a feature subset. A common metric involves the degree of consistency of a feature with the target concept (e.g., sample labels) in the training data; more sophisticated criteria concern how selected features interact with specific induction algorithms. 4) When to stop the search. Depending on which search and evaluation scheme is used, one can use thresholding or a significance test, or simply stop when performance stops improving. It should be clear that all the above design decisions must be made for a feature selection procedure, which leaves practitioners substantial freedom in designing their algorithms.

2.1   The Filter Methods

The filter model relies on general characteristics of the training data to select a feature subset, doing so without reference to the learning algorithm. Filter strategies range from sequentially evaluating each feature based on simple statistics from the empirical distribution of the training data to using an embedded learning algorithm (independent of the induction algorithm that uses its output) to produce a feature subset.

2.1.1   Discretization and Discriminability Assessment of Features

The measurements we obtained from microarrays are continuous values. In many situations in functional annotation (e.g., constructing regulatory networks) or data analysis (e.g. the information-theoretic-based filter technique we will discuss later), however, it is convenient to assume discrete values. One way to achieve this is to deduce the functional states of the genes based on their observed measurements.

A widely adopted empirical assumption about the activity of genes, and hence their expression, is that they generally assume distinct functional states such as "on" or "off". (We assume binary states for simplicity but generalization to more states is straightforward.) The combination of such binary patterns from multiple genes determines the sample phenotype. For concreteness, consider a particular gene i (feature F_i). Suppose that the expression levels of F_i in those samples where F_i is in the "on" state can be modeled by a probability distribution, such as a Gaussian distribution $N(x \mid \mu_1, \sigma_1)$ where μ_1 and σ_1 are the mean and standard deviation. Similarly, another Gaussian distribution $N(x \mid \mu_2, \sigma_2)$ can be assumed to model the expression levels of F_i in those samples where F_i is in the "off" state. Given the above assumptions, the marginal probability of any given expression level x_i of gene i can be modeled by a weighted sum of the two Gaussian probability functions corresponding to the two functional states of this gene (where the weights $\pi_{1/2}$ correspond to the prior probabilities of gene i being in the on/off states):

$$P(x_i) = \pi_1 N(x_i \mid \mu_1, \sigma_1) + \pi_2 N(x_i \mid \mu_2, \sigma_2) \tag{6.1}$$

Such a model is called a *univariate mixture model* with two components (which includes the degenerate case of a single component when either of the weights is zero). The histogram in Figure 6.1a gives the empirical marginal of gene 109, which clearly demonstrates the case of a two-component mixture distribution of the expression levels of this gene in the 72 leukemia samples (which indicates this gene can be either 'on' or 'off' in these samples), whereas Figure 6.1b is an example of a nearly uni-component distribution (which indicates that gene 1902 remains in the same functional state in all the 72 samples).

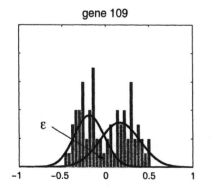 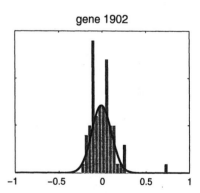

Figure 6.1. The histograms and estimated density functions of the expression profiles of two representative genes. The *x*-axes represent the normalized expression level.

For feature selection, if the underlying binary state of the gene does not vary between the two classes, then the gene is not discriminative for the classification problem and should be discarded. This suggests a heuristic procedure in which we measure the separability of the mixture components as an assay of the discriminability of the feature.

Given N microarray experiments for which gene i is measured in each experiment, the complete likelihood of all observations $X_i = \{x_{1i}, \ldots, x_{Ni}\}$ and their corresponding state indicator $Z_i = \{z_{1i}, \ldots, z_{Ni}\}$ is:

$$P_c(X_i, Z_i \mid \theta_i) = \prod_{n=1}^{N} \prod_{k=0}^{1} \left(\pi_{i,k} \left[\frac{1}{\sqrt{2\pi}\sigma_{i,k}} \exp\left\{ -\frac{(x_{ni}-\mu_{i,k})^2}{2(\sigma_{i,k})^2} \right\} \right] \right)^{z_{ni}^k} \qquad (6.2)$$

Random variable $z_{ni} \in \{0, 1\}$ indicates the underlying state of gene i in sample n (we omit sample index n in the subscript in the later presentation for simplicity) and is usually latent. We can fit the model parameters using the EM algorithm (Dempster et al., 1977). The solid curves in Figure 6.1(a) depict the density functions of the two Gaussian components fitted on the observed expression levels of gene 109. The curve in Figure 6.1(b) is the density of the single-component Gaussian distribution fitted on gene 1902. Note that each feature F_i is fitted independently based on its measurements in all N microarray experiments.

Suppose we define a decision $d(F_i)$ on feature F_i to be 0 if the posterior probability of $\{z_i = 0\}$ is greater than 0.5 under the mixture model, and let $d(F_i)$ equal 1 otherwise. We can define a mixture-overlap probability:

$$\varepsilon = P(z_i = 0)P(d(F_i) = 1 \mid z_i = 0) + P(z_i = 1)P(d(F_i) = 0 \mid z_i = 1). \qquad (6.3)$$

If the mixture model were a true representation of the probability of gene expression, then ε would represent the Bayes error of classification under this model (which equals to the area indicated by the arrow in Figure 6.1(a). We can use this probability as a heuristic surrogate for the discriminating potential of the gene.

The mixture model can be used as a quantizer, allowing us to discretize the measurements for a given feature. We can simply replace the continuous measurement f_i with the associated binary value $d(f_i)$.

2.1.2   Correlation-based Feature Ranking

The *information gain* is commonly used as a surrogate for approximating a conditional distribution in the classification setting (Cover and Thomas, 1991). Let the class labels induce a reference partition $S_1, \ldots, S_C$ (e.g. different types of cancers). Let the probability of this partition be the empirical proportions: $P(T) = |T| / |S|$ for any subset T. Suppose a test on feature F_i induces a partition of the training set into $E_1, \ldots, E_K$. Let $P(S_c \mid E_k) = P(S_c \cap E_k)/P(E_k)$. We define the information gain due to F_i with respect to the reference partition as:

$$I_g = H(P(S_1), \ldots, P(S_C)) - \sum_{k=1}^{K} P(E_k)H(P(S_1 \mid E_k), \ldots, P(S_C \mid E_k)) \qquad (6.4)$$

where H is the *entropy function*.[3] To calculate the information gain, we need to quantize the values of the features. This is achieved to the mixture model quantization discussed earlier. Back to the leukemia example: quantization of all the 72 measurement of gene 109 results in 36 samples in the "on" state (of which 20 are of type I leukemia and 16 type II) and 36 samples in the "off" state (27 type I and 9 type II). According to Equation 6.4, the information gain induced by gene 109 with respect to the original sample partition (47 type I and 25 type II) is:

$$I_g(F_{109}) = H(\tfrac{47}{72}, \tfrac{25}{72}) - (\tfrac{36}{72} H(\tfrac{20}{36}, \tfrac{16}{36}) + \tfrac{36}{72} H(\tfrac{27}{36}, \tfrac{9}{36})) = 0.0304.$$

The information gain reveals the degree of relevance of a feature to the reference partition. The greater the information gain, the more relevant the feature is to the reference partition.

2.1.3 Markov Blanket Filtering

If we have a large number of similar or redundant genes in a data set, all of them will score similarly in information gain.[4] This will cause undesirable dominance of the resulting classifier by a few gene families whose members have coherent expression patterns, or even by a group of replicates of genes. This will seriously compromise the predictive power of the classifier. To alleviate this problem, we turn to *Markov blanket filtering*, a technique due to Koller and Sahami (1996), which can screen out redundant features.

Let **G** be a subset of the overall feature set **F**. Let $\mathbf{f}_G$ denote the projection of **f** onto the variables in **G**. Markov blanket filtering aims to minimize the discrepancy between the conditional distributions $P(C\,|\,\mathbf{F}=\mathbf{f})$ and $P(C\,|\,\mathbf{G}=\mathbf{f}_G)$, as measured by a conditional entropy:

$$\Delta_{\mathbf{G}} = \sum_{\mathbf{f}} P(\mathbf{f}) D(P(C\,|\,\mathbf{F}=\mathbf{f}).\|P(C\,|\,\mathbf{G}=f_{\mathbf{G}})) \tag{6.5}$$

where

$$D(P\|Q) = \sum_{x} P(x)\log(P(x)/Q(x))$$

is the *Kullback-Leibler divergence*. The goal is to find a small set **G** for which $\Delta_{\mathbf{G}}$ is small.

---

[3] For discrete cases, the entropy of distribution $\{P_1, \ldots, P_c\}$ is given by $H = \sum_{i=1}^{c} -P_i \log P_i$.

[4] Such situations could either arise from true functional redundancy, or result from artifacts of the microarray (e.g., the probe of a particular gene is accidentally spotted k times and appears as k "similar genes" to a user who is unaware of the erroneous manufacturing process).

Intuitively, if a feature F_i is conditionally independent of the class label given some small subset of other features, then we should be able to omit F_i without compromising the accuracy of class prediction. Koller and Sahami formalize this idea using the notion of a Markov blanket.

Definition 6.1: *Markov blanket*

For a feature set **G** and class label C, the set $\mathbf{M}_i \subseteq \mathbf{G}$ ($F_i \notin \mathbf{M}_i$) is a Markov blanket of F_i ($F_i \in \mathbf{G}$) if: given $\mathbf{M}_i$, F_i is conditionally independent of $\mathbf{G} - \mathbf{M}_i - \{F_i\}$ and C.

Biologically speaking, one can view the Markov blanket $\mathbf{M}_i$ of gene i as a subset of genes that exhibit similar expression patterns as gene i in all the samples under investigation. Such a subset could correspond to genes of isozymes, coregulated genes, or even (erroneous) experimental/manufactural replicates of probes of the same gene in an array.

Theoretically, it can be shown that once we find a Markov blanket of feature F_i in a feature set **G**, we can safely remove F_i from **G** without increasing the divergence from the desired distribution (Xing et al., 2001). Furthermore, in a sequential filtering process in which unnecessary features are removed one by one, a feature tagged as unnecessary based on the existence of a Markov blanket $\mathbf{M}_i$ remains unnecessary in later stages when more features have been removed (Koller and Sahami, 1996).

In most cases, however, few if any features will have a Markov blanket of limited size. Hence, we must instead look for features that have an "approximate Markov blanket". For this purpose we define

$$\Delta(F_i \mid \mathbf{M}) = \sum_{f_\mathbf{M}, f_i} P(\mathbf{M} = f_\mathbf{M}, F_i = f_i)$$
$$D(P(C \mid \mathbf{M} = f_\mathbf{M}, F_i = f_i) \| P(C \mid \mathbf{M} = f_\mathbf{M})) \tag{6.6}$$

If **M** is a Markov blanket for F_i, then $\Delta(F_i \mid \mathbf{M}) = 0$ (following the definition of Markov blanket), which means all information carried by F_i about the sample is also carried by feature subset **M**. Since an exact zero is unlikely to occur, we relax the condition and seek a set **M** such that $\Delta(F_i \mid \mathbf{M})$ is small. It can be proved that those features that form an approximate Markov blanket of feature F_i are most likely to be more strongly correlated to F_i. We can construct a candidate Markov blanket of F_i by collecting the k features that have the highest correlations (defined by the Pearson correlations between the original feature vectors that are not discretized) with F_i, where k is a small integer. This suggests an easy heuristic way to search for features with approximate Markov blankets (Koller and Sahami, 1996):

Initialize
- $\mathbf{G} = \mathbf{F}$

Iterate
- For each feature $F_i \in \mathbf{G}$, let $\mathbf{M}_i$ be the set of k features $F_j \in \mathbf{G} - \{F_i\}$ for which the correlations between F_i and F_j are the highest.
- Compute $\Delta(F_i \mid \mathbf{M}_i)$ for each i.
- Choose the i that minimizes $\Delta(F_i \mid \mathbf{M}_i)$, and define $\mathbf{G} = \mathbf{G} - \{F_i\}$.

Figure 6.3. The Markov blanket filtering algorithm.

This heuristic method requires computation of quantities of the form $P(C \mid \mathbf{M} = \mathbf{f}_M, F_i = \mathbf{f}_i)$ and $P(C \mid \mathbf{M} = \mathbf{f}_M)$, which can be easily computed using the discretization technique described in Section 2.1. When working on a small data set, one should keep the Markov blankets small to avoid fragmenting the data.[5] The fact that in a real biological regulatory network the fan-in and fan-out will generally be small provides some justification for enforcing small Markov blankets.

Figure 6.2(a) shows the mixture overlap probability ε for the genes in the leukemia data set in ascending order. It can be seen that only a small percentage of the genes have an overlap probability significantly smaller than $\varepsilon \ll 0.5$, where 0.5 would constitute a random guessing under a Gaussian model if the underlying mixture components were construed as class labels. Figure 6.2(b) shows the information gain due to each individual gene with respect to the leukemia cancer labels. Indeed, only a very small fraction of the genes induce a significant information gain. One can rank all genes in the order of increasing information gain and select genes conservatively via a statistical significance test (Ben-Dor et al., 2000).

---

[5] This refers to the situation in which, given small number of samples, one has to estimate, for example, $P(C \mid \mathbf{M} = \mathbf{f}_M)$ for many different possible configurations of $\mathbf{f}_M$. When $\mathbf{M}$ is large, each $\mathbf{f}_M$ configuration is seen only in a very small number of samples, making estimation of the conditional probabilities based on empirical frequency very inaccurate.

Figure 6.2(c) depicts the result of the Markov blanket filtering for the leukemia data set.

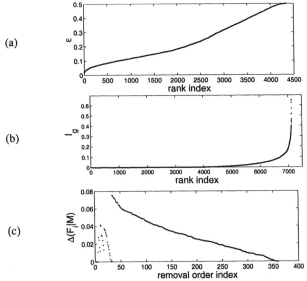

(a)

(b)

(c)

Figure 6.2. Feature selection using filter methods. (a) Genes ranked by mixture-overlap probability ε. Only 2-state genes (i.e. those whose distributions of expressions in all samples have two mixture components corresponding to the "on" and "off" states) are displayed. (b) Genes ranked by their information gains I_g with respect to the reference partition induced by the sample labels. (c) The $\Delta(F_i \mid M)$ of the last 360 genes removed during MB filter. (The *x*-axis indexes the inverse removal order of the genes. $\{x = 1\}$ refers to the gene that is removed last.)

2.1.4   Decision Tree Filtering

A *decision tree* is itself an induction algorithm and learns a decision rule (a Boolean function) mapping relevant attributes to the target concept (see also Chapter 10, Section 4). Since a decision tree typically contains only a subset of the features, those included in the final tree can be viewed as a relevant feature subset and fed into another classification algorithm of choice. Thus, we can use the decision-tree algorithm as an embedded selection scheme under the filter model.[6] This approach has worked well for some data sets, but does not have a guarantee of performance gain on an arbitrary classifier

---

[6] If at each tree-growing step, we choose to incorporate the feature whose information gain with respect to the target concept is the highest among all features not yet in the tree, then decision tree filtering is in a sense similar to information gain ranking mentioned previously. However, general decision tree learning algorithm can also use other criteria to choose qualified features (e.g., classification performance of the intermediate tree resulted from addition of one more feature), and usually a learned tree needs to be pruned and cross-validated. These differences distinguish decision tree filtering from information gain ranking.

since features that are good for a decision tree are not necessarily useful in other models. Essentially, a decision tree is itself a classifier (or an hypothesis), the features admitted to the learned tree inevitably bears *inductive bias*.[7] For high-dimensional microarray data, current methods of building decision trees may also suffer from data fragmentation and lack of sufficient samples. These shortcomings will result in a feature subset of possibly insufficient size. Nevertheless, if users have a strong prior belief that only a small number of genes are involved in a biological process of his/her interest, decision tree filtering could be a highly efficient way to pick them out.

2.2 The Wrapper Methods

The wrapper model makes use of the algorithm that will be used to build the final classifier to select a feature subset. Thus, given a classifier C, and given a set of features F, a wrapper method searches in the space of subsets of F, using cross-validation to compare the performance of the trained classifier C on each tested subset. While the wrapper model tends to be more computationally expensive, it also tends to find feature sets better suited to the inductive biases of the learning algorithm and tends to give superior performance.

A key issue of the wrapper methods is how to search the space of subsets of features. Note that when performing the search, enumeration over all 2^N possible feature sets is usually intractable for the high-dimensional problems in microarray analysis. There is no known algorithm for otherwise performing this optimization tractably. Indeed, the feature selection problem in general is NP-hard[8], but much work over recent years has developed a large number of heuristics for performing this search efficiently. A thorough review on search heuristics can be found in (Russel and Norvig, 1995).

It is convenient to view the search process as building up a search tree that is superimposed over the state space (which, in our case, means each node in the tree corresponds to a particular feature subset, and adjacent nodes correspond to two feature subsets that differ by one feature). The root of this tree is the initial feature set which could be full, empty, or randomly chosen. At each search step, the search algorithm chooses one leaf node in the tree to expand by applying an *operator* (i.e. adding, removing, or

---

[7] Any preference for one hypothesis over another, beyond mere consistency with the examples, is called an *inductive bias*. For example, over many possible decision trees that are consistent with all training examples, the learning algorithm may prefer the smallest one, but the features included in such a tree may be insufficient for obtaining a good classifier of another type, e.g., support vector machines.

[8] NP stands for *nondeterministic polynomial (time)*. In short, the NP-hard problems are a class of problems for which no polynomial-time solution is known.

replacing one of the features) to the feature subset corresponding to the node to produce a child. The first two search strategies described in the following can be best understood in this way.

2.2.1   Hill-Climbing Search

Hill-climbing search is one of the simplest search techniques also known as greedy search or steepest ascent. In fact, to perform this search one does not even need to maintain a search tree because all the algorithm does is to make the locally best changes to the feature subset. Essentially, it expands the current node and moves to the child with the highest accuracy based on cross-validation, terminating when no child improves over the current node. An important drawback of hill-climbing search is that it tends to suffer from the presence of local maxima, plateaux and ridges of the value surface of the evaluation function. Simulated annealing (occasionally picking a random expansion) provides a way to escape possible sub-optimality.

2.2.2   Best-First Search

Best-first search is a more robust search strategy than the hill-climbing search. Basically, it chooses to expand the best-valued leaf that has been generated so far in the search tree (for this purpose we need to maintain a record of the search tree to provide us the tree frontier). To explore the state space more thoroughly, we do not stop immediately when node values stop increasing, but keep on expanding the tree until no improvement (within ε error) is found over the last k expansions.

2.2.3   Probabilistic Search

For large search problems, it is desirable to concentrate the search in the regions of the search space that has appeared promising in the past yet still allow sufficient chance of exploration (in contrast to the greedy methods). A possible way to do so is to sample from a distribution of only the front-runners of the previously seen feature combinations. Define a random variable $z \in \{0, 1\}^n$: a string of n bits that indicates whether each of the n features is relevant. We can hypothesize a parametric probabilistic model, for example, a dependence tree or even a more elaborated Bayesian network, for random variable z and learn its distribution via an incremental procedure.

A dependence tree model is of the following form:

$$p(z) = p(z_r) \prod_{i \neq r} p(z_i \mid z_{\pi_i}) \qquad (6.7)$$

where z_r is the root node and π_i indexes the parent of node i. This tree should be distinguished from the search tree we mentioned earlier where a node represents a feature subset and the size of the tree grows during search up to 2^n. In a dependence tree each node corresponds to an indicator random variable concerning the inclusion or exclusion of a particular feature, and the size of the tree is fixed. Any particular composition of feature subset we may select is a *sample* from the distribution determined by this tree. Given a collection of previously tested feature subsets, we can use the Chow-Liu algorithm (Chow and Liu, 1968) to find the optimal tree model that fits the data (in the sense of maximizing the likelihood of the tested instances).[9] Then given the tree model, we can apply a depth first tree-traversal[10] that allows candidate feature subsets to be sampled from a concentrated subspace that is more likely to contain good solutions than mere random search. Figure 6.4 gives the pseudo-code of dependence-tree search. A detailed example of this algorithm can be found in (Baluja and Davies, 1997).

Initialize
- Generate N random bit-strings as candidate feature subsets

Iterate
- Evaluate each of the N candidate feature subsets by training the classifier on each feature subset and cross-validating.
- Collect the αN top-performing feature subsets (bit-strings), use them to update (with decay factor β) all pairwise mutual information between each pair of bits in the bit-string.
- Generate a maximum spanning tree for the bit-strings using Kruskal's algorithm .
- Generate N bit-strings based on joint probability encoded by the dependence tree (using depth first traversal).

if performance converges, **end** iteration

Figure 6.4. The dependence-tree search algorithm.

2.3   The ORDERED-FS Algorithm

For microarray data which have thousands of features, filter methods have the key advantage of significantly smaller computational complexity than wrapper methods. Therefore, these methods have been widely applied in the analysis of microarray data (Golub et al., 1999; Chow et al., 2002; Dudoit et al., 2000). But since a wrapper method searches for feature combinations that minimize classification error of a specific classifier, it can perform

---

[9] We skip the details of the Chow-Liu algorithm due to the space limit. Essentially, it constructs a maximum spanning tree from a complete graph of the feature nodes whose edges are weighted by the mutual information of the random variables connected by the edge.

[10] A strategy of touching every node in a tree by always visiting the child-node of the current node before going back to its parent-node and visit a sibling-node.

better than filter algorithms although at the cost of orders of magnitude of more computation time.

An additional problem with wrapper methods is that the repeated use of cross-validation on a single data set can potentially cause severe overfitting for problems with a few samples but very large hypothesis spaces, which is not uncommon for microarray data. While theoretical results show that exponentially many data points are needed to provide guarantees of choosing good feature subsets under the classic wrapper setting (Ng, 1998), Ng has recently described a generic feature selection methodology, referred to as ORDERED-FS, which leads to more optimistic conclusions (Ng, 1998). In this approach, cross-validation is used only to compare between feature subsets of different cardinality. Ng proves that this approach yields a generalization error that is upper-bounded by the logarithm of the number of irrelevant features.

Filter $(D = \{X_{N \times M}, C\})$
- Quantize each feature via mixture modeling (MM).
- Rank all features via information gain (IG) filter.
- Pick l features with highest IG, determine a removal order via Markov blanket (MB) filter.

Return an order π of the l features.

Wrapper (D, H, π)
   For $k = 1$ to l
- Train hypothesis $h_k \in H$ using the best k features.
- Leave-one-out cross-validation on h_k, compute ε_k.

   End
   $k^* = \arg \min_{k \in k}$
Return h_{k^*} (optimal hypothesis), k^* (optimal cardinality)

Figure 6.5. The ORDERED-FS algorithm.

Figure 6.5 presents an algorithmic instantiation of the ORDERED-FS approach in which filtering methods are used to choose best subsets for a given cardinality. We can use simple filter methods described earlier to carry out the major pruning of the hypothesis space, and use cross-validation for final comparisons to determine the optimal cardinality. This is in essence a hybrid of a filter and a wrapper method.

In Figure 6.6, we show training set and test set errors observed for the leukemia data when applying the ORDERED-FS algorithm.[11] Three different

---

[11] The 72 leukemia samples are split into two sets, with 38 (typeI / typeII = 27 / 11) serving as a training set and the remaining 34 (20 / 14) as a test set.

classifiers: a Gaussian quadratic classifier, a logistic linear classifier and a nearest neighbor classifier, are used (Xing et al., 2001). For all classifiers, after an initial coevolving trend of the training and testing curves for low-dimensional feature spaces, the classifiers quickly overfit the training data. For the logistic classifier and k-NN, the test error tops out at approximately 20 percent when the entire feature set of 7,130 genes is used. The Gaussian classifier overfits less severely in the full feature space. For all three classifiers, the best performance is achieved only in a significantly lower dimensional feature space.

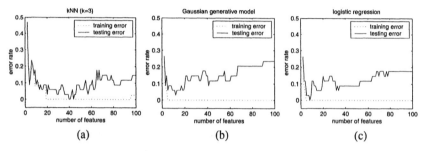

Figure 6.6. Classification in a sequence of different feature spaces with increasing dimensionality due to inclusion of gradually less qualified features. (a) Classification using k-NN classifier; (b) A quadratic Bayesian classifier given by a Gaussian generative model; (c) A linear classifier obtained from logistic regression. All three classifiers use the same 2-100 genes selected by the three stages of feature filtering.

Figure 6.6 shows that by an optimal choice of the number of features, it is possible to achieve error rates of 2.9%, 0%, and 0% for the Gaussian classifier, the logistic regression classifier, and k-NN, respectively. (Note that due to inductive bias, different types of classifiers admit different optimal feature subsets.) Of course, in actual diagnostic practice we do not have the test set available, so these numbers are optimistic. To choose the number of features in an automatic way, we make use of leave-one-out cross-validation on the training data.

The results of leave-one-out cross-validation are shown in Figure 6.7. Note that we have several minima for each of the cross-validation curves. Breaking ties by choosing the minima having the smallest cardinality, and running the resulting classifier on the test set, we obtain error rates of 8.8%, 0%, and 5.9% for the Gaussian classifier, the logistic regression classifier, and k-NN, respectively. The size of the optimal feature subsets determined hereby for the three classifiers are 6, 8 and 32, respectively.

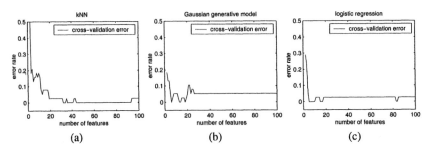

Figure 6.7. Plots of leave-one-out cross-validation error for the three classifiers.

3.   FEATURE WEIGHTING

Essentially, feature selection methods search in the combinatorial space of feature subsets, and pick an optimal subset of 'relevant' features as input to a learning algorithm. In contrast, feature weighting applies a weighting function to features, in effect assigning them a degree of perceived relevance and thereby performing feature selection implicitly during learning. In the following, we describe both a classic feature weighting scheme called WINNOW and a more general-purpose Bayesian learning technique that integrates feature weighting into the learning. For concreteness, we consider the generalized linear model (GLIM) for classification, where the input $x \in X$ (i.e. the measurement on the microarray) enters into the model via a linear combination $\xi = \theta^T x$ and the predictor, for example, the conditional distribution $p(y \mid x)$ of the corresponding label $y \in \{0, 1\}$ is characterized by an exponential family distribution with conditional mean $f(\xi)$, where f is known as a *response function*. Many popular classifiers belong to this family, for example, the logistic regression classifier:

$$P(y = 1 \mid x, \theta) = \frac{1}{1 + e^{-\theta^T x}} \tag{6.8}$$

3.1   The WINNOW Algorithm

The WINNOW algorithm is originally designed for learning Boolean monomials, or more generally, also k-DNF[12] formulas and r-of-k threshold functions[13], from noiseless data (Littlestone, 1988). Under these settings it enjoys worst-case loss logarithmic in the number of irrelevant features (i.e.

---

[12] A boolean formula is in *k-disjunctive normal form* (*k*-DNF) if it is expressed as a OR of clauses, each of which is the AND of *k* literals.

[13] For a chosen set of k ($k \le n$)\$ variables and a given number r ($1 \le r \le k$), an *r*-of-*k* threshold function is true if and only if at least r of the k relevant variables are true. The learning problem arises when both r and k are unknown.

the error rate is a function of the logarithm of the number of irrelevant features). For more realistic learning tasks encountered in microarray analysis, such as building a classifier from training set $\{(x^1, y^1),\ldots, (x^k, y^k)\}$, we can use the following multiplicative update rule for the weight of feature j: if the classifier misclassifies an input training vector x^i with true label y^i, then we update each component j of the weight vector w as:

$$w_j \leftarrow w_j \exp(\eta x_j^i y^i)$$

(6.9)

where η is a learning rate parameter, and the initial weight vector is set to $w_j = w_j, 0 > 0$. Where does w appear in the classifier? Back to the GLIM model, this simply means a slight change of the linear term ξ in the *response function*: $\xi = \theta^T (w \bullet x)$, where $w \bullet x$ means element-wise product of vectors w and x.

There are a number of variants of the WINNOW algorithm, such as normalized WINNOW, balanced WINNOW and large margin WINNOW. See (Zhang, 2000) and reference therein for more details.

3.2 Bayesian Feature Selection

Bayesian methods for feature selection have a natural appeal, because they model uncertainties present in the feature selection problems, and allow prior knowledge to be incorporated. In Bayesian feature selection, each feature is associated with a selection probability, and the feature selection process translates into estimating the posterior distribution over the feature-indicator variables. Irrelevant features quickly receive low albeit non-zero probability of being selected (Jebara and Jaakkola, 2000). This type of feature selection (which is carried out jointly with inductive learning) is most beneficial when the number of training examples is relatively small compared to their dimensionality.

Again, consider the classification of cancerous and non-cancerous samples measured on microarrays spanning n genes.

Following the representation introduced in Section 2.2, we can index each of the possible 2^n subsets of features by a random variable z, then the linear combination term ξ in the response function $f(\xi)$ essentially becomes

$$\xi = \sum_{i=1}^{n} \theta_i z_i x_i$$

(which obviates the effect of z_i as relevance indicator). Since the appropriate value of z is unknown, we can model the uncertainty underlying feature selection by a mixing prior:

$$P(\theta, z) = P_\theta(\theta) \prod_{i=1}^{n} P_z(z_i) \qquad (6.10)$$

where P_θ is a (conjugate) prior for the model parameters θ, and

$$P_z(z_i) = p_i^{z_i} (1 - p_i)^{1-z_i}, \qquad (6.11)$$

where p_i controls the overall prior probability of including feature i.

For a training set $D = \{X, Y\}$, the marginal posterior distribution $P(z \mid D)$ contains the information for feature selection, and the Bayesian optimal classifier is obtained by calculating:

$$P(y = 1 \mid x, D) = \sum_z \int_\theta p(y = 1 \mid x, \theta) P(\theta, z \mid D) d\theta \qquad (6.12)$$

For high dimensional problems and complex models we may encounter in microarray analysis, exact probabilistic computation of the posterior distribution $P(z, \theta \mid X, Y)$ as well as evaluation of the decision rule is intractable. Therefore, we need to use approximation techniques. George and McCulloch presented a detailed study of *Markov Chain Monte Carlo* methods such as *Gibbs sampler* or *Metropolis-Hasting algorithm* to explore the posterior distribution (George and McCulloch, 1997). Jebara and Jaakkola, on the other hand, took a *Maximum Entropy Discrimination* approach and derived a closed-form solution of the posterior distribution $P(z, \theta \mid X, Y)$ for some model families such as logistic regression and support vector machines (Jebara and Jaakola, 2000).

Recently, Ng and Jordan presented a *Voting Gibbs* classifier that solves the Bayesian feature selection problem in a surprisingly simple way (Ng and Jordan, 2001). Rather than taking Equation 6.11, they use a prior $P(\theta)$, assuming that the subset of relevant features is picked randomly according to the following procedure: **first**, sample the number r of relevant features uniformly from $\{0, 1, ..., n\}$; **then** a bit-string z in which r features are relevant is chosen randomly from one of the $\binom{n}{r}$ possible configurations. The prior $P(\theta)$ is constrained such that only the feature corresponding to an "on" bit in z has a non-zero prior. Thus, we have a parameter prior conditioned on z, $P_\Theta(\theta \mid z)$. Then we proceed to the usual Gibbs Voting classifier procedure where we sample N replicates of parameters θ from the posterior distribution $p(\theta \mid D)$, followed by N samples of y each from a particular $p(y = 1 \mid x, \theta)$. Finally, we vote for the result. A notable merit of this algorithm is its high tolerance to the presence of large number of irrelevant features. Ng and Jordan proved that their algorithm has sample complexity that is logarithmic in the number of irrelevant features.

4. FEATURE SELECTION FOR CLUSTERING

Clustering is another important type of analysis for microarray data. In contrast to classification, in this paradigm (known as unsupervised learning) a labeled training set is unavailable, and users are supposed to discover "meaningful" patterns (i.e. the existence of homogeneous groups that may correspond to particular macroscopic phenotypes such as clinical syndromes or cancer types) based on intrinsic properties of the data. Since microarrays usually measure thousands of genes for each sample, clustering a few hundred samples in such a high dimensional space may fail to yield a statistically significant pattern.

Eigenvector-based dimensionality reduction techniques such as *Multidimensional Scaling* (MDS) (Cox and Cox, 1994) and *Principal Component Analysis* (PCA) (Jolliffe, 1989; see Chapter 5 of the present book) handle this problem by trying to map the data onto a lower-dimensional space spanned by a small number of "virtual" features (e.g., the principal eigenvectors of the sample covariance matrix in case of PCA). However, microarray measurement is usually a highly noisy data source. Results from matrix stability theory suggest that even small perturbation may cause the eigenvector methods to pick a different set of eigenvectors (Ng et al., 2001). Moreover, in methods like PCA, the principal eigenvectors represent those directions in the original feature space along which data has the greatest variance, the presence of a few highly variable but not informative "noisy" genes tends to mislead the algorithm to a wrong set of discriminative eigenfeatures. Finally, identifiability remains an outstanding issue. In many situations we would like to explicitly recover genes that significantly contribute to the sample partition of interest. Eigenvector methods do not offer a convenient way to do so. (Each eigenvector from PCA is a linear combination of all the original features, eigenvectors from the Gram matrix in MDS even lack an explicit connection to the original features.)

Feature selection under the clustering paradigm is substantially more difficult than that for classification. The main difficulty lies in the absence of reference information for evaluating the relevance of features. Before concluding this chapter, we briefly introduce some of the recent attempts on this problem.

4.1 Category Utility

In the absence of class labels, one possible measure combining feature quality with the clustering performance is the average accuracy of predicting the value of each of the features in the data. The *category utility* metric is such a measure (Fisher, 1987). For a partition produced during clustering, the category utility is calculated as:

$$U = \frac{1}{K}[\sum_{k=1}^{K} P(C_k)\sum_{i=1}^{I}\sum_{j=1}^{J(i)} P(F_i = x_{ij} \mid C_k)^2 - \sum_{i=1}^{I}\sum_{j=1}^{J(i)} P(F_i = x_{ij})^2] \qquad (6.13)$$

where $P(F_i = x_{ij} \mid C_k)$ is the probability of feature F_i taking value x_{ij} conditional on class membership C_k, and $P(F_i = x_{ij})$ is the marginal probability of feature F_i taking value x_{ij} in the data set. Replacing the innermost summations with integration, category utility can be readily computed in the continuous domain for some distribution models (i.e. the mixture of two Gaussians we assumed in Section 2.1).

Devaney and Ram proposed a wrapper-like feature selection strategy using category utility as an evaluation function (Devaney and Ram, 1997). Essentially, any clustering algorithm can be used to evaluate the candidate feature subsets produced by a search heuristic based on this metric. The search terminates when category utility stops improving.

4.2 Entropy-Based Feature Ranking

Dash and Liu made an interesting empirical assumption on the relationship between the entropy and data distribution: two points belonging to the same cluster or in two different clusters will contribute less to the total entropy than if they were uniformly separated. They further reasoned that the former situation is more likely to happen if the *similarity* between the two points is either very high or low (rather than intermediate) (Dash and Liu, 2000). Then given distance measure (e.g., Euclidean distance or Pearson correlation computed using the selected features) $D_{i,j}$ between point i and j, we can compute the entropy of a data set as:

$$E = -\sum_{i \neq j} \sum_{j} (S_{i,j} \log S_{i,j} + (1 - S_{i,j}) \log(1 - S_{i,j})) \qquad (6.14)$$

where $S_{i,j} = \exp(-\alpha D_{i,j})$, and α is a scaling constant.

Based on this measure, one can rank features sequentially by discarding, one at a time, the feature whose removal results in minimum E (computed based on the remaining features). The optimal cardinality of the feature subset can be determined by an independent clustering algorithm (similar to the ORDERED-FS approach). However, the entropy assumption underlying this measure is only plausible when clusters are well separated and symmetric in shape. Under less ideal conditions, the performance is likely to break down.

4.3   The CLICK Algorithm

Xing and Karp proposed a strategy for feature selection in clustering that goes beyond the purely unsupervised feature evaluation techniques such as the entropy-based ranking or mixture-overlapping probability ranking (Xing and Karp, 2001). In their CLICK algorithm, they bootstrap an iterative feature selection and clustering process by using the most discriminative subset of features identified by the unsupervised mixture modeling to generate an *initial partition* of the samples. This partition is then used as an approximate reference for supervised feature selection based on information gain ranking and Markov blanket filtering, and then the algorithm alternates between computing a new reference partition given the currently selected features, and selecting a new set of features based on the current reference partition. It is hoped that at each iteration one can expect to obtain an approximate partition that is close to the target one, and thus allows the selection of an approximately good feature subset, which will hopefully draw the partition even closer to the target partition in the next iteration.

5.   CONCLUSION

At a conceptual level, one can divide the task of concept learning into the subtask of selecting a proper subset of features to use in describing the concept, and learning a hypothesis based on these features. This directly leads to a modular design of the learning algorithm which allows flexible combinations of explicit feature selection methods with model induction algorithms and sometimes leads to powerful variants. Many recent works, however, tend to take a more general view of feature selection as part of model selection and therefore integrate feature selection more closely into the learning algorithms (i.e. the Bayesian feature selection methods). Feature selection for clustering is a largely untouched problem, and there has been little theoretical characterization of the heuristic approaches we described in the chapter. In summary, although no universal strategy can be prescribed, for high-dimensional problems frequently encountered in microarray analysis, feature selection offers a promising suite of techniques to improve interpretability, performance and computation efficiency in learning.

ACKNOWLEDGMENTS

I thank Professor Richard Karp and Dr. Wei Wu for helpful comments on the manuscript.

REFERENCES

Baluja S. and Davies S. (1997). Using Optimal Dependency-Trees for Combinatorial Optimization: Learning the Structure of the Search Space, Proceedings of the Fourteenth International Conference on Machine Learning.

Ben-Dor A., Friedman N. and Yakhini Z. (2000). Scoring genes for relevance, Agilent Technologies Technical Report AGL-2000-19.

Blum A. and Langley P. (1997). Selection of Relevant Features and Examples in Machine Learning, Artificial Intelligence 97:245-271.

Chow M.L and Liu C. (1968). Approximating discrete probability distribution with dependency tree, IEEE Transactions on Information Theory 14:462-367.

Chow M.L., Moler E.J., Mian I.S. (2002). Identification of marker genes in transcription profiling data using a mixture of feature relevance experts, Physiological Genomics (in press).

Cover T. and Thomas J. (1991). Elements of Information Theory, Wiley, New York.

Cox T. and Cox M. (1994). Multidimensional Scaling, Chapman & Hall, London.

Dash M. and Liu H. (2000). Feature Selection for Clustering, PAKDD, 110-121.

Dempster A.P., Laird N.M., Revow M.(1977). Maximum likelihood from incomplete data via the EM algorithm, Journal of the Royal Statistical Society, B39(1):1-38.

Devaney M. and Ram A. (1997) Efficient feature selection in conceptual clustering, Proceedings of the Fourteenth International Conference on Machine Learning, Morgan Kaufmann, San Francisco, CA, 92-97.

Dudoit S., Fridlyand J., Speed T. (2000). Comparison of discrimination methods for the classification of tumors using gene expression data, Technical report 576, Department of Statistics, UC Berkeley.

Fisher D. H. (1987). Knowledge Acquisition via Incremental Conceptual Clustering, Machine Learning 2:139-172.

George E.I. and McCulloch R.E. (1997). Approaches for Bayesian variable selection, Statistica Sinica 7:339-373.

Golub T.R., Slonim D.K., Tamayo P., Huard C., Gaasenbeek M., Mesirov J.P., Coller H., Loh M.L., Downing J.R, Caligiuri M.A., Bloomfield C.D., Lander E.S. (1999). Molecular Classification of Cancer: Class Discovery and Class Prediction by Gene Expression Monitoring, Science 286:531-537.

Jebara T. and Jaakola T. (2000). Feature selection and dualities in maximum entropy discrimination, Proceedings of the Sixteenth Annual Conference on Uncertainty in Artificial Intelligence, Morgan Kaufman.

Jolliffe I.T. (1989). Principal Component Analysis, Springer-Verlag, New York.

Koller D. and Sahami M. (1996). Toward optimal feature selection, Proceedings of the Thirteenth International Conference on Machine Learning, ICML96, 284-292.

Littlestone N. (1988). Learning quickly when irrelevant attribute abound: A new linear-threshold algorithm, Machine Learning 2:285-318.

Ng A.Y. (1988). On feature selection: Learning with exponentially many irrelevant features as training examples, Proceedings of the Fifteenth International Conference on Machine Learning.

Ng A.Y. and Jordan M. (2001). Convergence rates of the voting Gibbs classifier, with application to Bayesian feature selection, Proceedings of the Eighteenth International Conference on Machine Learning.

Ng A.Y., Zheng A.X., Jordan M. (2001). Link analysis, eigenvectors, and stability, Proceedings of the Seventeenth International Joint Conference on Artificial Intelligence.

Russell S. and Norvig P. (1995). Artificial Intelligence, A Modern Approach, Prentice Hall, New Jersey.

Xing E.P., Jordan M., Karp R.M. (2001). Feature selection for high-dimensional genomic microarray data, Proceedings of the Eighteenth International Conference on Machine Learning.

Xing E.P. and Karp R.M. (2001). Cliff: Clustering of high-dimensional microarray data via iterative feature filtering using normalized cuts, Bioinformatics 1(1):1-9.

Zhang T. (2000). Large margin winnow methods for text categorization, KDD 2000 Workshop on Text Mining, 81-87.

Chapter 7

INTRODUCTION TO CLASSIFICATION IN MICROARRAY EXPERIMENTS

Sandrine Dudoit[1] and Jane Fridlyand[2]

[1]*Assistant Professor, Division of Biostatistics, School of Public Health, University of California, Berkeley, 140 Earl Warren Hall, # 7360, Berkeley, CA 94720-7360,*
e-mail: sandrine@stat.berkeley.edu.

[2]*Postdoctoral Scientist, Jain Lab, Comprehensive Cancer Center, University of California, San Francisco, 2340 Sutter St., # N412, San Francisco, CA 94143-0128,*
e-mail: janef@cc.ucsf.edu.

1. INTRODUCTION

1.1 Motivation: Tumor Classification Using Gene Expression Data

An important problem in microarray experiments is the classification of biological samples using gene expression data. To date, this problem has received the most attention in the context of cancer research; we thus begin this chapter with a review of tumor classification using microarray gene expression data. A reliable and precise classification of tumors is essential for successful diagnosis and treatment of cancer. Current methods for classifying human malignancies rely on a variety of clinical, morphological, and molecular variables. In spite of recent progress, there are still uncertainties in diagnosis. Also, it is likely that the existing classes are heterogeneous and comprise diseases that are molecularly distinct and follow different clinical courses. cDNA microarrays and high-density oligonucleotide chips are novel biotechnologies which are being used increasingly in cancer research (Alizadeh et al., 2000; Alon et al., 1999; Bittner et al., 2000; Chen et al., 2002; Golub et al., 1999; Perou et al., 1999; Pollack et al., 1999; Pomeroy et al., 2002; Ross et al., 2000; Sørlie et al., 2001). By allowing the monitoring of expression levels in cells for thousands of genes simultaneously, microarray experiments may lead to a more complete understanding of the molecular variations among tumors and hence to a finer and more reliable classification.

Recent publications on cancer classification using gene expression data have mainly focused on the cluster analysis of both tumor samples and genes, and include applications of hierarchical clustering (Alizadeh et al., 2000; Alon et al., 1999; Bittner et al., 2000; Chen et al., 2002; Perou et al., 1999; Pollack et al., 1999; Pomeroy et al., 2002; Ross et al., 2000; Sørlie et al., 2001) and partitioning methods such as self-organizing maps (Golub et al., 1999; Pomeroy et al., 2002). Alizadeh et al. used cDNA microarray analysis of lymphoma mRNA samples to identify two previously unrecognized and molecularly distinct subclasses of diffuse large B-cell lymphomas corresponding to different stages of B-cell differentiation (Alizadeh et al., 2000). One type expressed genes characteristic of germinal center B-cells (*germinal center B-like DLBCL* class) and the second type expressed genes normally induced during in vitro activation of peripheral blood B-cells (*activated B-like DLBCL* class). They also demonstrated that patients with the two subclasses of tumors had different clinical prognoses. Average linkage hierarchical clustering was used to identify the two tumor subclasses as well as to group genes with similar expression patterns across the different samples. Ross et al. used cDNA microarrays to study gene expression in the 60 cell lines from the National Cancer Institute's anti-cancer drug screen (NCI 60) (Ross et al., 2000). Hierarchical clustering of the cell lines based on gene expression data revealed a correspondence between gene expression and tissue of origin of the tumors. Hierarchical clustering was also used to group genes with similar expression patterns across the cell lines. Using acute leukemias as a test case, Golub et al. looked into both the cluster analysis and the discriminant analysis of tumors using gene expression data (Golub et al., 1999). For cluster analysis, or class discovery, self-organizing maps (SOMs) were applied to the gene expression data and the tumor groups revealed by this method were compared to known classes. For class prediction, Golub et al. proposed a weighted gene-voting scheme that turns out to be a variant of a special case of linear discriminant analysis, which is also known as naive Bayes' classification. More recently, Pomeroy et al. used Affymetrix oligonucleotide chips to study gene expression in embryonal tumors of the central nervous system (CNS) (Pomeroy et al., 2002). A range of unsupervised and supervised learning methods were applied to investigate whether gene expression data could be used to distinguish among new and existing CNS tumor classes and for patient prognosis. The recent studies just cited are instances of a growing body of research, in which gene expression profiling is used to, distinguish among known tumor classes, predict clinical outcomes such as survival and response to treatment, and identify previously unrecognized and clinically significant subclasses of tumors.

Microarray experiments in cancer research are not limited to monitoring transcript or mRNA levels. Other widely-used applications of the microarray technology are comparative genomic hybridization (CGH) (Jain et al., 2001) and methylation studies (Costello et al., 2000). CGH experiments measure DNA copy number across the genome, whereas methylation studies determine the methylation status of genes of interest. Similar classification questions arise in these other types of microarray experiments. In addition, cancer research is only one of the many areas of application of the microarray technology. In immunology, microarrays have recently been used to study the gene expression host response to infection by bacterial pathogens (Boldrick et al., 2002). Clinical implications include improved diagnosis of bacterial infections by gene expression profiling.

The above examples illustrate that class prediction is an important question in microarray experiments, for purposes of classifying biological samples and predicting clinical or other outcomes using gene expression data. A closely related issue is that of feature or variable selection, i.e. the identification of genes that characterize different tumor classes or have good predictive power for an outcome of interest.

1.2   Outline

The present chapter discusses statistical issues arising in the classification of biological samples using gene expression data from DNA microarray experiments. Section 2 gives further background on classification of microarray experiments and introduces the statistical foundations of classification. Section 3 provides an overview of traditional classifiers, such as linear discriminant analysis and nearest neighbor classifiers. The general issues of feature selection and classifier performance assessment are discussed in Sections 4 and 5, respectively.

The reader is referred to the texts of (Hastie et al., 2001; Mardia et al., 1979; McLachlan, 1992; Ripley, 1996) for general discussions of classification. Recent work on statistical aspects of classification in the context of microarray experiments includes: (Chow et al., 2001; Dettling and Buelmann, 2002; Golub et al., 1999; Pomeroy et al., 2002; Tibshirani et al., 2002; West et al., 2001). These articles have mostly focused on existing methods or variants thereof, and, in many cases, comparison studies have been limited and not always properly calibrated. Studies performed to date suggest that simple methods, such as nearest neighbor or naive Bayes classification, perform as well as more complex approaches, such as aggregated classification trees or Support Vector Machines (SVMs) (Dudoit et al., 2002; Dudoit and Fridlyand, 2002). Basic classifiers considered in these references are discussed in Section 3 of the present chapter, while SVMs and trees are discussed in Chapters 9 and 10, respectively. Feature selection is discussed in greater detail in Chapter 6.

2. STATISTICAL FOUNDATIONS OF CLASSIFICATION

2.1 Background on Classification

2.1.1 Unsupervised vs. Supervised Learning

In many situations, one is concerned with assigning objects to classes on the basis of measurements made on these objects. There are two main aspects to such problems: discrimination and clustering, or supervised and unsupervised learning. In *unsupervised learning* (also known as *cluster analysis*, *class discovery*, and *unsupervised pattern recognition*), the classes are *unknown a priori* and need to be discovered from the data. This involves estimating the number of classes (or clusters) and assigning objects to these classes (see Chapters 13 – 16). In contrast, in *supervised learning* (also known as *classification*, *discriminant analysis*, *class prediction*, and *supervised pattern recognition*), the classes are *predefined* and the task is to understand the basis for the classification from a set of labeled objects (learning set). This information is then exploited to build a classifier that will be used to predict the class of future unlabeled observations. In many situations, the two problems are related, as the classes which are discovered from unsupervised learning are often used later on in a supervised learning setting. Here, we focus on supervised learning, and use the simpler term classification.

2.1.2 Classification

Classification is a *prediction* or *learning* problem, in which the variable to be predicted assumes one of K unordered values, $\{c_1, c_2, ..., c_K\}$, arbitrarily relabeled as $\{1, 2, ..., K\}$ or $\{0, 1, ..., K-1\}$, and sometimes $\{-1, 1\}$ in binary classification. The K values correspond to K predefined classes, e.g., tumor class or bacteria type. Associated with each object is a *response* or *dependent variable* (class label), $Y \in \{1, 2, ..., K\}$ and a set of G measurements which form the *feature vector* or *vector of predictor variables*, $\mathbf{X} = (X_1, ..., X_G)$. The feature vector $\mathbf{X}$ belongs to a *feature space* X, e.g., the real numbers $\mathfrak{R}^G$. The task is to classify an object into one of the K classes on the basis of an observed measurement $\mathbf{X} = \mathbf{x}$, i.e. predict Y from $\mathbf{X}$.

A *classifier* or *predictor* for K classes is a mapping C from X into $\{1, 2,..., K\}$, $C : X \rightarrow \{1, 2,..., K\}$, where $C(\mathbf{x})$ denotes the predicted class for a feature vector $\mathbf{x}$. That is, a classifier C corresponds to a *partition* of the feature space X into K disjoint and exhaustive subsets, $A_1, ..., A_K$, such that a sample with feature vector $\mathbf{x} = (x_1,..., x_G) \in A_k$ has predicted class $\hat{y} = k$ (modifications can be made to allow doubt or outlier classes (Ripley, 1996)).

Classifiers are built from past experience, i.e. from observations which are known to belong to certain classes. Such observations comprise the *learning set* (LS), $\mathcal{L} = \{(\mathbf{x}_1, y_1),...,(\mathbf{x}_n, y_n)\}$. A classifier built from a learning set $\mathcal{L}$ is denoted by $C(\cdot; \mathcal{L})$. When the learning set is viewed as a collection of random variables, the resulting classifier is also a *random variable*. Intuitively, for a fixed value of the feature vector $\mathbf{x}$, as the learning set varies, so will the predicted class $C(\mathbf{x}; \mathcal{L})$. It is thus meaningful to consider distributional properties (e.g., bias and variance) of classifiers when assessing or comparing the performance of different classifiers (Friedman, 1996; Breiman, 1998).

2.1.3   Classification in Microarray Experiments

In the case of gene expression data from cancer DNA microarray experiments, features correspond to the expression measures of different genes and classes correspond to different types of tumors (e.g., nodal positive vs. negative breast tumors, or tumors with good vs. bad prognosis). There are three main types of statistical problems associated with tumor classification: (i) the identification of new tumor classes using gene expression profiles – *unsupervised learning*; (ii) the classification of malignancies into known classes – *supervised learning*; and (iii) the identification of marker genes that characterize the different tumor classes – *feature selection*. The present chapter focuses primarily on (ii) and briefly addresses the related issue of feature selection.

For our purpose, gene expression data on G genes for n tumor mRNA samples may be summarized by a $G \times n$ matrix $X = (x_{gi})$, where x_{gi} denotes the expression measure of gene (variable) g in mRNA sample (observation) i. The expression levels might be either absolute (e.g., Affymetrix oligonucleotide arrays) or relative to the expression levels of a suitably defined common reference sample (e.g., 2-color cDNA microarrays). When the mRNA samples belong to known classes (e.g., ALL and AML tumors), the data for each observation consist of a *gene expression profile* $\mathbf{x}_i = (x_{1i},..., x_{Gi})$ and a class label y_i, i.e. of predictor variables $\mathbf{x}_i$ and response y_i. For K tumor classes, the class labels y_i are defined to be integers ranging from 1 to K, and n_k denotes the number of learning set observations belonging to class k. Note that the expression measures x_{gi} are in general highly processed data: the raw data in a microarray experiment consist of image files, and important pre-processing steps include image analysis of these scanned images and normalization. Data from these new types of experiments present a so-called *large p, small n-problem*, that is, a very large number of variables (genes) relative to the number of observations (tumor samples). The publicly available datasets typically contain expression data on 5,000 – 10,000 genes for less than 100 tumor samples. Both numbers are expected to grow, the number of genes reaching on the order of 30,000, an estimate for the total number of genes in the human genome.

There are many different approaches for building a classifier for tumor samples using gene expression data. Basic methods such as naive Bayes and nearest neighbor classification have been found to perform very well and are described in Section 3. The important and closely related issue of gene selection is briefly discussed in Section 4. Different classifiers will clearly have varying performance, i.e. different classification error rates. In the context of tumor class prediction, errors could correspond to misdiagnosis and assignment to improper treatment protocol. Thus, an essential task is to assess the accuracy of the classifier. Performance assessment is discussed in Section 5.

2.2 Classification and Statistical Decision Theory

Classification can be viewed as a *statistical decision theory* problem. For each object, an observed feature vector $\mathbf{x}$ is examined to decide which of a fixed set of classes that object belongs to. Assume observations are independently and identically distributed (i.i.d.) from an unknown multivariate distribution. The class k *prior*, or proportion of objects of class k in the population of interest, is denoted as $\pi_k = p(Y = k)$. Objects in class k have feature vectors with *class conditional density* $p_k(\mathbf{x}) = p(\mathbf{x} \mid Y = k)$.

It will be useful to introduce the notion of a *loss function*. The loss function $L(h, l)$ simply elaborates the loss incurred if a class h case is erroneously classified as belonging to class l. The *risk function* for a classifier C is the expected loss when using it to classify, that is,

$$R(C) = E[L(Y, C(\mathbf{X}))] = \sum_{k=1}^{K} \int L(k, C(\mathbf{x})) p_k(\mathbf{x}) \pi_k, \qquad (7.1)$$

For an observation with feature vector $\mathbf{X}$ and true class Y, the predicted class is $C(\mathbf{X})$ and the corresponding loss is $L(Y, C(\mathbf{X}))$. The expected loss can be thought of as a weighted average of the loss function, where weights are given by the joint density of random variables $\mathbf{X}$ and Y. Thus, the risk is given by the double integral of the loss function times the joint density of $\mathbf{X}$ and Y. Since Y is a discrete random variable, taking on the values $1, \dots, K$, the integral is replaced by a sum $\sum_{k=1}^{K}$ over the possible values of Y.

Furthermore, for $Y = k$, we may express the joint density of $\mathbf{X}$ and Y as $p_k(\mathbf{x})\pi_k$. Typically $L(h, h)$, and in many cases the loss is symmetric with $L(h, l) = 1$, $h \neq l$ – making an error of one type is equivalent to making an error of a different type. Then, the risk is simply the *misclassification rate* (also often called *generalization error*),

$$p(C(\mathbf{X}) \neq Y) = \sum_k \int_{C(\mathbf{x}) \neq k} p_k(\mathbf{x})\pi_k .$$

However, for some important examples such as medical diagnosis, the loss function is not symmetric. Note that here the classifier is viewed as fixed, that is, probabilities are conditional on the learning set $\mathcal{L}$.

2.2.1 The Bayes Rule

In the unlikely situation that the class conditional densities $p_k(\mathbf{x}) = p(\mathbf{x} \mid Y = k)$ and class priors π_k are known, Bayes' Theorem may be used to express the posterior probability $p(k \mid \mathbf{x})$ of class k given feature vector $\mathbf{x}$ as

$$p(k \mid \mathbf{x}) = \frac{\pi_k p_k(\mathbf{x})}{\sum_{l=1}^{K} \pi_l p_l(\mathbf{x})} \tag{7.2}$$

The Bayes rule predicts the class of an observation $\mathbf{x}$ by that with highest posterior probability

$$C_B(\mathbf{x}) = \operatorname{argmax}_k p(k \mid \mathbf{x}) \tag{7.3}$$

The posterior class probabilities reflect the confidence in predictions for individual observations, the closer they are to one, the greater the confidence. The Bayes rule minimizes the total risk or misclassification rate under a symmetric loss function – Bayes risk. Note that the Bayes risk gives an upper bound on the performance of classifiers in the more realistic setting where the distributions of $\mathbf{X}$ and Y are unknown. For a general loss function, the classification rule which minimizes the total risk is

$$C_B(\mathbf{x}) = \operatorname{argmin}_l \sum_{h=1}^{K} L(h, l) p(h \mid \mathbf{x}) \tag{7.4}$$

In the special case when $L(h, l) = L_h I(h \neq l)$, that is, the loss incurred from misclassifying a class h observation is the same irrespective of the predicated class l, the Bayes rule is

$$C_B(\mathbf{x}) = \operatorname{argmax}_k L_k p(k \mid \mathbf{x}) \tag{7.5}$$

Suitable adjustments can be made to accommodate the doubt and outlier classes (Ripley, 1996).

Many classifiers can be viewed as versions of this general rule, with particular parametric or non-parametric estimates of $p(k \mid \mathbf{x})$. There are two general paradigms for estimating the class posterior probabilities $p(k \mid \mathbf{x})$: the density estimation and the direct function estimation paradigms (Friedman, 1996). In the density estimation approach, class conditional densities $p_k(\mathbf{x}) = p(\mathbf{x} \mid Y = k)$ (and priors π_k) are estimated separately for each class and Bayes' Theorem is applied to obtain estimates of $p(k \mid \mathbf{x})$. Classification procedures employing density estimation include: *Gaussian maximum likelihood discriminant rules*, a.k.a. discriminant analysis (chapter 3 in (Ripley, 1996), and Section 3.1 below); *learning vector quantization* (section 6.3 in (Ripley, 1996)); *Bayesian belief networks* (Chapter 8 in the present book and chapter 8 in (Ripley, 1996)). Another example is given by *naive Bayes methods* which approximate class conditional densities $p_k(\mathbf{x})$ by the product of their marginal densities on each feature variable. In the direct function estimation approach, class conditional probabilities $p(k \mid \mathbf{x})$ are estimated directly based on function estimation methodology such as regression. This paradigm is used by popular classification procedures such as: *logistic regression* (chapter 3 in (Ripley, 1996); *neural networks* (chapter in (Ripley, 1996), and Chapter 11 in the present book); *classification trees* (Chapter 10 in the present book and (Breiman et al., 1984)); *projection pursuit* (section 6.1 in (Ripley, 1996)); and *nearest neighbor classifiers* (section 6.2 in (Ripley, 1996), Chapter 12 in the present book, and Section 3.2 below).

2.2.2 Maximum Likelihood Discrininant Rules

The frequentist analogue of the Bayes rule is the maximum likelihood discriminant rule. For known class conditional densities $p_k(\mathbf{x}) = p(\mathbf{x} \mid Y = k)$, the *maximum likelihood* (ML) discriminant rule predicts the class of an observation $\mathbf{x}$ by that which gives the largest likelihood to $\mathbf{x}$: $C(\mathbf{x}) = \mathrm{argmax}_k p_k(\mathbf{x})$. In the case of equal class priors π_k, this amounts to maximizing the posterior class probabilities $p(k \mid \mathbf{x})$, i.e. the Bayes rule. Otherwise, the ML rule is not optimal, in the sense that it does not minimize the Bayes risk.

3. BASIC CLASSIFIERS

3.1 Linear and Quadratic Discriminant Analysis

Linear and quadratic (in the features $\mathbf{x}$) discriminant rules are classical and widely used classification tools. They arise as Bayes rules or maximum likelihood discriminant rules when features have Gaussian distributions within each class. For multivariate normal class densities, i.e. for $\mathbf{x} \mid y = k \sim N(\mu_k, \Sigma_k)$ (here, μ_k and Σ_k denote respectively the expected value and the $G \times G$ covariance matrix of the feature vector in class k), the Bayes rule is

$$C(\mathbf{x}) = \operatorname{argmin}_k \left\{ (\mathbf{x} - \mu_k) \Sigma_k^{-1} (\mathbf{x} - \mu_k)' + \log |\Sigma_k| - 2 \log \pi_k \right\}. \qquad (7.6)$$

In general, this is a quadratic discriminant rule – *Quadratic Discriminant Analysis* – QDA. The main quantity in the discriminant rule is $(\mathbf{x} - \mu_k) \Sigma_k^{-1} (\mathbf{x} - \mu_k)'$, the squared Mahalanobis distance from the observation $\mathbf{x}$ to the class k mean vector μ_k. Interesting special cases are described below for homogeneous priors, i.e. for π_k constant in k.

Linear Discriminant Analysis (LDA). When the class densities have the same covariance matrix, $\Sigma_k = \Sigma$, the discriminant rule is based on the square of the Mahalanobis distance and is linear in $\mathbf{x}$ and given by

$$C(\mathbf{x}) = \operatorname{argmin}_k (\mathbf{x} - \mu_k) \Sigma^{-1} (\mathbf{x} - \mu_k)'.$$

Diagonal Quadratic Discriminant Analysis (DQDA). When the class densities have diagonal covariance matrices, $\Delta_k = \operatorname{diag}(\sigma_{k1}^2, ..., \sigma_{kG}^2)$, the discriminant rule is given by additive quadratic contributions from each gene, that is,

$$C(\mathbf{x}) = \operatorname{argmin}_k \sum_{g=1}^{G} \left\{ \frac{(x_g - \mu_{kg})^2}{\sigma_{kg}^2} + \log \sigma_{kg}^2 \right\}.$$

Diagonal Linear Discriminant Analysis (DLDA). When the class densities have the same diagonal covariance matrix $\Delta = \operatorname{diag}(\sigma_1^2, ..., \sigma_G^2)$, the discriminant rule is linear and given by

$$C(\mathbf{x}) = \operatorname{argmin}_k \sum_{g=1}^{G} \frac{(x_g - \mu_{kg})^2}{\sigma_g^2}.$$

Note that DLDA and DQDA correspond to naive Bayes rules for Gaussian class conditional densities. As with any classifier explicitly estimating the Bayes rule, class posterior probabilities may be used to assess the confidence in the predictions for individual observations. For the sample Bayes or ML discriminant rules, the population mean vectors and covariance matrices are estimated from a learning set L, by the sample mean vectors and covariance matrices, respectively: $\hat{\mu}_k = \overline{\mathbf{x}}_k$ and $\hat{\Sigma}_k = S_k$. For the constant covariance matrix case, the pooled estimate of the common covariance matrix is used:

$$\hat{\Sigma} = S = \sum\nolimits_k (n_k - 1) S_k / (n - K).$$

The weighted gene voting scheme of (Golub et al., 1999) turns out to be a variant of sample DLDA for $K = 2$ classes, with the variance σ_g^2 replaced by a sum of standard deviations $\sigma_{g1} + \sigma_{g2}$, thus resulting in the wrong units for the discriminant rule (Dudoit et al., 2002; Dudoit and Fridlyand, 2002).

The above simple classifiers may be modified easily to allow unequal class priors; estimates of the priors may be obtained from the sample class proportions $\hat{\pi}_k = n_k / n$. Biased sampling of the classes (e.g., oversampling of a class that is rare in the population) and differential misclassification costs (e.g., higher cost for misclassifying a diseased person as healthy than for the reverse error) can be handled similarly by imposing different weights or cutoffs on the posterior class probabilities. These and other extensions to linear discriminant analysis are summarized in (Dudoit and Fridlyand, 2002).

3.2 Nearest Neighbor Classifiers

Nearest neighbor methods are based on a *distance function* for pairs of samples, such as the Euclidean distance or one minus the correlation of their gene expression profiles. The basic *k-nearest neighbor* (*k*-NN) rule proceeds as follows to classify a new observation on the basis of the learning set: (i) find the k closest samples in the learning set, and (ii) predict the class by majority vote, i.e. choose the class that is most common among those k neighbors. Nearest neighbor classifiers were initially proposed by Fix and Hodges (1951) as consistent non-parametric estimates of maximum likelihood discriminant rules.

Number of neighbors *k*. Although classifiers with $k = 1$ are often quite successful, the number of neighbors k can have a large impact on the performance of the classifier and should be chosen carefully. A common approach for selecting the number of neighbors is *leave-one-out cross-validation*. Each sample in the learning set is treated in turn as if its class were unknown: its distance to all of the other learning set samples (except itself) is computed and it is classified by the nearest neighbor rule. The classification for each learning set observation is then compared to the truth to produce the cross-validation error rate. This is done for a number of k's (e.g., $k \in \{1, 3, 5, 7\}$) and the k for which the cross-validation error rate is smallest is retained.

The nearest neighbor rule can be refined and extended to deal with unequal class priors, differential misclassification costs, and feature selection. Many of these refinements involve some form of *weighted voting* for the neighbors, where weights reflect priors and costs. Feature selection may be performed so that the relevance of each variable is estimated locally for each new case (Friedman, 1994). The reader is referred to section 6.2 in (Ripley, 1996), sections 13.3 – 13.5 in (Hastie et al., 2001), and (Dudoit and Fridlyand, 2002) for a more detailed discussion of the nearest neighbor rule and its extensions.

4. FEATURE SELECTION

Feature selection is one of the most important issues in classification; it is particularly relevant in the context of microarray datasets with thousands of features, most of which are likely to be uninformative. Some classifiers like classification trees (Breiman et al., 1984) perform automatic feature selection and are relatively insensitive to the variable selection scheme. In contrast, standard LDA and nearest neighbor classifiers do not perform feature selection; all variables, whether relevant or not, are used in building the classifier. For many classifiers, it is thus important to perform some type of feature selection; otherwise performance could degrade substantially with a large number of irrelevant features. Feature selection may be performed *explicitly*, prior to building the classifier, or *implicitly*, as an inherent part of the classifier building procedure, for example using modified distance functions. In the machine learning literature, these two approaches are referred to as *filter* and *wrapper* methods, respectively (see Chapter 6 for greater detail).

Filter methods. The simplest approaches are *one-gene-at-a-time* approaches, in which genes are ranked based on the value of univariate test statistics such as: t- or F-statistics (Dudoit et al., 2002); ad hoc signal-to-noise statistics (Golub et al., 1999; Pomeroy et al., 2002); non-parametric Wilcoxon statistics (Dettling and Buelmann, 2002); p-values. Possible meta-parameters for feature selection include the number of genes G or a p-value cut-off. A formal choice of these parameters may be achieved by cross-validation or bootstrapping. More refined feature selection procedures consider the *joint* distribution of the gene expression measures, and include forward variable selection (Bø and Jonassen, 2002) and selection based on prediction accuracy (Breiman, 1999).

Wrapper methods. Feature selection may also be performed implicitly by the classification rule itself. In this case, different approaches to feature selection will be used by different classifiers. In classification trees (e.g., CART (Breiman et al., 1984)), features are selected at each step based on reduction in impurity and the number of features used (or size of the tree) is determined by pruning the tree using cross-validation. Thus, feature selection is an inherent part of tree building and pruning deals with the issue of overfitting. Suitable modifications of the distance function in nearest neighbor classification allow automatic feature selection.

The importance of taking feature selection into account when assessing the performance of the classifier cannot be stressed enough (see West et al., 2001, and Section 5 below). Feature selection *is* an aspect of building the predictor, whether done explicitly or implicitly. Thus, when using for example cross-validation to estimate generalization error, feature selection should *not* be done on the entire learning set, but separately for each cross-validation training sample used to build the classifier. Leaving out feature selection from cross-validation or other resampling-based performance assessment methods results in overly optimistic error rates.

5. PERFORMANCE ASSESSMENT

Different classifiers clearly have different accuracies, i.e. different misclassification rates. In certain medical applications, errors in classification could have serious consequences. For example, when using gene expression data to classify tumor samples, errors could correspond to misdiagnosis and assignment to improper treatment protocol. It is thus essential to obtain reliable estimates of the classification error $p(C(\mathbf{X}) = Y)$ or of other measures of performance. Different approaches are reviewed next. For a more detailed discussion of performance assessment and of the bias and variance properties of classifiers, the reader is referred to section 2.7 in (Ripley, 1996), (Friedman, 1996), (Breiman, 1998), and chapter 7 in (Hastie et al., 2001).

5.1 Resubstitution Estimation

In this naive approach, known as resubstitution error rate estimation or training error rate estimation, the same dataset is used to build the classifier and assess its performance. That is, the classifier is *trained* using the entire learning set $\mathcal{L}$ and an estimate of the classification error rate is obtained by running the *same* learning set $\mathcal{L}$ through the classifier and recording the number of observations with discordant predicted and actual class labels. Although this is a simple approach, the resubstitution error rate can be severely biased downward. Consider the trivial and extreme case when the feature space is partitioned into n sets, each containing a learning set observation. In this extreme overfitting situation, the resubstitution error rate is zero. However, such a classifier is unlikely to generalize well, that is, the classification error rate (as estimated from a test set) is likely to be high. In general, as the complexity of the classifier increases (i.e. the number of training cycles or epochs increases), the training set error decreases. In contrast, the true generalization error initially decreases but subsequently increases due to overfitting.

5.2   Test Set Estimation

Suppose that a test set of labeled observations sampled independently from the same population as the learning set is available. In such a case, an unbiased estimate of the classification error rate may be obtained by running the test set observations through the classifier built from the learning set and recording the proportion of test cases with discordant predicted and actual class labels.

In the absence of a genuine test set, cases in the learning set $\mathcal{L}$ may be divided into two sets, a *training set* $\mathcal{L}_1$ and a *validation set* $\mathcal{L}_2$. The classifier is built using $\mathcal{L}_1$, and the error rate is computed for $\mathcal{L}_2$. It is important to ensure that observations in $\mathcal{L}_1$ and $\mathcal{L}_2$ can be viewed as i.i.d. samples from the population of interest. This can be achieved in practice by randomly dividing the original learning set into two subsets. In addition, to reduce variability in the estimated error rates, this procedure may be repeated a number of times (e.g., 50) and error rates averaged (Breiman, 1998). A general limitation of this approach is that it reduces effective sample size for training purposes. This is an issue for microarray datasets, which have a limited number of observations. There are no widely accepted guidelines for choosing the relative size of these artificial training and validation sets. A possible choice is to leave out a randomly selected 10% of the observations to use as a validation set. However, for comparing the error rates of different classifiers, validation sets containing only 10% of the data are often not sufficiently large to provide adequate discrimination. Increasing validation set size to one third of the data provides better discrimination in the microarray context.

5.3   Cross-Validation Estimation

In *V-fold cross-validation* (CV), cases in the learning set $\mathcal{L}$ are randomly divided into V sets $\mathcal{L}_v$, $v = 1,\ldots, V$ of as nearly equal size as possible. Classifiers are built on *training sets* $\mathcal{L} - \mathcal{L}_v$, error rates are computed for the *validation sets* $\mathcal{L}_v$, and averaged over v. There is a *bias-variance trade-off* in the selection of V: small V's typically give a larger bias, but a smaller variance and mean squared error.

A commonly used form of CV is *leave-one-out cross-validation* (LOOCV), where $V = n$. LOOCV often results in low bias but high variance estimates of classification error. However, for stable (low variance) classifiers such as k-NN, LOOCV provides good estimates of generalization error rates. For large learning sets, LOOCV carries a high computational burden, as it requires n applications of the training procedure.

5.4   General Issues in Performance Assessment

The use of cross-validation (or any other estimation method) is intended to provide accurate estimates of classification error rates. It is important to note that these estimates relate *only* to the experiment that was (cross-) validated. There is a common practice in microarray classification of doing feature selection using *all* of the learning set and then using cross-validation only on the classifier-building portion of the process. In that case, inference can only be applied to the latter portion of the process. However, in most cases, the important features are unknown and the intended inference includes feature selection. Then, CV estimates as above tend to suffer from a downward bias and inference is not warranted. Features should be selected only on the basis of the samples in the training sets $\mathcal{L} - \mathcal{L}_v$ for CV estimation. This applies to other error rate estimation methods (e.g., test set error and out-of-bag estimation), and also to other aspects of the classifier training process, such as variable standardization and parameter selection. Examples of classifier parameters that should be included in cross-validation are: the number of predictor variables, the number of neighbors k for k-nearest neighbor classifiers, and the choice of kernel for SVMs. The issue of "honest" cross-validation analysis is discussed in (West et al., 2001).

The approaches described above can also be extended to reflect differential misclassification costs; in such situations, performance is assessed based on the general definition of risk in Equation 7.1. In the case of unequal representation of the classes, some form of stratified sampling may be needed to ensure balance across important classes in all subsamples. In addition, for complex experimental designs, such as factorial or time-course designs, the resampling mechanisms used for computational inference should reflect the design of the experiment.

Finally, note that in machine learning, a frequently employed alternative to simple accuracy-based measures is the *lift*. The lift of a given class k is computed from a test set as the proportion of correct class k predictions divided by the proportion of class k test cases, i.e.

$$lift_k = \frac{\dfrac{\#\ \text{test cases } correctly \text{ predicted in class } k}{\#\ \text{test cases predicted in class } k}}{\dfrac{\#\ \text{test cases actually belonging to class } k}{\#\ \text{test cases}}} \tag{7.7}$$

In general, the greater the lift, the better the classifier.

6.   DISCUSSION

Classification is an important question in microarray experiments, for purposes of classifying biological samples and predicting clinical or other outcomes using gene expression data. In this chapter, we have discussed the statistical foundations of classification and described two basic classification approaches, nearest neighbor and linear discriminant analysis (including the special case of DLDA, also known as naive Bayes classification). We have addressed the important issues of feature selection and honest classifier performance assessment, which takes into account gene screening and other training decisions in error rate estimation procedures such as cross-validation.

The reader is referred to (Dudoit et al., 2002) and (Dudoit and Fridlyand, 2002) for a more detailed discussion and comparison of classification methods for microarray data. The classifiers examined in these two studies include linear and quadratic discriminant analysis, nearest neighbor classifiers, classification trees, and SMVs. Resampling methods such as bagging and boosting were also considered, including random forests and LogitBoost for tree stumps. Simple methods such as nearest neighbors and naive Bayes classification were found to perform remarkably well compared to more complex approaches, such as aggregated classification trees or SVMs. Dudoit and Fridlyand (2002) also discussed the general questions of feature selection, standardization, distance function, loss function, biased sampling of classes, and binary vs. polychotomous classification. Decisions concerning all these issues can have a large impact on the performance of the classifier; they should be made in conjunction with the choice of classifier and included in the assessment of classifier performance.

Although classification is by no means a new subject in the statistical literature, the large and complex multivariate datasets generated by microarray experiments raise new methodological and computational challenges. These include building accurate classifiers in a "large p, small n" situation and obtaining honest estimates of classifier performance. In particular, better predictions may be obtained by inclusion of other predictor variables such as age or sex. In addition to accuracy, a desirable property of a classifier is its ability to yield insight into the predictive structure of the data, that is, identify individual genes and sets of interacting genes that are related to class distinction. Further investigation of the resulting genes may improve our understanding of the biological mechanisms underlying class distinction and eventually lead to marker genes to be used in a clinical setting for predicting outcomes such as survival and response to treatment.

ACKNOWLEDGMENTS

We are most grateful to Leo Breiman for many insightful conversations on classification. We would also like to thank Robert Gentleman for valuable discussions on classification in microarray experiments while designing a short course on this topic. Finally, we have appreciated the editors' careful reading of the chapter and very helpful suggestions.

REFERENCES

Alizadeh A. A., Eisen M. B., Davis R. E., Ma C., Lossos I. S., Rosenwald A., Boldrick J. C., Sabet H., Tran T., Yu X., Powell J. I., Yang L., Marti G. E., Moore T., Jr J. H., Lu L., Lewis D. B., Tibshirani R., Sherlock G., Chan W. C., Greiner T. C., Weisenburger D. D., Armitage J. O., Warnke R., Levy R., Wilson W., Grever M. R., Byrd J. C., Botstein D., Brown P. O., and Staudt L. M. (2000). Distinct types of diffuse large B-cell lymphoma identified by gene expression profiling. Nature, 403:503-511.

Alon U., Barkai N., Notterman D. A., Gish K., Ybarra S., Mack D., and Levine A. J. (1999). Broad patterns of gene expression revealed by clustering analysis of tumor and normal colon tissues probed by oligonucleotide arrays. Proc. Natl. Acad. Sci., 96:6745-6750.

Bittner M., Meltzer P., Chen Y., Jiang Y., Seftor E., Hendrix M., Radmacher M., Simon R., Yakhini Z., Ben-Dor A., Sampas N., Dougherty E.,Wang E.,Marincola F., Gooden C., Lueders J., Glatfelter A., Pollock P., Carpten J., Gillanders E., Leja D., Dietrich K., Beaudry C., Berens M., Alberts D., Sondak V., Hayward N., and Trent J. (2000). Molecular classification of cutaneous malignant melanoma by gene expression profiling. Nature, 406:536-540.

Bø T. H. and Jonassen I. (2002). New feature subset selection procedures for classification of expression profiles. Genome Biology, 3(4):1-11.

Boldrick J. C., Alizadeh A. A., Diehn M., Dudoit S., Liu C. L., Belcher C. E., Botstein D., Staudt L. M., Brown P. O., and Relman D. A. (2002). Stereotyped and specific gene expression programs in human innate immue responses to bacteria. Proc. Natl. Acad. Sci., 99(2):972-977.

Breiman L. (1998). Arcing classifiers. Annals of Statistics, 26:801-824.

Breiman L. (1999). Random forests - random features. Technical Report 567, Department of Statistics, U.C. Berkeley.

Breiman L., Friedman, J.H. Olshen, R., and Stone C.J. (1984). Classification and regression trees. The Wadsworth statistics/probability series. Wadsworth International Group.

Chen X., Cheung S.T., So S., Fan S.T., Barry C., Higgins J., Lai K.-M., Ji J., Dudoit S., Ng I. O. L., van de Rijn M., Botstein D., and Brown P.O. (2002). Gene expression patterns in human liver cancers. Molecular Biology of the Cell, 13(6):1929-1939.

Chow M.L., Moler E.J., and Mian I.S. (2001). Identifying marker genes in transcription profiling data using a mixture of feature relevance experts. Physiological Genomics, 5:99-111.

Costello J. F., Fruehwald M.C., Smiraglia D.J., Rush, L. J., Robertson G.P., Gao X., Wright F.A., Feramisco J.D., Peltomki P., Lang J.C., Schuller D.E., Yu L., Bloomfeld C.D., Caligiuri M.A., Yates A., Nishikawa R., Huang H.J.S., Petrelli N.J., Zhang X., O'Dorisio

M.S., Held W.A., Cavenee W.K., and Plass C. (2000). Aberrant CpG-island methylation has non-random and tumour-typespecific patterns. Nature Genetics, 24:132-138.

Dettling M. and Buelmann P. (2002). How to use boosting for tumor classification with gene expression data. Available at http://stat.ethz.ch/~dettling/boosting.

Dudoit S. and Fridlyand J. (2002). Classification in microarray experiments. In Speed, T. P., editor, Statistical Analysis of Gene Expression Microarray Data. Chapman & Hall/CRC. (To appear).

Dudoit S., Fridlyand J., and Speed T P. (2002). Comparison of discrimination methods for the classification of tumors using gene expression data. Journal of the American Statistical Association, 97(457):77-87.

Fix E. and Hodges J. (1951). Discriminatory analysis, nonparametric discrimination: consistency properties. Technical report, Randolph Field, Texas: USAF School of Aviation Medicine.

Friedman J.H. (1994). Flexible metric nearest neighbor classification. Technical report, Department of Statistics, Stanford University.

Friedman J.H. (1996). On bias, variance, 0/1-loss, and the curse-of-dimensionality. Technical report, Department of Statistics, Stanford University.

Golub T.R., Slonim D.K., Tamayo P., Huard C., Gaasenbeek M., Mesirov J.P., Coller H., Loh M., Downing J.R., Caligiuri M.A., Bloomfield C.D., and Lander E.S. (1999). Molecular classification of cancer: class discovery and class prediction by gene expression monitoring. Science, 286:531-537.

Hastie T., Tibshirani R., and Friedman J.H. (2001). The Elements of Statistical Learning : Data Mining, Inference, and Prediction. Springer Verlag.

Jain A.N., Chin K., Børresen-Dale A., Erikstein B.K., Eynstein L.P., Kaaresen R., and Gray J.W. (2001). Quantitative analysis of chromosomal CGH in human breast tumors associates copy number abnormalities with p53 status and patient survival. Proc. Natl. Acad. Sci., 98:7952-7957.

Mardia K.V., Kent J.T., and Bibby J.M. (1979). Multivariate Analysis. Academic Press, Inc., San Diego.

McLachlan G.J. (1992). Discriminant analysis and statistical pattern recognition. Wiley, New York.

Perou C.M., Jeffrey S.S., van de Rijn M., Rees C.A., Eisen M.B., Ross D.T., Pergamenschikov A., Williams C.F., Zhu S.X., Lee J.C.F., Lashkari D., Shalon D., Brown, P.O., and Botstein D. (1999). Distinctive gene expression patterns in human mammary epithelial cells and breast cancers. Proc. Natl. Acad. Sci., 96:9212-9217.

Pollack J.R., Perou C.M., Alizadeh A.A., Eisen M.B., Pergamenschikov A., Williams C.F., Jeffrey S.S., Botstein D., and Brown P.O. (1999). Genome-wide analysis of DNA copy-number changes using cDNA microarrays. Nature Genetics, 23:41-46.

Pomeroy S.L., Tamayo P., Gaasenbeek M., Sturla L.M., Angelo M., McLaughlin M.E., Kim J.Y., Goumnerova L.C., Black P.M., Lau C., Allen J.C., Zagzag D., Olson J., Curran T., Wetmore C., Biegel J.A., Poggio T., Mukherjee S., Rifkin R., Califano A., Stolovitzky G., Louis D.N., Mesirov J.P., Lander E.S., and Golub T.R. (2002). Prediction of central nervous system embryonal tumour outcome based on gene expression. Nature, 415(24):436-442. (and supplementary information).

Ripley, B. D. (1996). Pattern recognition and neural networks. Cambridge University Press, Cambridge, New York. Ross, D. T., Scherf, U., Eisen, M. B., Perou, C. M., Spellman, P., Iyer, V., Jeffrey, S. S., de Rijn, M. V., Waltham, M., Pergamenschikov, A., Lee, J. C. F., Lashkari, D., Shalon, D., Myers, T. G., Weinstein, J. N., Botstein, D., and Brown, P. O. (2000). Systematic variation in gene expression patterns in human cancer cell lines. Nature Genetics, 24:227-234.

Sørlie, T., Perou, C. M., Tibshirani, R., Aas, T., Geisler, S., Johnsenb, H., Hastie,T., Eisen,M. B., van deRijn,M., Jeffrey, S. S.,Thorsen, T., Quist, H., Matese, J. C., Brown, P. O., Botstein,D., Lønningg, P. E., and Børresen-Dale, A. L. (2001). Gene expression patterns of breast carcinomas distinguish tumor subclasses with clinical implications. Proc. Natl. Acad. Sci., 98(19):10869-10874.

Tibshirani R., Hastie T., and G. Chu B.N. (2002). Diagnosis of multiple cancer types by shrunken centroids of gene expression. Proc. Natl. Acad. Sci. 99:6567-6572.

West M., Blanchette C., Dressman H., Huang E., Ishida S., Spang R., Zuzan H., Marks J.R., and Nevins J.R. (2001). Predicting the clinical status of human breast cancer using gene expression profiles. Proc. Natl. Acad. Sci., 98:11462-11467.

Chapter 8

BAYESIAN NETWORK CLASSIFIERS FOR GENE EXPRESSION ANALYSIS

Byoung-Tak Zhang and Kyu-Baek Hwang

Biointelligence Laboratory, School of Computer Science and Engineering, Seoul National University, Seoul 151-742, Korea
e-mail: {btzhang, kbhwang}@bi.snu.ac.kr

1. INTRODUCTION

The recent advent of DNA chip technologies has made it possible to measure the expression level of thousands of genes in the cell population. The parallel view on gene expression profiles offers a novel opportunity to broaden the knowledge about various life phenomena. For example, the microarray samples from normal and cancer tissues are accumulated for the study of differentially expressed genes in the malignant cell (Golub et al., 1999; Alon et al., 1999; Slonim et al., 2000; Khan et al., 2001). The eventual knowledge acquired by such a study could aid in discriminating between carcinoma cells and normal ones based on the gene expression pattern. One of the main objectives of machine learning is to build a discriminative (classification) model from data, automatically.

There exist various kinds of machine learning models deployed for the classification task i.e. *k-nearest neighbor* (*k*NN) models (Li et al., 2002), *decision trees* (Dubitzky et al., 2002), *artificial neural networks* (Khan et al., 2001; Dubitzky et al., 2002), and *Bayesian networks* (Hwang et al., 2002). These models differ mostly in their way of representing the learned knowledge. The *k*NN methods just lay aside learning examples in computer memory. When a new example without class label is encountered, a set of *k* similar examples are retrieved from memory and used for classification. Decision trees represent the learned knowledge in the form of a set of 'if-then' rules. Neural networks learn the functional relationships between the class variable and input attributes. Bayesian networks represent the joint probability distribution over the variables of interest. The *k*NN model is the simplest among the above classification models. The Bayesian network might be the most complicated and flexible one. In general, the more complicated model requires the more elaborate and complex learning techniques. Nonetheless, each of the above classification models could achieve the classification performance comparable to each other, regardless of its representation power. Then, what is the reason for using more complex models? The answer might be that they enable the acquisition of more flexible and comprehensive knowledge. And the Bayesian network is probably the most suitable model for such purposes. Thus, it has been employed for the sample classification (Hwang et al., 2002) as well as for the genetic network analysis (Friedman et al., 2000; Hartemink et al., 2001) with microarray data.

This chapter deals with the Bayesian network for the classification of microarray data and is organized as follows. In Section 2, we give a simple explanation of the Bayesian network model. Methods of data preprocessing and learning Bayesian networks from data are provided in Section 3. In Section 4, the advantages of the Bayesian networks as well as the difficulties in applying them to the classification task are described. Some techniques for improving the classification performance of the Bayesian network are also presented. In Section 5, we compare the classification accuracy of the Bayesian network with other state-of-the-art techniques on two microarray data sets. The use of Bayesian networks for knowledge discovery is also illustrated. Finally, we give some concluding remarks in Section 6.

2. BAYESIAN NETWORKS

The Bayesian network (Heckerman, 1999; Jensen, 2001) is a kind of *probabilistic graphical model*, which represents the joint probability distribution over a set of variables of interest. [1] In the framework of probabilistic graphical models, the *conditional independence* is exploited for the efficient representation of the joint probability distribution. For three sets of variables **X**, **Y**, and **Z**,[2] **X** is conditionally independent from **Y** given the value of **Z**, if $P(\mathbf{x} \mid \mathbf{y}, \mathbf{z}) = P(\mathbf{x} \mid \mathbf{z})$ for all **x**, **y**, and **z** whenever $P(\mathbf{y}, \mathbf{z}) > 0$. The Bayesian network structure encodes various conditional independencies among the variables. Formally, a Bayesian network assumes a *directed-*

---

[1] When applying Bayesian networks to microarray data analysis, each gene or the experimental condition is regarded as a variable. The value of the gene variable corresponds to its expression level. The experimental conditions include the characteristics of tissues, cell cycles, and others.

[2] Following the standard notation, we represent a random variable as a capital letter (e.g., *X*, *Y*, and *Z*) and a set of variables as a boldface capital letter (e.g., **X**, **Y**, and **Z**). The corresponding lowercase letters denote the instantiation of the variable (e.g., *x*, *y*, and *z*) or all the members of the set of variables (e.g., **x**, **y**, and **z**), respectively.

acyclic graph (DAG) structure where a node corresponds to a variable[3] and an edge denotes the direct probabilistic dependency between two connected nodes. The DAG structure asserts that each node is independent from all of its non-descendants conditioned on its parent nodes. By these assertions, the Bayesian network over a set of N variables, $\mathbf{X} = \{X_1, X_2, ..., X_N\}$, represents the joint probability distribution as

$$P(\mathbf{X}) = \prod_{i=1}^{N} P(X_i \mid \mathbf{Pa}(X_i)), \qquad (8.1)$$

where $\mathbf{Pa}(X_i)$ denotes the set of parents of X_i. $P(X_i \mid \mathbf{Pa}(X_i))$ is called the *local probability distribution* of X_i. The local probability distribution describes the conditional probability distribution of each node given the values of its parents. The appropriate local probability distribution model is chosen according to the variable type. When all the variables are discrete, the multinomial model is used. When all the variables are continuous, the *linear Gaussian model*[4] can be used.[5]

Figure 8.1 is an example Bayesian network for cancer classification.

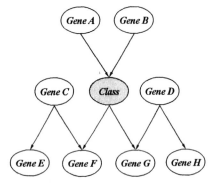

Figure 8.1. A Bayesian network consisting of eight gene nodes ('*Gene A*' to '*Gene H*') and '*Class*' node. The '*Class*' node represents the type of cancer. The Bayesian network structure encodes various conditional independencies among '*Class*' variable and eight gene variables. Each node has its local probability distribution although omitted in this figure. This Bayesian network represents the joint probability distribution over nine variables

This Bayesian network represents the joint probability distribution over eight gene variables and '*Class*' variable. Unlike other machine learning models

---

[3] Because each node in the Bayesian network is one-to-one correspondent to a variable, 'node' and 'variable' denote the same object in this paper. We use both terms interchangeably according to the context.

[4] In the linear Gaussian model, a variable is normally distributed around a mean that depends linearly on the values of its parent nodes. The variance is independent of the parent nodes.

[5] The hybrid case, in which the discrete variables and the continuous variables are mixed, could also exist. Such a case is not dealt with in this paper.

for classification, the Bayesian network does not discriminate between the class variable and the input attributes. The class variable is simply regarded as one of the data attributes. When classifying a sample without class label using Bayesian networks in Figure 8.1, we calculate the conditional probability of '*Class*' variable given the values of eight gene variables as follows:

$$P(Class \mid Gene\ A, Gene\ B, Gene\ C, Gene\ D, Gene\ E, Gene\ F, Gene\ G, Gene\ H)$$
$$= \frac{P(Class, Gene\ A, Gene\ B, Gene\ C, Gene\ D, Gene\ E, Gene\ F, Gene\ G, Gene\ H)}{\sum_{Class} P(Class, Gene\ A, Gene\ B, Gene\ C, Gene\ D, Gene\ E, Gene\ F, Gene\ G, Gene\ H)} \quad (8.2)$$

where the summation is taken over all the possible states of '*Class*' variable. Among the possible cancer class labels, the one with the highest conditional probability value might be selected as an answer.[6] The joint probability in the numerator and denominator of Equation (2) can be decomposed into a product of the local probability of each node based on the DAG structure in Figure 8.1.[7] In addition to the classification, the Bayesian network represents the probabilistic relationships among variables in a comprehensible DAG format. For example, the Bayesian network in Figure 8.1 asserts that the expression of '*Gene D*' might affect the expression of both '*Gene G*' and '*Gene H*' ('*Gene D*' is the common parent of '*Gene G*' and '*Gene H*').[8]

3.   APPLYING BAYESIAN NETWORKS TO THE CLASSIFICATION OF MICROARRAY DATA

Figure 8.2 shows the overall procedure of applying Bayesian networks to the classification of microarray data. First, an appropriate number of genes are selected and the expression level of each gene is transformed into the discrete value[9]. After the discretization and selection process, a Bayesian network is learned from the reduced microarray data set which has only the selected genes and the '*Class*' variable as its attributes. Finally, the learned

---

[6] In the case of a tie, the answer can be selected randomly or just 'unclassified'.

[7] The local probability distribution of each node is estimated from the data in the procedure of Bayesian network learning.

[8] An edge in the Bayesian network just means the probabilistic dependency between two connected nodes. In Figure 8.1, '*Gene D*' depends on '*Gene G*' and vice versa. The probabilistic dependency does not always denote the causal relationship but the possibility of its existence.

[9] The discretization of gene expression level is related to the choice of the local probability distribution model for each gene node. It is not compulsory in microarray data analysis with Bayesian networks.

Bayesian network is used for the classification of microarray samples and for the knowledge discovery.

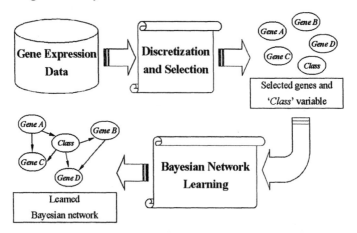

Figure 8.2. The overall procedure of building Bayesian network classifiers from microarray data. The learned Bayesian network classifier can also be used for the knowledge discovery.

3.1   Discretization and Selection

Discretization means to categorize the gene expression levels into several regions, e.g., 'over-expressed' and 'under-expressed'. The discretization of gene expression levels before learning Bayesian networks has its own benefits and drawbacks. The multinomial model for discrete variables could represent more diverse and complex relationships than the linear Gaussian model for continuous variables because the latter could only represent the linear relationships among variables. Nevertheless, the discretization step must incur some information loss. There are various methods for discretization (Dougherty et al., 1995) and one simple method is to divide the expression level of a gene based on its mean value across the experiments. The selection step is necessary because there exist a large amount of genes that are not related to the sample classification. Generally, considering all of these genes increases the dimensionality of the problem, presents computational difficulties, and introduces unnecessary noise. So, it is often helpful to select more relevant or predictive genes for the classification task. For the selection of genes, *mutual information* (Cover and Thomas, 1991), *P*-metric (Slonim et al., 2000), or other statistical methods can be used. The mutual information between two random variables X and Y, $I(X;Y)$, measures the amount of information that X contains about Y and is calculated as

$$I(X;Y) = \sum_{X,Y} P(X,Y) \log \frac{P(X,Y)}{P(X)P(Y)}. \tag{8.3}$$

Here, $P(\cdot)$ denotes the empirical probability estimated from the data. The summation is taken over all the possible X and Y values.[10] In order to select the genes much related to the class variable, the mutual information between the class variable and each gene is calculated. And all genes are ranked according to the corresponding mutual information value. Then, we can select appropriate numbers of genes from this gene list. Other measures are also applied in a similar way.

3.2 Learning Bayesian Networks from Data

The Bayesian network learning procedure consists of two parts. The first is to learn the DAG structure of Bayesian network as outlined in Section 2. The other is to learn the parameters for each local probability distribution under the fixed structure. Parameter learning is to estimate the most likely parameter values based on the training data. As an example, consider the Bayesian network structure in Figure 8.1. We assume that each variable is binary. The value of '*Class*' variable is either 'class 0 (0)' or 'class 1 (1)'. Each gene node has the value of either 'under-expressed (0)' or 'over-expressed (1)'. Then, the local probability distribution of '*Gene G*' could be represented as depicted in Table 8.1.

Table 8.1. The local probability distribution of '*Gene G*' node in Figure 8.1. '*Class*' and '*Gene D*' are the parents of '*Gene G*'. The parameter θ_1 denotes the conditional probability, $P(Gene\ G = 0 \mid Class = 0, Gene\ D = 0)$. Note that $P(Gene\ G = 1 \mid Class = 0, Gene\ D = 0)$ is $1 - \theta_1$.

Gene G	(Class, Gene D)			
	(0, 0)	(0, 1)	(1, 0)	(1, 1)
0	θ_1	θ_2	θ_3	θ_4

The local probability distribution model of '*Gene G*' has four parameters $(\theta_1, \theta_2, \theta_3, \theta_4)$. These parameters could be simply estimated from the data. For example, the *maximum-likelihood* value of θ_1 is calculated as follows:

$$\theta_1 = \frac{\#\text{ of cases where } Class = 0, Gene\ D = 0, \text{ and } Gene\ G = 0}{\#\text{ of cases where } Class = 0 \text{ and } Gene\ D = 0} \tag{8.4}$$

---

[10] Equation 8.3 is for the case of discrete variables. If X and Y are continuous variables, then the summation in this equation can be replaced by the integral.

When the data is not complete (some cases have missing values), the *expectation-maximization* (EM) *algorithm* [11] (Dempster et al., 1977) is generally used for the maximum-likelihood estimation of the parameter values (Heckerman, 1999).

Structure learning corresponds to searching for the plausible network structure based on the training data. The fitness of the network structure is measured by some scoring metrics. Two popular such metrics are the *minimum description length* (MDL) score and the *Bayesian Dirichlet* (BD) score (Heckerman et al., 1995; Friedman and Goldszmidt, 1999). The MDL score and the logarithm of the BD score assume the similar form and asymptotically have the same value with opposite sign. They can be decomposed into two terms i.e. the *penalizing term* about the complexities of the network structure and the log likelihood term. The scoring metric for the Bayesian network structure G with N variables and the training data D consisting of M cases $\{x_1, x_2, ..., x_M\}$ can be expressed as:

$$
\begin{aligned}
Score(G;D) &= penalizing\ term + log\ likelihood \\
&= penalizing\ term + \sum_{t=1}^{M} \log P_G(\mathbf{x}_t) \\
&= penalizing\ term + \sum_{t=1}^{M} \sum_{i=1}^{N} \log P(X_i = x_{ti} \mid \mathbf{Pa}_G(X_i) = \mathbf{pa}_{ti}),
\end{aligned}
\tag{8.5}
$$

where $P_G(\mathbf{x}_t)$ is the joint probability of $\mathbf{x}_t$ represented by the Bayesian network G. The value of X_i is given by x_{ti} and $\mathbf{pa}_{ti}$ is the configuration of $\mathbf{Pa}_G(X_i)$ at $\mathbf{x}_t$. However, the number of possible structures of Bayesian networks consisting of N variables is super-exponential in N. And it is known to be an *NP-hard problem*[12] to find the best structure (Chickering, 1996). Hence, several search heuristics, such as *greedy hill-climbing*, are used to find good structures in a general way. The greedy search algorithm for learning the Bayesian network structure proceeds as follows.

- Generate the initial Bayesian network structure G_0.
- For $m = 1, 2, 3, ...$ until convergence.
  - Among all the possible local changes (insertion of an edge, reversal of an edge, and deletion of an edge) in G_{m-1}, the one that leads to the largest improvement in the score is performed. The resulting graph is G_m.

---

[11] The EM algorithm proceeds as follows. First, all the parameter values are assigned randomly. Then, the following two steps are iterated until convergence. The expectation step is to estimate the necessary sufficient statistics based on the present parameter values and given incomplete data. The maximization step is to calculate the maximum-likelihood parameter values based on the sufficient statistics estimated in the expectation step.

[12] The complex class of decision problems which are intrinsically harder than those that can be solved by a non-deterministic Turing machine in polynomial time.

The algorithm stops when the score of G_{m-1} is equal to the score of G_m. The greedy search algorithm does not guarantee to find the best solution because it will get stuck at the local maximum. Nevertheless, this algorithm has shown acceptable performance in many applications. In some situations, the *greedy search algorithm with random restarts* is used for escaping from the local maxima. In this algorithm, when a local maximum is found, the network structure is randomly perturbed and the greedy search procedure is applied again.

4. PROS AND CONS OF THE BAYESIAN NETWORK CLASSIFIER

4.1 The Advantages of the Bayesian Network Classifier

Most advantages of the Bayesian network classifier come from its representation power. Other predictive classification models are basically focusing on learning only the relationships between the class label and input attributes. In contrast, Bayesian networks represent the joint probability distribution over the class label and input attributes. One advantage of the Bayesian network classifier is that it can predict the class label when only partial information about the input attributes is available. For example, consider again the Bayesian network in Figure 8.1. Assume that we are given a sample consisting of only the values of '*Gene A*' and '*Gene F*'. Then, this sample could also be classified by calculating the conditional probability, P(*Class* | *Gene A*, *Gene F*) = P(*Class*, *Gene A*, *Gene F*) / P(*Gene A*, *Gene F*) from the Bayesian network. Of course, the calculation of this conditional probability is not straightforward because it requires the summation over all the possible configurations for unknown variables. If all the variables are binary, we should enumerate $2^7 (= 128)$ possible configurations for the summation. [13] Calculation of the conditional probabilities in the Bayesian network is often called *probabilistic inference*. Although the probabilistic inference of arbitrary conditional probabilities from arbitrary Bayesian networks is known to be NP-hard (Cooper, 1990), there exist several algorithms applicable to the special type of network structures (Pearl, 1988; Spirtes et al., 2000; Jensen, 2001). One example is Pearl's message passing scheme (Pearl, 1988).

Another advantage of the Bayesian network classifier is that it can be used as a hypothesis generator about the domain. The Bayesian network structure learned from microarray data represents various probabilistic relationships among gene expressions and the class label in a

---

[13] When the number of unknown variables or the range of possible values of the discrete variable is large, this problem becomes serious.

comprehensible graph format. Such relationships might be used as a hypothesis and be verified by further biological experiments. Of course, other predictive models such as decision trees could represent the relationships between the class label and input attributes in the form of comprehensible 'if-then' rules. However, the representation power of the Bayesian network is superior to that of other predictive models.

4.2   Difficulties in Using Bayesian Networks for Classification

Although the Bayesian network has some advantages, it requires special tuning techniques to achieve the good classification accuracy of other predictive models in practice. In principle, this comes from the fact that Bayesian networks try to represent the joint probability distribution. When calculating the conditional probability of the class variable in the Bayesian network, only the Markov blanket (Pearl, 1988; also refer to Chapter 6 of this volume) of the class variable affects the results. For a set of $N-1$ input attributes, $\mathbf{A} = \{A_1, A_2, ..., A_{N-1}\}$ and the class variable C, the Markov blanket of C, $\mathbf{MB}(C)$, is the subset of $\mathbf{A}$ which satisfies the following equation.

$$P(C \mid \mathbf{A}) = P(C \mid \mathbf{MB}(C)) \qquad (8.6)$$

In other words, C is conditionally independent of $\mathbf{A} - \mathbf{MB}(C)$ given the values of all the members of $\mathbf{MB}(C)$.[14] Given a Bayesian network structure, determination of the Markov blanket of a node is straightforward. By the conditional independencies asserted by the network structure, the Markov blanket of a variable C consists of all the parents of C, all the children of C, and all the spouses of C. In Figure 8.1, the Markov blanket of '*Class*' node consists of six gene nodes i.e. '*Gene A*', '*Gene B*', '*Gene C*', '*Gene D*', '*Gene F*', and '*Gene G*'. Because only the members of the Markov blanket of the class variable participate in the classification process [15], the construction of the accurate Markov blanket structure around the class variable is most important for the classification performance. However, the nature of the scoring metrics used in general Bayesian network learning is not favorable to this point. Consider learning the Bayesian network consisting of one class variable C and $N-1$ input variables, $\{A_1, A_2, ..., A_{N-1}\}$. Then, the *log likelihood* term in Equation 8.5 can be decomposed into two components as

---

[14] Several Markov blankets could exist for a variable. The minimal Markov blanket is also called Markov boundary (Pearl, 1988). In our paper, Markov blanket always denotes the Markov boundary.

[15] This situation occurs when all the input attribute values are given in the sample.

$$log \; likelihood = \sum_{t=1}^{M} \log P_G(C = c_t \mid A_1 = a_{t1}, A_2 = a_{t2}, ..., A_{N-1} = a_{t(N-1)})$$
$$+ \sum_{t=1}^{M} \log P_G(A_1 = a_{t1}, A_2 = a_{t2}, ..., A_{N-1} = a_{t(N-1)}), \tag{8.7}$$

where c_t is the value of C and a_{ti} is the value of A_i ($1 \leq i \leq N-1$) in the tth training example. Because only the first term of Equation 8.7 is related with the classification accuracy, maximizing the second term might mislead the search for the Bayesian network structure as a good classifier. In the greedy search procedure, the essential variable for the classification might be eliminated from the Markov blanket of the class variable. More details on this issue could be found in (Friedman et al., 1997). In the next subsection, some methods for improving classification accuracy of the Bayesian networks are briefly presented.

4.3 Improving the Classification Accuracy of the Bayesian Network

There are various criteria for the classification accuracy including the total rate of correctly classified samples, the *sensitivity* and *specificity*, and the *receiver operating characteristic* (ROC) curve. In this chapter, we rely on the simple and intuitive measure, the total rate of correctly classified cases in the test data set.

One simple solution for the problem discussed in Section 4.2 is to fix the structure as appropriate for the classification task. The *naive Bayes classifier* (Mitchell, 1997) is a typical example, where all the input variables are the children of the class variable and are conditionally independent from each other given the class label. The classification performance of the naive Bayes classifier is reported to be comparable to other state-of-the-art classification techniques in many cases. However, the strong restriction on the network structure hides one advantage of the Bayesian network, that is to say, the ability of exploratory data analysis. Friedman et al. (1997) suggested the *tree-augmented naive Bayes classifier* (TAN) model. The TAN model assumes a little more flexible structure than the naive Bayes classifier. Here, the correlations between input variables can be represented in some restricted forms. This approach outperforms the naive Bayes classifier in some cases. Bang and Gillies (2002) deployed the hidden nodes for capturing the correlations among the input attributes. This approach has also shown the better classification accuracy although the experiments were confined to only one classification problem. Zhang et al. (2002) proposed to use the ensemble of heterogeneous Bayesian networks in order to improve the classification accuracy. This approach is based on the concept of committee machines (Haykin, 1999) and showed the improved performance applied to the classification of microarray data. In the next section, the experimental results of the ensemble of Bayesian network classifiers (Zhang et al., 2002) on two microarray data sets are presented.

5.   EXPERIMENTS: CANCER CLASSIFICATION

5.1   The Microarray Data Sets

We demonstrate the classification performance of the ensemble of Bayesian networks (Zhang et al., 2002) on two microarray data sets. These two microarray data sets are as follows.

Leukemia data: This data set is the collection presented by Golub et al. (1999).[16] The data set contains 72 acute leukemia samples which consist of 25 samples of acute myeloid leukemia (AML) and 47 samples of acute lymphoblastic leukemia (ALL). 38 leukemia samples (11 AML and 27 ALL) were derived from bone marrow taken before treatment and are used as a training set in (Golub et al., 1999; Slonim et al., 2000). Additional 34 samples (14 AML and 20 ALL) were obtained as a test set among which 25 samples were derived from bone marrow and 9 were from peripheral blood. Each sample consists of 7,129 gene expression measurements obtained by a high-density oligonucleotide array. The classification task is to discriminate between AML and ALL.

Colon cancer data: This data set was presented by Alon et al. (1999) and contains 62 colon tissue samples. 40 tumor samples were collected from patients and paired 22 normal tissues were obtained from some of the patients.[17] More than 6,500 human gene expressions were analyzed with an Affymetrix oligonucleotide array. Among them, 2,000 genes with highest minimal intensity across the samples were chosen (Alon et al., 1999). Finally, each sample is represented by expression levels of 2,000 genes. The classification task is to discriminate between normal tissue and cancer tissue.

5.2   Experimental Settings

The *P*-metric (Slonim et al., 2000) was used to select 50 genes from each data set. Each gene expression level was discretized into two values, 'over-expressed' and 'under-expressed' based on its mean value across the training examples. As the scoring metric for the Bayesian network learning, BD score (Heckerman et al., 1995) with the following penalizing term was used:

---

[16] The leukemia data set is available at http://www.genome.wi.mit.edu/MPR.

[17] The colon cancer data set is available at http://microarray.princeton.edu/oncology/affydata.

$$-\sum_{i=1}^{N}\left\{\left(\log N+\log\left(\frac{N}{|\,\mathbf{Pa}(X_i)\,|}\right)\right)+\frac{1}{2}\|\,\mathbf{Pa}(X_i)\,\|\,(\|\,X_i\,\|-1)\log M\right\}. \qquad (8.8)$$

Here, N (= 51) is the number of nodes in the Bayesian network, M is the sample size, $|\cdot|$ denotes the size of a set of nodes, and $\|\cdot\|$ denotes the number of possible configurations of a node or a set of nodes. And the ensemble machines consisting of 5, 7, 10, 15, and 20 Bayesian networks were constructed from two cancer data sets, respectively (Zhang et al., 2002).

5.3　Experimental Results

Due to the small number of data samples in two microarray data sets used, we applied the *leave-one-out cross validation* (LOOCV) (Mitchell, 1997) to assess the classification performance. In the case of the leukemia data, the ensemble of seven Bayesian networks achieved the best classification accuracy (97.22%) among five ensemble machines. In the case of the colon cancer data, the ensemble machine of five Bayesian networks showed the best accuracy (85.48%) among five ensemble machines (Zhang et al., 2002).

For comparative studies, we also show the classification accuracy of other classification techniques, including weighted voting[18] (Golub et al., 1999; Slonim et al., 2000), C4.5 decision trees, naive Bayes classifiers, multilayer perceptrons (MLPs), and support vector machines (SVMs) (Ben-Dor et al., 2000; Chapter 9 of this volume). For the weighted voting scheme, the original value of gene expression level was used. The decision trees were applied to the discretized data sets. The naive Bayes classifiers, the MLPs, and the SVMs were run on both the discretized and original data sets. Table 8.2 summarizes the best classification accuracy of each method on two cancer data sets. Among all of these approaches, the SVM achieved the best classification accuracy. However, the SVM used the original gene expression values. When using discretized gene expression levels, the ensemble of Bayesian networks also shows the best performance like the SVM (in the case of leukemia data) or the MLP as well as the naive Bayes classifier (in the case of colon cancer data).

---

[18] In this approach, the classification of a new sample is based on the "weighted voting" of a set of informative genes. Each gene votes for the class depending on the distance between its expression level in the new sample and the mean expression level in each class. For more details, refer to (Golub et al., 1999; Slonim et al., 2000).

Table 8.1. Comparison of the classification accuracy of the ensemble of Bayesian networks and other state-of-the-art classification techniques on two microarray data sets.

Classifiers	Data attributes	Classification accuracy (LOOCV)	
		Leukemia data	Colon cancer data
Weighted voting	Real	95.83%	87.10%
Decision trees (C4.5)	Binary	95.83%	83.87%
Naive Bayes classifiers	Binary	95.83%	**85.48%**
	Real	97.22%	85.48%
Multilayer perceptrons	Binary	95.83%	**85.48%**
	Real	97.22%	**88.71%**
Support vector machines	Binary	**97.22%**	83.87%
	Real	**98.61%**	**88.71%**
Ensemble of Bayesian networks	Binary	**97.22%**	**85.48%**

Figure 8.3 shows the part around the class variable of a member Bayesian network which belongs to the ensemble machine learned from the leukemia data and the colon cancer data.

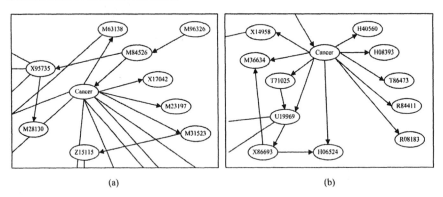

(a)   (b)

Figure 8.3. Parts of the member Bayesian network of the ensemble machines, learned from leukemia data (a) and colon cancer data (b). 'Cancer' node denotes the class variable. Gene nodes are represented by the respective GenBank accession number.

These graph structures generate the hypotheses on the relationships among the cancer class and gene expression profiles. These hypotheses could further guide biological experiments for the verifications. We illustrate some of them. In Figure 8.3(a), M96326, M31523, and X17042 are closely related through M84526 and 'Cancer' node. The E2A locus (M31523) is a frequent target of chromosomal translocations in B-cell ALL (Khalidi et al., 1999). E2A encodes two products, E12 and E47, that are part of the basic helix-loop-helix (bHLH) family of transcription factors. The disruption of E2A allele contributes to leukemogenesis (Herblot et al., 2002). Accidentally, azurocidin (M96326), adipsin (M84526), and E2A (M31523) are located in the same chromosomal region of 19p13.3 that is known to be the site of recurrent abnormalities in ALL and AML. Among these proteins, special attention should be made to azurocidin, also known as heparin-binding protein (HBP) or CAP37, that has antibacterial properties and chemotactic activity toward monocytes (Ostergaard and Flodgaard, 1992). Azurocidin released from human neutrophils binds to endothelial cell-surface proteoglycans. A significant fraction of proteoglycan-bound azurocidin is taken up by the cells. And the internalized azurocidin markedly reduces growth factor deprivation-induced caspase-3 activation and protects endothelial cells from apoptosis (Olofsson et al., 1999). This kind of reaction might affect the behavior of leukemia in the context of cell proliferation. It might be investigated whether adipsin (M84526) plays any role in the interaction between azurocidin (M96326) and hematopoetic proteoglycan core protein (X17042).

6. CONCLUDING REMARKS

We presented the Bayesian network method for the classification of microarray data. The Bayesian network is a probabilistic graphical model which represents the joint probability distribution over a set of variables. In the microarray data analysis, the variables denote the gene expression levels or the experimental conditions such as the characteristics of tissues and cell cycles. The Bayesian network consisting of the gene variables and the class variable can be learned from the microarray data set and used for the classification of new samples. For classification in the Bayesian network, the conditional probability of the class variable given the values of input attributes, is calculated from the joint probability representation. One of the most interesting points of the Bayesian network as a classifier compared to other predictive classification models is that it does not discriminate between the class label and the input attributes but tries to show the probabilistic dependency among arbitrary attributes. This enables the Bayesian network to represent correlations among input attributes as well as between the class variable and the input variables. Due to these features, the Bayesian network learned from microarray data could aid in broadening the knowledge about the domain by representing these relations in a comprehensible graph format. However, this flexibility of the Bayesian network makes it harder to learn the model with high classification accuracy as other classification models. This problem could be partially resolved by other techniques such as the ensemble of Bayesian networks. In our experiments on the gene expression analysis of leukemias and colon cancers, we showed that the ensemble of Bayesian classifiers could achieve the competitive classification accuracy compared to the other state-of-the-art techniques. We also demonstrated that the analysis of the learned Bayesian network could generate some interesting hypotheses which could guide further biological experiments.

ACKNOWLEDGMENTS

This work was supported by the BK-21 Program from Korean Ministry of Education and Human Resources Development, by the IMT-2000 Program and the NRL Program from Korean Ministry of Science and Technology.

REFERENCES

Alon U., Barkai N., Notterman D.A., Gish K., Ybarra S., Mack D., Levine A.J. (1999). Broad patterns of gene expression revealed by clustering analysis of tumor and normal colon tissues probed by oligonucleotide arrays. Proc Natl Acad Sci USA 96(12):6745-50

Bang J.-W., Gillies D. Using Bayesian networks with hidden nodes to recognize neural cell morphology (2002). Proc of the Seventh Pacific Rim International Conference on Artificial Intelligence; 2002 August 18 - 22; Tokyo. Heidelberg: Springer-Verlag (to appear).

Ben-Dor A., Bruhn L., Friedman N., Nachman I., Schummer M., Yakhini Z. (2000). Tissue classification with gene expression profiles. J Comput Biol 7(3/4):559-84

Chickering D.M. (1996). "Learning Bayesian Networks is NP-Complete." In *Learning from Data: Artificial Intelligence and Statistics V*, Doug Fisher, Hans-J. Lenz, eds. New York, NY: Springer-Verlag.

Cooper G.F. (1990). The computational complexity of probabilistic inference using Bayesian belief networks. Artificial Intelligence 42(2-3):393-405

Cover T.M., Thomas J.A. (1991). *Elements of Information Theory*. NY: John Wiley & Sons.

Dempster A.P., Laird N.M., Rubin D.B. (1977). Maximum likelihood from incomplete data via the EM algorithm (with discussion). J R Stat Soc Ser B 39(1): 1-38

Dougherty J., Kohavi R., Sahami M. (1995). Supervised and unsupervised discretization of continuous features. Proc of the Twelfth International Conference on Machine Learning; 1995 July 9 - 12; Tahoe City. San Francisco: Morgan Kaufmann Publishers.

Dubitzky W., Granzow M., Berrar D. (2002). "Comparing Symbolic and Subsymbolic Machine Learning Approaches to Classification of Cancer and Gene Identification." In *Methods of Microarray Data Analysis*, Simon M. Lin, Kimberly F. Johnson, eds. Norwell, MA: Kluwer Academic Publishers.

Friedman N., Geiger D., Goldszmidt M. (1997). Bayesian network classifiers. Machine Learning 29(2/3):131-63.

Friedman N., Goldszmidt M. (1999.) "Learning Bayesian Networks with Local Structure." In *Learning in Graphical Models*, Michael I. Jordan, ed. Cambridge, MA: MIT Press, 1999.

Friedman N., Linial M., Nachman I., Pe'er D (2000). Using Bayesian networks to analyze expression data. J Comput Biol 7(3/4):601-20.

Golub T.R., Slonim D.K., Tamayo P., Huard C., Gaasenbeek M., Mesirov J.P., Coller H., Loh M.L., Downing J.R., Caligiuri M.A., Bloomfield C.D., Lander E.S. (1999). Molecular classification of cancer: class discovery and class prediction by gene expression monitoring. Science 286(5439):531-7.

Hartemink A.J., Gifford D.K., Jaakkola T.S., Young R.A. (2001). Combining location and expression data for principled discovery of genetic regulatory network models. Proc. Seventh Pacific Symposium on Biocomputing; 2002 January 3 - 7; Lihue. Singapore: World Scientific Publishing.

Haykin S. (1999). *Neural Networks – A Comprehensive Foundation*, 2nd edition. NJ: Prentice-Hall.

Heckerman D., Geiger D., Chickering D.M. (1995). Learning Bayesian networks: the combination of knowledge and statistical data. Machine Learning 20(3):197-243

Heckerman D. (1999). "A Tutorial on Learning with Bayesian Networks." In *Learning in Graphical Models*, Michael I. Jordan, ed. Cambridge, MA: MIT Press.

Herblot S., Aplan P.D., Hoang T. (2002). Gradient of E2A activity in B-cell development. Mol Cell Biol 22:886-900

Hwang K.-B., Cho D.-Y., Park S.-W., Kim S.-D., Zhang B.-T. (2002). "Applying Machine Learning Techniques to Analysis of Gene Expression Data: Cancer Diagnosis." In *Methods of Microarray Data Analysis*, Simon M. Lin, Kimberly F. Johnson, eds. Norwell, MA: Kluwer Academic Publishers.

Jensen F.V. *Bayesian Networks and Decision Graphs*. NY: Springer-Verlag, 2001.

Khalidi H.S., O'Donnell M.R., Slovak M.L., Arber D.A. (1999). Adult precursor-B acute lymphoblastic leukemia with translocations involving chromosome band 19p13 is associated with poor prognosis. Cancer Genet Cytogenet 109(1):58-65.

Khan J., Wei J.S., Ringnér M., Saal L.H., Ladanyi M., Westermann F., Berthold F., Schwab M., Antonescu C.R., Peterson C., Meltzer P.S. (2001). Classification and diagnostic prediction of cancers using gene expression profiling and artificial neural networks. Nat Med 7(6):673-9.

Li L., Pederson L.G., Darden T.A., Weinberg C.R. (2002). "Computational Analysis of Leukemia Microarray Expression Data Using the GA/KNN Method." In *Methods of Microarray Data Analysis*, Lin S.M., Johnson K.F., eds., Norwell, MA, Kluwer Academic Publishers.

Mitchell T.M. (1997). *Machine Learning*. NY: McGraw-Hill Companies.

Olofsson A.M., Vestberg M., Herwald H., Rygaard J., David G., Arfors K.-E., Linde V., Flodgaard H., Dedio J., Muller-Esterl W., Lundgren-Åkerlund E. (1999). Heparin-binding protein targeted to mitochondrial compartments protects endothelial cells from apoptosis. J Clin Invest 104(7):885-94

Ostergaard E., Flodgaard H. (1992). A neutrophil-derived proteolytic inactive elastase homologue (hHBP) mediates reversible contraction of fibroblasts and endothelial cell monolayers and stimulates monocyte survival and thrombospondin secretion. J Leukoc Biol 51(4):316-23

Pearl J. (1988). *Probabilistic Reasoning in Intelligent Systems: Networks of Plausible Inference*. CA: Morgan Kaufmann Publishers.

Slonim D.K., Tamayo P., Mesirov J.P., Golub T.R., Lander E.S. (2000). Class prediction and discovery using gene expression data. Proc. of the Fourth Annual International Conference on Computational Molecular Biology; 2000 April 8 - 11; Tokyo. New York: ACM Press.

Spirtes P., Glymour C., Scheines R. (2000). *Causation, Prediction, and Search*, 2nd edition. MA: MIT Press.

Zhang B.-T., Hwang K.-B., Chang J.-H., Augh S.J. Ensemble of Bayesian networks for gene expression-based classification of cancers. Artif Intell Med (submitted) 2002.

Chapter 9

CLASSIFYING MICROARRAY DATA USING SUPPORT VECTOR MACHINES

Sayan Mukherjee

PostDoctoral Fellow: MIT/Whitehead Institute for Genome Research and Center for Biological and Computational Learning at MIT,

e-mail: sayan@mit.edu

1. INTRODUCTION

Over the last few years the routine use of DNA microarrays has made possible the creation of large data sets of molecular information characterizing complex biological systems. Molecular classification approaches based on machine learning algorithms applied to DNA microarray data have been shown to have statistical and clinical relevance for a variety of tumor types: Leukemia (Golub et al., 1999), Lymphoma (Shipp et al., 2001), Brain cancer (Pomeroy et al., 2002), Lung cancer (Bhattacharjee et al., 2001) and the classification of multiple primary tumors (Ramaswamy et al., 2001).

One particular machine learning algorithm, *Support Vector Machines* (SVMs), has shown promise in a variety of biological classification tasks, including gene expression microarrays (Brown et al., 2000, Mukherjee et al., 1999). SVMs are powerful classification systems based on regularization techniques with excellent performance in many practical classification problems (Vapnik, 1998, Evgeniou et al., 2000).

This chapter serves as an introduction to the use of SVMs in analyzing DNA microarray data. An informal theoretical motivation of SVMs both from a geometric and algorithmic perspective, followed by an application to leukemia classification, is described in Section 2. The problem of gene selection is described in Section 3. Section 4 states some results on a variety of cancer morphology and treatment outcome problems. Multiclass classification is described in Section 5. Section 6 lists software sources and rules of thumb that a user may find helpful. The chapter concludes with a brief discussion of SVMs with respect to some other algorithms.

2. SUPPORT VECTOR MACHINES

In binary microarray classification problems, we are given l experiments $\{(x_1, y_1),\ldots, (x_l, y_l)\}$. This is called the *training set*, where x_i is a vector corresponding to the expression measurements of the ith experiment or sample (this vector has n components, each component is the expression measurement for a gene or EST) and y_i is a binary class label, which will be ± 1. We want to estimate a multivariate function from the training set that will accurately label a new sample, that is $f(x_{new}) = y_{new}$. This problem of learning a classification boundary given positive and negative examples is a particular case of the problem of approximating a multivariate function from sparse data. The problem of approximating a function from sparse data is *ill-posed*. Regularization is a classical approach to making the problem *well-posed* (Tikhonov and Arsenin, 1977).

A problem is *well-posed* if it has a solution, the solution is unique, and the solution is stable with respect to perturbations of the data (small perturbations of the data result in small perturbations of the solution). This last property is very important in the domain of analyzing microarray data where the number of variables, genes and ESTs measured, is much larger than the number of samples. In statistics when the number of variables is much larger that the number of samples one is said to be facing the *curse of dimensionality* and the function estimated may very accurately fit the samples in the training set but be very inaccurate in assigning the label of a new sample, this is referred to as over-fitting. The imposition of stability on the solution mitigates the above problem by making sure that the function is smooth so new samples similar to those in the training set will be labeled similarly, this is often referred to as the *blessing of smoothness*.

2.1 Mathematical Background of SVMs

We start with a geometric intuition of SVMs and then give a more general mathematical formulation.

2.1.1 A Geometrical Interpretation

A geometric interpretation of the SVM illustrates how this idea of smoothness or stability gives rise to a geometric quantity called the margin which is a measure of how well separated the two classes can be. We start by assuming that the classification function is linear

$$f(x) = w \cdot x = \sum_{i=1}^{n} w_i x_i \qquad (9.1)$$

where x_i and w_i are the ith elements of the vectors x and w, respectively. The operation $w \cdot x$ is called a *dot product*. The label of a new point x_{new} is the sign of the above function, $y_{new} = \text{sign} [f(x_{new})]$. The classification boundary, all values of x for which $f(x) = 0$, is a *hyperplane*[1] defined by its normal vector w (see Figure 9.1).

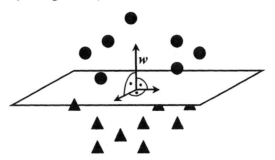

Figure 9.1. The hyperplane separating two classes. The circles and the triangles designate the members of the two classes. The normal vector to the hyperplane is the vector w.

Assume we have points from two classes that can be separated by a hyperplane and x is the closest data point to the hyperplane, define x_0 to be the closest point on the hyperplane to x. This is the closest point to x that satisfies $w \cdot x = 0$ (see Figure 9.2). We then have the following two equations:

$$w \cdot x = k \quad \text{for some } k, \text{ and}$$
$$w \cdot x = 0.$$

Subtracting these two equations, we obtain $w \cdot (x - x_0) = k$.
Dividing by the norm of w (the norm of w is the length of the vector w), we obtain:[2]

$$\frac{w}{\|w\|} \cdot (x - x_0) = \frac{k}{\|w\|}$$

---

[1] A *hyperplane* is the extension of the two or three dimensional concepts of lines and planes to higher dimensions. Note, that in an n-dimensional space, a hyperplane is an $n-1$ dimensional object in the same way that a plane is a two dimensional object in a three dimensional space. This is why the *normal* or *perpendicular* to the hyperplane is always a vector.

[2] The notation "$|a|$" refers to the absolute value of the variable a. The notation "$\|a\|$" refers the length of the vector a.

where $\|w\| = \sqrt{\sum_{i=1}^{n} w_i^2}$. Noting that $w / \|w\|$ is a unit vector (a vector of length 1), and the vector $x - x_0$ is parallel to w, we conclude that

$$\|x - x_0\| = \frac{|k|}{\|w\|} .$$

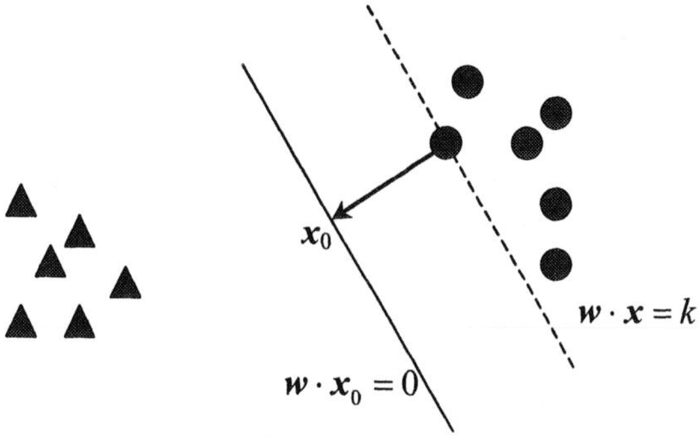

Figure 9.2. The black line is the hyperplane separating the triangles from the circles defined by its normal vector w. The circle on the dashed line is the point x closest to the hyperplane, and x_0 the closest point to x on the hyperplane.

Our objective is to maximize the distance between the hyperplane and the closest point, with the constraint that the points from the two classes fall on opposite sides of the hyperplane. The following optimization problem satisfies the objective:

$$\max_{w} \min_{x_i} \frac{y_i(w \cdot x_i)}{\|w\|} \quad \text{subject to} \quad y_i(w \cdot x) > 0 \text{ for all } x_i \tag{9.2}$$

Note that $y(w \cdot x) = |k|$ when the point x is the circle closest to the hyperplane in Figure 9.2.

For technical reasons, the optimization problem stated above is not easy to solve. One difficulty is that if we find a solution w, then cw for any positive constant c is also a solution. This is because we have not fixed a scale or unit to the problem. So without any loss of generality, we will require that for the point x_i closest to the hyperplane $k = 1$. This fixes a scale and unit to the problem and results in a guarantee that $y_i(w \cdot x_i) \geq 1$ for all x_i. All other points are measured with respect to the closest point, which is distance 1 from the optimal hyperplane. Therefore, we may equivalently solve the problem

$$\max_{w} \min_{x_i} \frac{y_i(w \cdot x_i)}{\|w\|} \quad \text{subject to} \quad y_i(w \cdot x_i) \geq 1 \tag{9.3}$$

An equivalent, but simpler problem (Vapnik, 1998) is

$$\min_{w} \tfrac{1}{2}\|w\|^2 \quad \text{subject to} \quad y_i(w \cdot x_i) \geq 1 \tag{9.4}$$

Note that so far, we have considered only hyperplanes that pass through the origin. In many applications, this restriction is unnecessary, and the standard separable (i.e. the hyperplane can separate the two classes) SVM problem is written as

$$\min_{w,b} \tfrac{1}{2}\|w\|^2 \quad \text{subject to} \quad y_i(w \cdot x_i + b) \geq 1 \tag{9.5}$$

where b is a free threshold parameter that translates the optimal hyperplane relative to the origin. The distance from the hyperplane to the closest points of the two classes is called the margin and is $1/\|w\|^2$. SVMs find the hyperplane that maximize the margin. Figure 9.3 illustrates the advantage of a *large margin*.

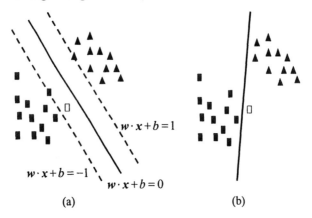

(a)   (b)

Figure 9.3.(a) The maximum margin hyperplane separating two classes. The solid black line is the hyperplane ($w \cdot x + b = 0$). The two dashed lines are those for the points in the two classes closest to the hyperplane ($w \cdot x + b = \pm 1$). A new point, the blank rectangle, is classified correctly in (a). Note, the larger the margin the greater the deviation allowed or margin for error. (b) A non-maximum margin hyperplane separating the two classes. Note, that the same new point is now classified incorrectly. There is less margin for error.

In practice, data sets are often not linearly separable. To deal with this situation, we add slack variables that allow us to violate our original distance constraints. The problem becomes now:

$$\min_{w,b,\xi} \tfrac{1}{2}\|w\|^2 + C\sum_i \xi_i \quad \text{subject to} \quad y_i(w \cdot x_i + b) \geq 1 - \xi_i \qquad (9.6)$$

where $\xi_i \geq 0$ for all i. This new formulation trades off the two goals of finding a hyperplane with large margin (minimizing $\|w\|$), and finding a hyperplane that separates the data well (minimizing the ξ_i). The parameter C controls this trade-off. This formulation is called the soft margin SVM. The parameter C controls this trade-off. It is no longer simple to interpret the final solution of the SVM problem geometrically. Figure 9.4 illustrates the *soft margin* SVM.

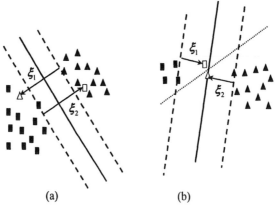

(a)   (b)

Figure 9.4.(a) The data points are not linearly separable. The solid black line is the SVM solution. The white triangle and the white rectangle are misclassified. The slack variables designate the distance of these points from the dashed lines for the corresponding classes. (b) The classes are separable. The dotted line is the solution when the tradeoff parameter C is very large (e.g., infinite), and this gives us the *maximum margin classifier* for the separable case. If the tradeoff parameter is small, then one allows errors (given by the two slack variables), but one gets a much larger margin.

SVMs can also be used to construct nonlinear separating surfaces. The basic idea here is to nonlinearly map the data to a feature space of high or possibly infinite dimensions, $x \rightarrow \phi(x)$. We then apply the linear SVM algorithm in this feature space. A linear separating hyperplane in the feature space corresponds to a nonlinear surface in the original space. We can now rewrite Equation 9.6 using the data points mapped into the feature space, and we obtain Equation 9.7.

$$\min_{w,b,\xi} \tfrac{1}{2}\|w\|^2 + C\sum_i \xi_i \qquad (9.7)$$

$\xi \geq 0$ for all i, where the vector w has the same dimensionality as the feature space and can be thought of as the normal of a hyperplane in the feature space. The solution to the above optimization problem has the form

$$f(x) = w \cdot \phi(x) + b = \sum_{i=1}^{\ell} c_i \phi(x_i) \cdot \phi(x) + b \tag{9.8}$$

since the normal to the hyperplane can be written as a linear combination of the training points in the feature space,

$$w = \sum_{i=1}^{l} c_i \phi(x_i).$$

For both the optimization problem (Equation 9.7) and the solution (Equation 9.8), the dot product of two points in the feature spaces needs to be computed. This dot product can be computed without explicitly mapping the points into feature space by a *kernel function*, which can be defined as the dot product for two points in the feature space:

$$K(x_i, x_j) \equiv \phi(x_i) \cdot \phi(x_j) \tag{9.9}$$

So our solution to the optimization problem has now the form:

$$f(x) = \sum_{i=1}^{l} c_i K(x, x_i) + b \tag{9.10}$$

Most of the coefficients c_i will be zero; only the coefficients of the points closest to the maximum margin hyperplane in the feature space will have nonzero coefficients. These points are called the *support vectors*. Figure 9.5 illustrates a nonlinear decision boundary and the idea of support vectors.

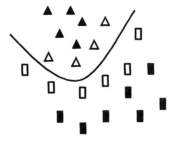

Figure 9.5 The curve is the nonlinear decision boundary given by the SVM. The white triangles and the white rectangles are the support vectors. Only these points contribute in defining the nonlinear decision boundary.

The following example illustrates the connection between the mapping into a feature space and the kernel function. Assume that we measure the expression levels of two genes, TrkC and SonicHedghog (SH). For each sample, we have the expression vector $x = (x_{SH}, x_{TrkC})$. We use the following mapping $\phi(x)$:

$$\phi : x \rightarrow \left\{ x_{SH}^2, x_{TrkC}^2, \sqrt{2} x_{SH} x_{TrkC}, x_{SH}, x_{TrkC}, 1 \right\}$$

If we have two samples x and z, then we obtain:

$$K(x,z) \equiv \phi(x) \cdot \phi(z) = (x \cdot z + 1)^2$$
$$= x_{SH}^2 z_{SH}^2 + x_{TrkC}^2 z_{TrkC}^2 + 2 x_{SH} x_{TrkC} z_{SH} z_{TrkC} +$$
$$x_{SH} z_{SH} + x_{TrkC} z_{TrkC} + 1$$

which is called a *second order polynomial kernel*. Note, that this kernel uses information about both expression levels of individual genes and also expression levels of pairs of genes. This can be interpreted as a model that incorporates co-regulation information. Assume that we measure the expression levels of 7,000 genes. The feature space that the second order polynomial would map into would have approximately 50 million elements, so it is advantageous that one does not have to explicitly construct this map.

The following two kernels, the polynomial and Gaussian kernel, are commonly used:

$$K(x, z) = (x \cdot z + 1)^P \text{ and } K(x, z) = \exp(-\|x - z\| / 2\sigma^2).$$

2.1.2 A Theoretical Motivation for SVMs and Regularization

SVMs can also be formulated as an algorithm that finds a function f that minimizes the following functional[3] (note that this functional is basically Equation 9.7, rewritten in a more general way):

$$\min_f \frac{1}{l} \sum_{i=1}^{l} \left(1 - y_i f(x_i) \right)_+ + \frac{1}{C} \|f\|_K^2 \qquad (9.11)$$

where the first term, the *hinge loss function*, is used to measure the error between our estimate $f(x_i)$ and y_i, and $(a)_+ = \min(a, 0)$. The expression $\|f\|_K^2$ is a measure of smoothness, and the margin is $1 / \|f\|_K^2$. The variable l is the number of training examples, and $1 / C$ is the regularization parameter that trades off between smoothness and errors. The first term ensures that the estimated function has a small error on the training samples and the second term ensures that this function is also smooth. This functional is a particular

---

[3] A *functional* is a function that maps other functions to real numbers.

instance of the *Tikhonov regularization principle* (Tikhonov and Arsenin, 1977).

Given a data set $S = \{(x_1, y_1), \ldots, (x_l, y_l)\}$, the SVM algorithm takes this training set and finds a function f_S. The error of this function on the training set is called the *empirical error*, and we can measure it using Equation 9.12:

$$I_{emp}[f_S] = \frac{1}{l}\sum_{i=1}^{l}(1 - y_i f(x_i))_+ \qquad (9.12)$$

However, what we really care about is how accurate we will be given a new data sample, which is $(1 - y_{new} f(x_{new}))_+$. In general, we want to weight this error by the probability of drawing the sample (x_{new}, y_{new}), and average this over all possible data samples. This weighted measure is called the *expected* or *generalization error*:

$$I_{exp}[f_S] = \int (1 - yf(x))_+ p(x, y)dxdy \qquad (9.13)$$

where $p(x, y)$ is the distribution which the data is drawn from.

For algorithms that implement Tikhonov regularization, one can say with high probability (Bousquet and Elisseeff, 2002, Mukherjee et al., 2002):

$$\left| I_{emp}[f_s] - I_{exp}[f_s] \right| \leq \Phi(l, \|f_S\|) \qquad (9.14)$$

where the function Φ decreases as $\|f\|$ decreases (or margin increases) and l increases. This tells us that if our error rate on the training set is low and the margin is large, then the error rate on average for a new sample will also be low. This is a theoretical motivation for using regularization algorithms such as SVMs. Note, that for the number of samples typically seen in microarray expression problems, plugging in values of l and f_S into Φ will not yield numbers small enough to serve as practical error bars.

2.2   An Application of SVMs

One of the first cancer classification studies was discriminating acute myeloid leukemia (AML) from acute lymphoblastic leukemia (ALL) (Golub et al., 1999). In this problem, a total of 38 training samples belong to the two classes, 27 ALL cases vs. 11 AML cases. The accuracy of the trained classifier was assessed using 35 test samples. The expression levels of 7,129 genes and ESTs were given for each sample. A linear SVM trained on this data accurately classified 34 of 35 test samples (see Figure 9.6.)

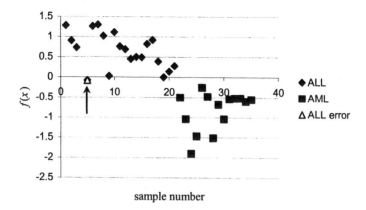

Figure 9.6 The signed distance, $f(x)$, from the optimal separating hyperplane for the test samples. The diamonds are the correctly labeled ALL samples. The squares indicate the correctly labeled AML samples. The triangle marks the misclassified ALL case (see arrow).

From Figure 9.6 an intuitive argument can be formulated: the larger the absolute value of the signed distance, $|f(x)|$, the more confident we can be in the classification. There exist approaches to convert the real-valued $f(x)$ into confidence values $p(y = \pm 1 \mid x)$ (Mukherjee et al., 1999, Platt, 1999). Thus, we can classify samples that have a confidence value larger than a particular threshold, whereas the classification of samples with confidence values below this threshold will be rejected. We applied this methodology in (Mukherjee et al., 1999) to the above problem, with a positive threshold of $|f(x)| = 0.1$ and a negative threshold of $|f(x)| = -0.2$.

A simple rule of thumb value to use as threshold is $|f(x)| = 1$. Function values greater than this threshold are considered *high confidence*. This value is in general too large to use for rejections.

Polynomial or Gaussian kernels did not increase the accuracy of the classifier (Mukherjee et al., 1999). However, when "important" genes were removed, the polynomial classifier did improve performance. This suggests that correlation information between genes can be helpful. Therefore, we removed 10 to 1,000 of the most "informative" genes from the test and training sets according to the signal-to-noise (S2N) criterion (see Section 3.1). We then applied linear and polynomial SVMs on this data set and reported the error rates (Mukherjee et al., 1999). Although the differences are not statistically significant, it is suggestive that until the removal of 300 genes, the polynomial kernel improves the performance, which suggests that modeling correlations between genes helps in the classification task.

3.   GENE SELECTION

It is important to know which genes are most relevant to the binary classification task and select these genes for a variety of reasons: removing noisy or irrelevant genes might improve the performance of the classifier, a candidate list of important genes can be used to further understand the biology of the disease and design further experiments, and a clinical device recording on the order of tens of genes is much more economical and practical than one requiring thousands of genes.

The gene selection problem is an example of what is called *feature selection* in machine learning (see Chapter 6 of this volume). In the context of classification, feature selection methods fall into two categories *filter methods* and *wrapper methods*. Filter methods select features according to criteria that are independent of those criteria that the classifier optimizes. On the other hand, wrapper methods use the same or similar criteria as the classifier. We will discuss two feature selection approaches: *signal-to-noise* (S2N, also known as *P-metric*) (Golub et al., 1999, Slonim et al., 2000;), and *recursive feature elimination* (RFE) (Guyon et al., 2002). The first approach is a filter method, and the second approach is a wrapper method.

3.1   Signal-to-Noise (S2N)

For each gene j, we compute the following statistic:

$$S(j) = \frac{\mu_+(j) - \mu_-(j)}{\sigma_+(j) + \sigma_-(j)}, \qquad (9.15)$$

where $\mu_+(j)$ and $\mu_-(j)$ are the means of the classes $+1$ and -1 for the jth gene. Similarly, $\sigma_+(j)$ and $\sigma_-(j)$ are the standard deviations for the two classes for the jth gene. Genes that give the most positive values are most correlated with class $+1$, and genes that give the most negative values are most correlated with class -1. One selects the most positive $m/2$ genes and the most negative $m/2$ genes, and then uses this reduced dataset for classification. The question of estimating m is addressed in Section 3.3.

3.2   Recursive Feature Elimination (RFE)

The method recursively removes features based upon the absolute magnitude of the hyperplane elements. We first outline the approach for linear SVMs. Given microarray data with n genes per sample, the SVM outputs the normal to the hyperplane, w, which is a vector with n components, each corresponding to the expression of a particular gene. Loosely speaking, assuming that the expression values of each gene have similar ranges, the absolute magnitude of each element in w determines its importance in classifying a sample, since the following equation holds:

$$f(x) = w \cdot x + b = \sum_{i=1}^{n} w_i x_i + b$$

The idea behind RFE is to eliminate elements of w that have small magnitude, since they do not contribute much in the classification function. The SVM is trained with all genes; then we compute the following statistic for each gene:

$$S(j) = |w_j| \tag{9.16}$$

Where w_j the value of the jth element of w. We then sort S from largest to smallest value and we remove the genes corresponding to the indices that fall in the bottom 10% of the sorted list S. The SVM is retrained on this smaller gene expression set, and the procedure is repeated until a desired number of genes, m, is obtained. When a nonlinear SVM is used, the idea is to remove those features that affect the margin the least, since maximizing the margin is the objective of the SVM (Papageorgiou et al., 1998). The nonlinear SVM has a solution of the following form:

$$f(x) = \sum_{i=1}^{l} c_i K(x, x_i) + b \tag{9.17}$$

Let M denote the margin. Then we obtain Equation 9.18:

$$\frac{1}{M} = \sum_{p,r=1}^{l} c_p c_r K(x_p, x_r) \tag{9.18}$$

So for each gene j, we compute to which extent the margin changes using the following statistic:

$$S(j) = \left| \frac{\partial (1/M)}{\partial x_j} \right| \tag{9.19}$$

where x_j is the jth element of a vector of expression values x. We then sort S from the largest to the smallest value, and we remove the genes corresponding to the indices that fall in the bottom 10% of the sorted list S. The SVM is retrained and the procedure is repeated just as in the linear case.

3.3   How Many Genes To Use?

A basic question that arises for all feature selection algorithms is how many genes the classifier should use. One approach to answer this question is using hypothesis and permutation testing (Golub et al., 1999). The null hypothesis is that the S2N or RFE statistic for each gene computed on the training set comes from the same distribution as that for a *random data set*. A random data set is the training set with its labels randomly permuted.

In detail, the permutation test procedure for the S2N or RFE statistic is as follows:

(1) Generate the statistic for all genes using the actual class label and sort the genes accordingly.
(2) Generate 100 or more random permutations of the class labels. For each case of randomized class labels, generate the statistics for all genes and sort the genes accordingly.
(3) Build a histogram from the randomly permuted statistics using various numbers of genes. We call this number k. For each value of k, determine different percentiles (1%, 5%, 50% etc.) of the corresponding histogram.
(4) Compare the actual signal-to-noise scores with the different significance levels obtained for the histograms of permuted class labels for each value of k (see Figure 9.7 for an illustration).

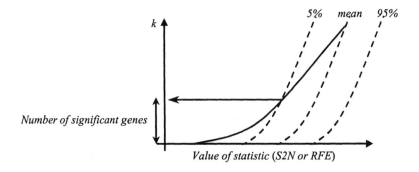

Figure 9.7. The solid curve is the S2N or RFE statistic rank ordered computed on the training set. The three dashed lines are the 5th, 50th, and 95th percentiles of the same rank ordered statistic as computed from the random data. The number of statistical genes is designated as the value of k, where the solid curve crosses the 5th percentile curve.

4.   ERROR RATES FOR MORPHOLOGY AND TREATMENT OUTCOME PREDICTION

We examined the error rate for SVMs and two other algorithms, *Weighted Voting Average* (WVA), and k-nearest neighbors (*k*NN) on seven binary cancer classification problems. The problems are as follows: discriminating acute myeloid leukemia (AML) from acute lymphoblastic leukemia (ALL), and discriminating B-cells from T-cells for acute lymphoblastic leukemia (Golub et al., 1999), discriminating follicular (FSC) lymphoma from diffuse large cell lymphoma (DLCL) and discriminating high risk from low risk lymphoma patients (Shipp et al., 2001), discriminating glioblastomas (GL) from meduloblastomas (MD), and discriminating high risk from low risk patients with medulloblastoma (Pomeroy et al., 2002). See Table 9.1 for number of samples in each class for the data sets.

Error rates for all data sets except for AML vs. ALL were measured using leave-one-out cross validation[4]. For AML vs. ALL, the test/train split described in (Golub et al., 1999) was used. S2N was used for feature selection for WVA and kNN algorithms. The SVM used the radius-margin ratio as a gene selection methodology. The errors for both outcome prediction problems were much larger than those for the morphology prediction problems. The errors are reported in Table 9.2.

The number of genes used in each classification task for each algorithm was determined using cross-validation. In general, SVMs required more genes in the classification tasks.

Table 9.1. The number of samples in the various data sets.

Data set	# of Samples	Class −1	Class +1
AML vs. ALL (*train*)	38	27 ALL	11 AML
AML vs. ALL (*test*)	35	21 ALL	14 AML
B-cell vs. T-cell	23	15 B-cell	8 T-cell
FSC vs. DLCL	77	19 FSC	58 DLCL
GL vs. MD	41	14 GL	27 MD
Lymphoma outcome	58	20 Low risk	38 High risk
Medullo outcome	50	38 Low risk	12 High risk

---

[4] See Chapter 7 of this volume for more details on *leave-one-out cross-validation*.

Table 9.2. Absolute number of errors for the various data sets.

Data set	Method	Errors Total	Class 1	Class −1	# of genes used
AML vs. ALL	WVA	2	1	1	50
	kNN	3	1	2	10
	SVM	0	0	0	40
B-cell vs. T-cell	WVA	0	0	0	9
	kNN	0	0	0	10
	SVM	0	0	0	10
FSC vs. DLCL	WVA	6	1	5	30
	kNN	3	1	2	200
	SVM	4	2	2	250
GL vs. MD	WVA	1	1	0	3
	kNN	0	0	0	5
	SVM	1	1	0	100
Lymphoma outcome	WVA	15	5	10	12
	KNN	15	8	7	15
	SVM	13	3	10	100
Medullo outcome	WVA	13	6	7	6
	kNN	10	6	4	5
	SVM	7	6	1	50

5.   MULTICLASS CLASSIFICATION

Ramaswamy et al. investigated whether the diagnosis of multiple adult malignancies could be achieved purely by molecular classification, using DNA microarray gene expression profiles (Ramaswamy et al., 2001). In total, 218 tumor samples, spanning 14 common tumor types, and 90 normal tissue samples were subjected to oligonucleotide microarray gene expression analysis. These tumor types/localizations are: breast (BR), prostate (PR), lung (LU), colorectal (CO), lymphoma (L), bladder (BL), melanoma (ME), uterus (UT), leukemia (LE), renal (RE), pancreas (PA), ovary (OV), mesothelioma (MS), and central nervous system (CNS). The expression levels of 16,063 genes and ESTs were used to train and evaluate the accuracy of a multiclass classifier based on the SVM algorithm. The overall classification accuracy was 78%, far exceeding the accuracy of random classification (9%). Table 9.3 shows the number of training and test samples per tumor class.

Table 9.3. The number of samples in the training set (train) and the test set (test).

	BR	PR	LU	CO	L	BL	ME	UT	LE	RE	PA	OV	MS	CNS
train	8	8	8	8	16	8	8	8	24	8	8	8	8	16
test	3	2	3	5	6	3	2	2	6	3	3	3	3	4

Multiple class prediction is intrinsically more difficult than binary prediction because the classification algorithm has to learn to construct a greater number of separation boundaries or relations. In binary classification, an algorithm can "carve out" the appropriate decision boundary for only one of the classes; the other class is simply the complement. In multiclass classification problems, each class has to be defined explicitly. A multiclass problem can be decomposed into a set of binary problems, and then combined to make a final multiclass prediction.

The basic idea behind combining binary classifiers is to decompose the multiclass problem into a set of easier and more accessible binary problems. The main advantage in this divide-and-conquer strategy is that any binary classification algorithm can be used. Besides choosing a decomposition scheme and a classifier for the binary decompositions, one also needs to devise a strategy for combining the binary classifiers and providing a final prediction. The problem of combining binary classifiers has been studied in the computer science literature (Hastie and Tibshirani, 1998, Allwein et al., 2000, Guruswami and Sahai, 1999) from a theoretical and empirical perspective. However, the literature is inconclusive, and the best method for combining binary classifiers for any particular problem is open.

Standard modern approaches for combining binary classifiers can be stated in terms of what is called *output coding* (Dietterich and Bakiri, 1991). The basic idea behind output coding is the following: given k classifiers trained on various partitions of the classes, a new example is mapped into an output vector. Each element in the output vector is the output from one of the k classifiers, and a *codebook* is then used to map from this vector to the class label (see Figure 9.7). For example, given three classes, the first classifier may be trained to discriminate classes 1 and 2 from 3, the second classifier is trained to discriminate classes 2 and 3 from 1, and the third classifier is trained to discriminate classes 1 and 3 from 2.

Two common examples of output coding are the *one-versus-all* (OVA) and all-pairs (AP) approaches. In the OVA approach, given k classes, k independent classifiers are constructed where the ith classifier is trained to separate samples belonging to class i from all others. The codebook is a diagonal matrix, and the final prediction is based on the classifier that produces the strongest confidence:

$$class = \arg\max_{i=1..k} f_i \qquad (9.20)^5$$

where f_i is the signed confidence measure of the ith classifier (see Figure 9.8). In the all-pairs approach, $\frac{1}{2} k(k-1)$ classifiers are constructed, with each classifier trained to discriminate between a class pair i and j. This can be

---

[5] The procedure arg max$f(x)$ simply selects the value or argument of x that maximizes $f(x)$.

thought of as a $k \times k$ matrix, where the ij^{th} entry corresponds to a classifier that discriminates between classes i and j. The codebook, in this case, is used to simply sum the entries of each row and select the row for which this sum is maximal:

$$class = \arg\max_{i=1..k}\left[\sum_{j=1}^{k} f_{ij}\right] \qquad (9.21)$$

where f_{ij} is the signed confidence measure for the ij^{th} classifier.

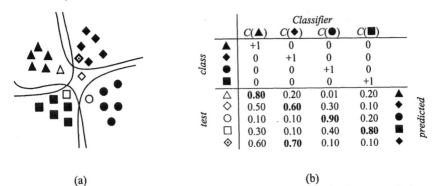

		Classifier				
		$C(▲)$	$C(◆)$	$C(●)$	$C(■)$	
class	▲	+1	0	0	0	
	◆	0	+1	0	0	
	●	0	0	+1	0	
	■	0	0	0	+1	
test	△	**0.80**	0.20	0.01	0.20	▲
	◇	0.50	**0.60**	0.30	0.10	◆
	○	0.10	0.10	**0.90**	0.20	●
	□	0.30	0.10	0.40	**0.80**	■
	◈	0.60	**0.70**	0.10	0.10	◆

(predicted)

(a)   (b)

Figure 9.8. OVA classification. (a) Four binary classifiers are trained. The curves designate the non-linear decision boundaries. The first classifier discriminates between the ▲-class and all other classes; the second classifier discriminates between the ◆-class and all other classes; the third classifier discriminates between the ●-class and all other classes, and the fourth classifier discriminates between the ■-class and all other classes. The following five cases are the test cases: △, ◇, ○, □, and ◈. (b) The codebook for OVA classification is represented in the upper part of the table. The numbers in the table are the ideal outputs of the classifiers for the cases of the four classes. For example, the classifier $C(▲)$, which has learnt to discriminate ▲-cases from all other cases, ideally outputs +1 for ▲-cases and 0 otherwise. The lower part of the table shows the outputs of the classifiers for the test cases. The classifier that outputs the highest number determines the class. For example, $C(●)$ yields 0.90 for the case ○; consequently, ○ is classified as member of class ●.

Intuitively, there is a tradeoff between the OVA and AP approaches. The discrimination surfaces that need to be learned in the all-pairs approach are, in general, more natural and, theoretically, should be more accurate. However, with fewer training examples, the empirical surface constructed may be less precise. The actual performance of each of these schemes, or others such as random codebooks, in combination with different classification algorithms is problem dependent. The OVA approach gave the best results on this dataset, and these results are reported in Table 9.4. The train/test split in the table is the same as that in (Ramaswamy et al., 2001). The error rate on the training set was measured using leave-one-out cross validation. A sample classification was considered *high confidence* if max $f_i \geq 1$, and low confidence otherwise.

Table 9.4. Error rates for the multiclass data set.

Data set	Validation Method	Samples	Total accuracy	Confidence			
				High		Low	
				Fraction	Accuracy	Fraction	Accuracy
Train	cross-validation	144	78%	80%	90%	20%	28%
Test	train/test	54	78%	78%	83%	22%	58%

6. SOFTWARE SOURCES AND RULES OF THUMB

The SVM experiments described in this chapter were performed using a modified version of the SvmFu package, which can be downloaded from http://www.ai.mit.edu/projects/cbcl/). Another available SVM package is SVMTorch (http://www.idiap.ch/learning/SVMTorch.html). SvmFu has some advantages in that it has both multiclass and leave-one-out cross-validation built in. SVMTorch is the only software of the above that does SVM regression as well as classification.

The following are some rules of thumb when using SVMs:

(1) *Normalizing your data*: in general it is a good idea to rescale your data such that all kernel values fall between -100 and 100; a simple way to do this is by normalizing all entries of the microarray such that they fall between $-10(n)^{\frac{1}{2}}$ and $+10(n)^{\frac{1}{2}}$, where n is the number of expression values per sample;

(2) *Choosing the regularization parameter C*: given the above normalization, the regularization parameter usually does not have much effect, so set it somewhere between $1-100$;

(3) *Choosing the kernel*: for microarray applications, a linear kernel is usually sufficient; you can use polynomial kernels if you want to examine correlations between genes, but it will in general not greatly improve the performance; if the linear kernel does not give good performance, it is worth trying the Gaussian kernel;

(4) *Choosing the variance of the Gaussian kernel*: set σ such that the average distance between two training points

$$K(x_1, x_2) = \exp(\|x_1 - x_2\|^2 / 2\sigma^2) = 1/l,$$

where l is the number of samples.

(5) *Multiclass problems*: for a multiclass problem, use the OVA and AP decompositions; in general, more complicated coding systems do not help. When there are very few (5-10) training samples per class, OVA will in general give the best results.

7.   DISCUSSION

The theoretical advantage of SVMs is that the idea of margin or stability mitigates the problem of overfitting the training data. This is crucial in the DNA microarray domain, which is characterized by thousands of genes/variables and very few samples. Unlike most other algorithms (for example, *k*NN or WVA), the SVM performs very well even without feature selection, using all 7,000-16,000 expression values. The multiclass example illustrates that the SVM is especially helpful when there are very few training samples. Other algorithms such as *k*NN or WVA are made stable by removing the noisy genes and reducing the number of features to 10-100. One can loosely conclude that the *curse of dimensionality* is overcome in SVMs using the *blessing of smoothness*, and simply by reducing the dimensionality in the other algorithms. A practical result of this is that for some classification problems, the SVM will tend to use more genes to make a classification that *k*NN or WVA.

ACKNOWLEDGEMENTS

I would like to acknowledge my colleagues at Cancer Genomics Group at the Whitehead/MIT Center for Genome Research and at the Center for Biological and Computational Learning at MIT. The author is supported by the Office of Science (BER), US Department of Energy, Grant No. DE-FG02-01ER63185.

REFERENCES

Allwein E.L., Schapire R.E., Singer Y. (2000). Reducing multiclass to binary: a unifying approach for margin classifiers. Journal of Machine Learning Research 1:113-141.

Bhattacharjee A., Richards W.G, Staunton J., Li C., Monti S., Vasa P., Ladd C., Beheshti J., Bueno R., Gillette M., Loda M., Weber G., Mark E.F., Lander E.S., Wong W., Johnson B.E., Golub T.R., Sugarbaker D.J., Meyerson M. (2001). Classification of human lung carcinomas by mRNA expression profiling reveals distinct adenocarcinoma subclasses. Proc. Natl. Acad. Sci. USA 98:13790–13795.

Bousquet O. and Elisseeff A. (2002). Stability and Generalization. Journal of Machine Learning Research 2, 499-526.

Brown M.P.S., Grundy W.N., Lin D., Cristianini N., Sugnet C., Furey T.S., Ares M., Jr., Haussler D. (2000). Knowledge-based analysis of microarray gene expression data using support vector machines. Proc. Natl. Acad. Sci. USA 97(1):262-267.

Dietterich T.G. and Bakiri G. (1991). Error-correcting output codes: A general method for improving multiclass inductive learning programs. Proc. of the Ninth National Conference on Artificial Intelligence, AAAI Press, 572-577.

Evgeniou T., Pontil M., Poggio T. (2000). Regularization networks and support vector machines. Advances in Computational Mathematics 13:1-50.

Golub T.R., Slonim D.K., Tamayo P., Huard C., Gaasenbeek M., Mesirov J.P., Coller H., Loh M.L., Downing J.R., Caligiuri M.A., Bloomfield C.D., Lander E.S. (1999). Molecular classification of cancer: class discovery and class prediction by gene expression monitoring. Science 286(5439):531-7.

Guruswami V. and Sahai A. 1999. Multiclass learning, boosting and error-correcting codes. Proc. of the Twelfth Annual Conference on Computational Learning Theory, ACM Press, 145-155.

Guyon I., Weston J., Barnhill S., Vapnik V. (2002). Gene selection for cancer classification using support vector machines. Machine Learning 46:389-422.

Hastie T.J. and Tibshirani R.J. (1998). Classification by pairwise coupling. In Jordan M.I., Kearnsa M.J., Solla S.A., eds., Advances in Neural Information Processing Systems, volume 10, MIT Press.

Mukherjee S., Rifkin R., Poggio T. (2002). Regression and Classification with Regularization. Proc. of Nonlinear Estimation and Classification, MSRI, Berkeley, Springer-Verlag.

Mukherjee S., Tamayo P., Slonim D., Verri A., Golub T., Mesirov J.P., Poggio T. (2000). Support vector machine classification of microarray data. Technical Report, Artificial Intelligence Laboratory, Massachusetts Institute of Technology.

Papageorgiou C., Evgeniou T., Poggio T. (1998). A trainable pedestrian detection system. Intelligent Vehicles, pp. 241-246.

Platt J.C. 1999. Probabilistic Outputs for Support Vector Machines and Comparisons to Regularized Likelihood methods. Advances in Large Margin Classifiers, MIT Press.

Pomeroy S.L., Tamayo P., Gaasenbeek M., Sturla L.M., Angelo M., McLaughlin M.E., Kim J.Y., Goumnerova L.C., Black P.M., Lau C., Allen J.C., Zagzag D., Olson J., Curran T., Wetmore C., Biegel J.A., Poggio T., Mukherjee S., Rifkin R., Califano A., Stolovitzky G., Louis D.N., Mesirov J.P., Lander E.S., and Golub T.R. (2002). Prediction of central nervous system embryonal tumour outcome based on gene expression. Nature, 415(24):436-442. (and supplementary information).

Ramaswamy S., Tamayo P., Rifkin R., Mukherjee S., Yeang C.H., Angelo M. Ladd C., Reich M., Latulippe E., Mesirov J.P., Poggio T., Gerald W., Loda M., Lander E.S., Golub. T.R.: Multiclass cancer diagnosis using tumor gene expression signatures, Proc. Natl. Acad. Sci. USA. 98(26):15149-15154.

Shipp MA, Ross KN, Tamayo P, Weng AP, Kutok JL, Aguiar RC, Gaasenbeek M, Angelo M, Reich M, Pinkus GS, Ray TS, Koval MA, Last KW, Norton A, Lister TA, Mesirov J, Neuberg DS, Lander ES, Aster JC, Golub TR. (2002). Diffuse large B-cell lymphoma outcome prediction by gene expression profiling and supervised machine learning. Nat Med 8(1):68-74.

Slonim D., Tamayo P., Mesirov J., Golub T., Lander E. (2000). Class prediction and discovery using gene expression data. In Proc. of the 4th Annual International Conference on Computational Molecular Biology (RECOMB), Universal Academy Press, pp. 263-272, Tokyo, Japan.

Tikhonov A.N. and Arsenin V.Y. (1977). Solutions of Ill-posed Problems. W.H. Winston, Washington D.C.

Vapnik V.N. (1998). Statistical Learning Theory. John Wiley & Sons, New York.

Chapter 10

WEIGHTED FLEXIBLE COMPOUND COVARIATE METHOD FOR CLASSIFYING MICROARRAY DATA

A Case Study of Gene Expression Level of Lung Cancer

Yu Shyr[1] and KyungMann Kim[2]

[1]*Division of Biostatistics, Department of Preventive Medicine, Vanderbilt University, 571 Preston Building, Nashville, TN 37232-6848, USA,*
e-mail: yu.shyr@vanderbilt.edu

[2]*Department of Biostatistics and Medical Informatics, University of Wisconsin-Madison, 600 Highland Ave, Box 4675, Madison, WI 53792, USA,*
e-mail: kmkin@biostat.wisc.edu

1. INTRODUCTION

Just as in any biomedical investigations, the foundation of the microarray data analysis is based on the primary objective(s) of the experiment. In general, there are three objectives in the microarray experiments: class discovery (non-supervised method), class comparison (supervised method) and class prediction (supervised method). The investigator may be interested in one or more of these three objectives, but the statistical considerations for design and experiment and statistical data analysis should correspond to the objective(s) of the study. For example, the hierarchical clustering techniques may be useful in class discovery but may not be appropriate in class comparison. The objective for class discovery can be "Are there more than three histological groups, i.e. large cell, adenocarcinoma and squamous cell, in non-small cell lung cancer?" The objective for class comparison can be "Is there a set of genes that are differentially expressed between patients responding and non-responding to a therapy?" The objective for class prediction can be "Is there a set of genes that can predict the clinical features?" In this chapter, we will discuss the class comparison and prediction methods with an example from Vanderbilt lung cancer SPORE.

Lung cancer causes deaths of more men and women in the United States than the next four most common types of cancer combined. The five-year survival rate of all lung cancer is only 14 percent (Landis et al., 1999). Recent advances in microarray technology enabled us to get detailed simultaneous expression profiles of thousands of genes expressed in a tissue. In order to better understand the biological diversity in human lung cancer, we have begun to use this technology to determine the RNA profiles of the genes expressed in freshly resected human lung cancers and to correlate these with clinical and histological parameters.

In an investigation at Vanderbilt-Ingram Cancer Center, total RNAs were extracted from 26 tumors (24 lung tumors, 1 breast and 1 sarcoma) and 3 normal tissues. As a control RNA, six RNAs were pooled from different lung cancer cell lines representing the common histological types of lung cancer. To obtain clean RNA, two step extraction using Trizol (GibcoBRL) followed by RNeasy kit (Qiagen) was used. The 5,088 cDNAs were prepared from sequence-verified clones of Research Genetics to represent 4,749 unique genes: 2,162 were expressed sequence tags (ESTs) and 2,587 were non-ESTs. The cDNAs were amplified by polymerase chain reaction (PCR) and spotted onto poly-L-lysine coated glass slides using a Stanford-type microarrayer robot at the Vanderbilt Microarray Shared Resource.

Labeled cDNAs were made from total RNAs through an oligo-dT primed reverse transcriptase reaction in the presence of Cy3 or Cy5 dCTP. Cy3- and Cy5-labeled cDNAs were mixed and applied onto 5k human cDNA arrays on glass slides. After overnight hybridization and post-hybridization washing, the images on the glass slides were captured using a dual-laser confocal microscopy scanner. Using the analysis software package (GenePix Pro3), output was expressed as the ratio of intensity for the probe tumor RNA compared to the control RNA for each cDNA spot on the slide. For each sample, two reciprocal hybridizations were performed on different arrays (switching the dyes) to account for dye bias.

The primary statistical challenges of analysis of RNA expression patterns are to identify a set of genes that are differentially expressed between different classes and to develop predictive models of the statistical relationships between multivariate RNA expression data and the clinical features. In this chapter, the *mutual-information scoring* (Info Score), *weighted gene analysis* (WGA), *significance analysis of microarrays* (SAM), and permutation *t-test* or *F-test* have been reviewed for identifying the genes that are differentially expressed between different classes. One of the challenges for these analyses is the genes identified by different methods may not be the same. The class prediction model may be used to examine the significances of the genes. Hedenfalk et al. (2001) successfully applied the *t*-test based *compound covariate method* to class prediction analysis for BRCA1+ vs. BRCA1−. We reviewed a recently proposed *weighted flexible compound covariate method* (*WFCCM*) based on Info Score, WGA, SAM, and permutation test. We also reviewed the basic concepts of the classification tree methods. We conclude with suggestions for general ideas of analyzing RNA expression data.

2. CLASS COMPARISON - VARIABLE SELECTION

Most classification and pattern recognition methodologies are limited by the number of independent variables that can be assessed simultaneously. Thus, a useful subset of independent variables must be selected. This is referred to as the feature or variable subset selection problem. The mutual-information scoring (Info Score), weighted gene analysis (WGA), significance analysis of microarrays (SAM), and a permutation *t*-test or F-test were successfully applied in several RNA expression profile data for determining the genes that are differentially expressed between classes (Hedenfalk et al., 2001; Tusher et al., 2001). These methods can provide the significance, the score or the rank of the genes of the multi-dimensional data, and the subset selection can be based on the resulting significance, the score or the rank.

2.1   Information-Theoretic Score

The Info Score is an information-theoretic score, which was introduced by Ben-Dor, Friedman and Yakhini (2000). It uses a ranking-based scoring system and combinatorial permutation of sample labels to produce a rigorous statistical benchmarking of the overabundance of genes whose differential expression pattern correlates with sample type, e.g., tumor (+) vs. normal (−). Let N denote the number of tissues, consisting of p tissues from class P, e.g., tumor tissues, with g_{jp} expression level for gene (j), and q tissues from Class Q, e.g., normal tissues, with g_{jq} expression level. We define the rank vector v_j of g_j to be a vector of $\{+, -\}$ with a $N \times l$ dimension, where g_j is a $N \times l$ vector of expression level and $N = p + q$.

$$v_j = \begin{cases} + & \text{if } g_j \in P \\ - & \text{if } g_j \in Q \end{cases} \qquad (10.1)$$

For example, if the expression levels for gene j are $\{1, 2, 3, 5, 6, 7, 11, 14\}$ in class P and $\{4, 8, 9, 10, 12, 13, 15\}$ in class Q, then $v_j = \{+, +, +, -, +, +, +, -, -, -, +, -, -, +, -\}$. Note that the rank vector v_j captures the essence of the differential expression profile of g_j. If g_j is underexpressed in class P, then the positive entries of v_j are concentrated in the left hand side of the vector, and the negative entries are concentrated at the right hand side.

The Info score of a rank vector v_j is defined as

$$Info(v_j) = \min_{x, y = v} \{(|x|/|v|) \, Ent(x) + (|y|/|v|) \, Ent(y)\} \quad (10.2)$$

where $Ent(\bullet)$ is the entropy of $\bullet$ defined by

$$Ent(\bullet) = H(\phi) = -\phi \log_2(\phi) - (1 - \phi) \log_2(1 - \phi) \quad (10.3)$$

and ϕ denotes the fraction of positive entries in $\bullet$. An entropy can be viewed as measurement of the degree of disorder or uncertainty in a system. Using the same example above, the best partition with respect to the Info Score for gene j is

$$Info(v_j) = \frac{7}{15} H\left(\frac{6}{7}\right) + \frac{8}{15} H\left(\frac{2}{8}\right) = 0.71 \quad (10.4)$$

The range of the Info Score is between 0 and 1, and the smaller Info Scores indicate the stronger evidence of the different expression profiles of two classes.

In summary, the Info Score uses a rank-based scoring system and combinatorial permutation of sample labels to produce a rigorous statistical benchmarking of the overabundance of genes whose differential expression pattern correlates with sample type. Genes may be ranked on the basis of Info Score.

2.2 Weighted Gene Analysis

Weighted Gene Analysis (WGA) is described by Bittner and Chen (2001). It can be defined as follows: For a given two-class setting, a discriminative weight (score) for each gene j can be evaluated by

$$w_j = d_B /(f_1 d_{wP} + f_2 d_{wQ} + \alpha) \quad (10.5)$$

where d_B is the center-to-center Euclidean distance between the two classes (P and Q), d_{wg} is the average Euclidean distance among all sample pairs within class g, $g = P, Q$. L is the number of sample pairs in class P, and M is the number of sample pairs in class Q, e.g., tumor and normal. $f_1 = L /(L + M)$ and $f_2 = M /(L + M)$. α is a small constant, e.g., 0.01, to prevent zero denominator case.

The range of the w_j is between 0 and ∞, and the bigger w_j scores indicate the stronger evidence of the different expression profiles of two classes.

2.3   Significance Analysis of Microarrays

Significance Analysis of Microarrays (SAM) (Tusher et al., 2001) is a statistical technique for finding significant genes in a set of microarray experiments. It is a method for identifying genes on a microarray with statistically significant changes in expression, developed in the context of an actual biological experiment. SAM assigns a score to each gene on the basis of change in gene expression relative to the standard deviation of repeated measurements to estimate the percentage of gene identified by chance, the false positive rate (FPR). It is based on the analysis of random fluctuations in the data. In general, the signal-to-noise ratio is decreased with decreasing gene expression. The "relative difference" $d(j)$ in gene expression is:

$$d(j) = \frac{\overline{x}_P(j) - \overline{x}_Q(j)}{s(j) + s_0} \qquad (10.6)$$

where $\overline{x}_P(j)$ and $\overline{x}_Q(j)$ are defined as the average levels of expression for gene j in class P and Q, respectively.

$$s(j) = \sqrt{a\left(\sum_m [x_m(j) - \overline{x}_P(j)]^2 + \sum_n [x_n(j) - \overline{x}_Q(j)]^2\right)} \qquad (10.7)$$

where Σ_m and Σ_n are summation of the expression measurements in class P and Q, respectively, $a = (1/m + 1/n)(m + n - 2)$, and m and n are the numbers of measurements in class P and Q, respectively.

The distribution of $d(j)$ should be independent of the level of gene expression. At low expression levels, variance in $d(j)$ can be high because of small values of $s(j)$. To ensure that the variance of $d(j)$ is independent of gene expression, SAM adds a small positive constant s_0 to the denominator.

The coefficient of variation (CV) of $d(j)$ can be computed as a function of $s(j)$ in moving windows across the data. The value of s_0 can be chosen to minimize the coefficient of variation. The bigger $|d(j)|$ scores indicate the stronger evidence of the different expression profiles of two classes.

2.4   Permutation *t*-Test

Permutation *t*-test (Radmacher and Simon, 2001) is a strategy for establishing differences in gene expression pattern between classes. In permutation *t*-test, the standard *t* statistic is computed on the log-expression ratios of each gene in order to analyze the variation between classes, e.g., tumor vs. normal. Then, labels (i.e. tumor and normal) are randomly permuted among the specimens and the t statistic for each gene (j) in the permuted data set is computed. This process was repeated 10,000 or more times (Hedenfalk et al., 2001). Finally, a critical value of the t statistic, e.g., .999, is determined for each gene based on the empirical distribution of t from permuted data sets for the gene (j). If the t statistic for a gene (j) in the original labeling of specimens is larger than its critical value, the gene is deemed differentially expressed between the two groups and is considered as the significant difference.

2.5 Inconsistence of the Variable Selection

We have reviewed four methods for the variable selection process in class comparison. There are several other existing methods that can do similar job (see Chapter 6 for more feature selections), e.g., REML-based *mixed model* (Wolfinger et al., 2001), *Threshold Number of Misclassification Score* (*TNoM*) (Ben-Dor et al., 2000). There are also more new methods coming out for the class comparison purpose of the microarray data, for example, the *P-value for Identifying Differentially Expressed genes* (*PIDEX*) method (Ge et al., unpublished manuscript). This method combines the fold change, change in the absolute intensity measurements and data reproducibility. PIDEX produces p-values for identifying differentially expressed genes. The genes may be ranked on the basis of p-value. More new methods will be developed as the microarray researches advance.

Each of these methods can generate a list of genes based on their significances or scores. The question is "Do the results from these methods agree with each other?" The answer is "No!" Based on the nature of the development of each method, the results form each method will not agree with each other totally. Figure 10.1 shows the results from the Vanderbilt lung cancer SPORE study based on four methods – permutation t-test, SAM, WGA, and Info Score. The investigators would like to generate a set of genes that performed differently between non-small cell lung cancer tumors and normal tissues. We picked top 30 genes that performed most differently between these two classes from each of these four statistical methods. There are 49 "winner" genes based on the union of all four methods. Only 13 genes or 26.5% of genes were selected as "winners" by all four methods. The results of the selection were influenced by sample size, variation of the data, percent of missing data, as well as the outliers/extreme values of the data within each class. In practice, it is not a bad idea to use more than one method to generate the gene list. The investigators may focus on the genes selected by all methods first.

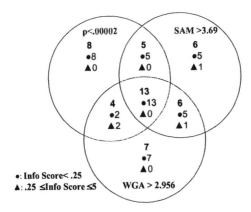

Figure 10.1. Variable selection results of Vanderbilt lung cancer SPORE study.

3. CLASS PREDICTION

Class prediction methods can be used to examine the goodness of the set of genes identified in the class comparison step. There are two types of the class prediction methods: (1) based on the "training" data set and (2) based on the "test" data set. It is highly recommended that the class prediction models be applied to these two types of the data sets for the microarray data analysis because of the class prediction model may easily over-fit the "training" data set. In addition, the sample size of the blinded/test data set probably should be comparable to the sample size of the training data set if the sample size of the training data set is very small. On the other hands, if the sample size of the training data set is large, i.e. several hundred samples, a training set: test set ratio of $k : 1$, where $k > 1$, may be appropriate.

3.1 Compound Covariate Method (CCM)

Hedenfalk et al. (2001) successfully applied the compound covariate method (CCM) (Tukey, 1993) to class prediction analysis for BRCA1+ vs. BRCA1-. This predictor is built in two steps. First, a standard two-sample *t*-test is performed to identify genes with significant differences (at level α, Hedenfalk et al. picked $\alpha = 0.0001$) in log-expression ratios between the two tissue classes. Second, the log-expression ratios of differentially expressed genes are combined into a single compound covariate for each tissue sample; the compound covariate is used as the basis for class prediction. The compound covariate for tissue sample i is defined as

$$c_i = \sum_j t_j x_{ij} \qquad (10.8)$$

where t_j is the *t*-statistic for the two-group comparison of classes with respect to gene *j*, x_{ij} is the log-ratio measured in tissue sample *i* for gene *j* and the sum is over all differentially expressed genes.

The CCM reduces the data dimension from $N \times J$ to $N \times 1$, where *N* is the total number of the samples, *J* is total number of study genes. We can view CCM as the "overall score" of each tissue sample, which combines information of all important genes from one statistical method.

3.2 Weighted Flexible Compound Covariate Method

The *weighted flexible compound covariate method* (WFCCM) (Shyr, 2002) is an extension of the compound covariate method, which allows considering more than one statistical analysis methods in the compound covariate. Before we apply the WFCCM, it is important to make sure that the "sign" of each statistical method is consistent. For example, the sign of the *t*-statistic and SAM are always consistent, but WGA scores are always positive since its scoring system is based upon Euclidean distance. Therefore, multiplying a (-1) to the WGA scores for all the genes that have negative scores in SAM or *t*-statistic is the first step for applying the WFCCM. The second step is to select the "winners" with all statistical methods. We may arbitrarily pick genes from each statistical method, e.g., top 1% or top 100 genes, or we may use some significant information to select genes from the statistical methods, e.g., p-value < 0.0001 for *t*-statistic, p-value < 0.01 for REML-based mixed effect models, or SAM > 3.5.

The WFCCM for tissue sample *i* is defined as

$$WFCCM(i) = \sum_j \left[\sum_k \left(ST_{jk} W_k \right) \right] \left[W_j \right] x_{ij} \qquad (10.9)$$

where x_{ij} is the log-ratio measured in tissue sample *i* for gene *j*. ST_{jk} is the standardized statistic/score of gene *j*, e.g., standardized SAM score, for statistical analysis method *k*. W_k is the weight of method *k*, which can be determined as

$$W_k = (1 - CCM\ misclassification\ rate_k) \qquad (10.10)$$

where "*CCM misclassification rate_k*" stands for the misclassification rate of the compound covariate method for statistical analysis method *k*. W_j is the weight of gene *j*, which can be determined as

$$W_j = \sum_k V_{jk} / K \qquad (10.11)$$

where $V_{jk} = 1$, if the gene j is selected as the "winner" in method k; $V_{jk} = 0$, if the gene j is not selected as the "winner" in method k. If gene j is selected by all methods then $W_j = 1$.

The W_k and W_j can be determined by other methods, too. For example, we may assign $W_k = 1$ for all K methods used in variable selection stage if we believe they perform equally well. We may also modify W_j as

$$W_j = \left[\left(\sum_k X_{jk}/K\right)\left(1 - \textit{Info Score}_j\right)\right] \qquad (10.12)$$

In this case, if gene j is selected by all methods and the *Info Score$_j$* = 0, then $W_j = 1$.

The WFCCM also reduces the data dimension from $N \times P$ to $N \times 1$. We can certainly view WFCCM as the "overall score" of each tissue sample, which combines all information of all important genes from several statistical methods.

3.3   Leave-One-Out Cross-Validated Class Prediction Model

The misclassification rate can be assessed using leave-one-out cross-validated (LOOCV) class prediction method. Cross-validation is a method for estimating generalization error based on resampling. LOOCV is one specific type of cross-validation. The LOOCV is processed in four steps in the Vanderbilt lung cancer SPORE study. First, apply the WFCCM to calculate the single compound covariate for each tissue sample based on the significant genes. Second, one tissue sample is selected and removed from the dataset, and the distance between two tissue classes for the remaining tissue samples is calculated. Third, the removed tissue sample is classified based on the closeness of the distance of two tissue classes, e.g., k-nearest neighbor approach, which k=2, or using the midpoint of the means of the WFCCM for the two classes as the threshold. Fourth, repeat step 2 and 3 for each tissue sample. To determine whether the accuracy for predicting membership of tissue samples into the given classes (as measured by the number of correct classifications) is better than the accuracy that may be attained for predicting membership into random grouping of the tissue samples, we may create 2,000-5,000 random data sets by permuting class labels among the tissue samples. Cross-validated class prediction is performed on the resulting data sets and the percentage of permutations that results in as few or fewer misclassifications as for the original labeling of samples can be reported. If less than 5% of the permutations result in as few or fewer misclassifications, the accuracy of the prediction of the given classes is considered significant. Therefore, this rate may be considered as the "p-value" for the class prediction model. We recently have succeeded in applying the WFCCM class prediction analysis method to the preliminary data generated by the Vanderbilt lung cancer SPORE study.

The perfect WFCCM class-prediction models based on 54, 62, 77 and 27 differentially expressed genes were found to classify tumor tissue samples vs. normal tissue samples, primary lung cancer tissue samples vs. non-primary lung cancer tissue samples, non-small cell lung cancer (NCLS) tissue samples vs. normal tissue samples as well as adenocarcinoma tissue samples vs. squamous cell carcinoma tissue samples. Table 10.1 shows the results from the Vanderbilt lung cancer study. WGA, SAM, Info-Score, and permutation *t*-test were applied in the analysis. The cut-off points were 3.0, 3.7 and p < 0.0001 for WGA, SAM, and Permutation *t*-test respectively. We selected $W_k = 1$ for all methods, and $W_j = [(\Sigma_k X_{jk} / k) (1 - \textit{Info Score}_j)]$.

Table 10.1. WFCCM class prediction model in training data set.

Classification (sample size)	# of diff. expressed genes	# of misclassified samples	Prob. of random permutations with misclassifications
All samples			
Normal lung (3) vs. Tumor (26)	54	0	< 0.001
Normal lung and metastatic lung tumor(5) vs. Primary lung tumor (24)	62	0	< 0.0001
Normal lung (3) vs. NSCLC (23)	77	0	< 0.001
Non-small cell lung cancer			
Adeno (8) vs. Non-adeno (15)	6	1 (Non-adeno)	< 0.001
Squamous (8) vs. Non-squamous (15)	10	1 (Squamous)	< 0.001
Large (7) vs. Non-large (16)	2	1 (Large)	= 0.001
Adeno (8) vs. Squamous (8)	27	0	< 0.0001

We also applied the WFCCM to a set of blinded/test samples. Table 10.2 shows the results of the analyses. In general, the model performed reasonably well (the average correct prediction rate in the blinded/test data set was 93%) except for predicting large cell vs. non-large cell tissues. Because there were only two genes reached the selection criteria in WFCCM for comparing large cell tissues with non-large cell tissues, this result was not a surprise.

Table 10.2. WFCCM class prediction model in test data set

Classification (sample size)	# of diff. expressed genes	# of misclassified samples	Percent of correct prediction rate
All samples			
Normal lung (0) vs. Tumor (13)	54	0	100%
Normal lung and metastatic lung tumor(4) vs. Primary lung tumor (9)	62	1 (Metastatic)	92%
Normal lung (0) vs. NSCLC (9)	77	0	100%
Non-small cell lung cancer			
Adeno (4) vs. Non-adeno (5)	6	0	100%
Squamous (4) vs. Non-squamous (5)	10	1 (Squamous)	89%
Large (1) vs. Non-large (8)	2	3 (non-large)	67%
Adeno (4) vs. Squamous (4)	27	0	100%

Figure 10.2 shows the results from the agglomerative hierarchical clustering algorithm for clustering adenocarcinoma and squamous cell carcinomas. The average linkage algorithm was applied to calculate the distance between the clusters. All the adenocarcinoma tissues clustered together, so did squamous cell tissues. The results looked very promising but we might only use these results to reconfirm the genes performed differently between two classes. Having a perfect or near perfect cluster result was expected if we applied the cluster analysis after we selected the genes that performed differently using any of the supervised methods. It is important to know that we could not apply these results in any class discovery conclusion!

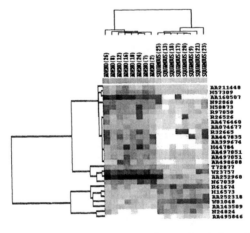

Figure 10.2. Results of the agglomerative hierarchical clustering algorithm.

4.   CLASSIFICATION TREE METHODS

We have reviewed several class comparison methods and class prediction methods separately in this chapter. In this section, we would like to discuss some basic concepts of the classification tree methods since constructing classification trees may be seen as a type of variable selection while at the same time a predictive model is developed. Figure 10.3 shows the diagram illustrating a (simplified) decision tree for lung cancer risk evaluation. Classification trees are the structures that rigorously define and link choices and possible outcomes (Reggia, 1985).

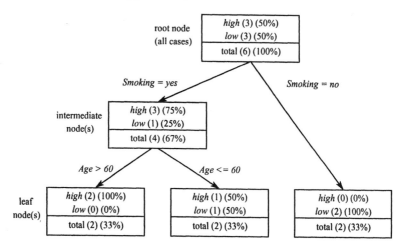

Rule 1: if *smoking* = *yes* and *Age* > *60*, then *lung cancer risk* = *high*.
Rule 2: if *smoking* = *no*, then *lung cancer risk* = *low*.

Figure 10.3. Decision tree analysis of lung cancer risk (simplified example).

The ideal situation for using classification tree methods is when there is a partition of the space Δ that will correctly classify all observations, and the task is to find a tree to describe it succinctly. In some cases the distributions of the classes over Δ overlap, so there is no partition that completely describes the classes. Then, for each cell of the partition, there will be a probability distribution over the classes, and the Bayes decision rule will choose the class with highest probability. The training set idea described above can also apply to classification tree methods. The misclassification rate can be calculated by the proportion of the training set that is misclassified, and the generalization ability is estimated. With "noisy" data, such as microarray data, it is quite possible to construct a tree which fits the training set well, but which has adapted too well to features of that particular subset of Δ. In other words, it can be too elaborate and over-fit the training data set. Overfitting is always a problem in any classifier.

Ji and Kim (unpublished manuscript) successfully applied the classification tree methods to classify toxic chemicals based on gene expression levels. The approach chooses the predictive genes as well as determines the classification rule. Three classification tree methods investigated are described below.

CART (Breiman et al., 1984) which stands for "Classification and Regression Trees" is biased when there are categorical predictors and when there are many missing values. With gene expression data, neither is the case, so CART gives a good fit. However, since it uses exhaustive search for variable selection, it often causes model overfitting. The CART algorithm is likely to give high accuracy in classification, but the genes selected may not be the most predictive ones.

QUEST (Loh and Shih, 1997) which stands for "Quick, Unbiased, Efficient Statistical Trees" is a program for tree-structured classification. The main strengths of QUEST are unbiased variable selection and fast computational speed. Also it is sensitive in identifying predictive genes since it uses statistical tests instead of exhaustive search. A common criticism of classification trees is that the construction of a classification tree can be extremely time-consuming. In addition, QUEST has options to perform CART-style exhaustive search and cost-complexity cross-validation pruning.

CRUISE (Kim and Loh, 2001), which stands for "Classification Rule with Unbiased Interaction Selection and Estimation", consists of several algorithms for the construction of classification trees. It provides the features of unbiased variable selection via bootstrap calibration, multi-way splits for more than two classes, missing value treatment by global imputation or by node-wise imputation, and choice of tree pruning by cross-validation or by a test sample. CRUISE differs from QUEST and CART in that it allows multiway splits, which is natural when there are multiple classes.

Different classification tree methods may identify different sets of genes that have the same or similar misclassification rates. The concept of the WFCCM may be applied in this situation for combining all the possible "winners" genes from different classification tree methods.

5.   CONCLUSION

The statistical class comparison and class prediction analyses for the microarray data may focus on the following steps: (1) Selecting the important gene patterns that perform differently among the study groups, (2) Using the class prediction model based upon the Weighted Flexible Compound Covariate Method (WFCCM), classification tree methods, or other methods to verify if the genes selected in step one have the statistical significant prediction power on the training samples, (3) Applying the prediction model generated from step two to a set of test samples for examining the prediction power on the test samples, and (4) Employing the agglomerative hierarchical clustering algorithm to investigate the pattern among the significant discriminator genes as well as the biologic status.

The selection of important gene patterns may be based on different methods, such as Significance Analysis of Microarrays (SAM), Weighted Gene Analysis (WGA), and the permutation *t*-test. The cutoff points may be determined based on the significance as well as the prediction power of each method. The genes will be on the final list if they are selected by at least one of the methods.

The weighted flexible compound covariate method may be employed for the class-prediction model based on the selected genes. This method was designed to combine the most significant genes associated with the biologic status from each analysis method. The WFCCM is an extension of the compound covariate method, which allows considering more than one statistical analysis method into the compound covariate. The class prediction model can be applied to determine whether the patterns of gene expression could be used to classify tissue samples into two or more classes according to the chosen parameter, e.g., normal tissue vs. tumor tissue. We reviewed the leave-one-out cross-validated class prediction method based on the WFCCM to estimate the misclassification rate. The random permutation method may be applied to determine whether the accuracy of prediction is significant.

Applying the results of WFCCM from the training data set to the test samples is highly recommended. The test sample can be classified based on the closeness of the distance of two tissue classes, which is determined using the WFCCM in the training data set.

The classification tree methods have the features of variable selection and class prediction. Applying the concept of WFCCM for combining different "winners' genes from different tree classification methods may be necessary.

ACKNOWLEDGEMENTS

This work was supported in part by Lung Cancer SPORE (Special Program of Research Excellence) (P50 CA90949) and Cancer Center Support Grant (CCSG) (P30 CA68485) for Shyr and by CCSG (P30 CA14520) for Kim. The authors thank Dr. David Carbone, PI of Vanderbilt Lung Cancer SPORE, for permission to use the study data for the illustration. The authors also thank Dr. Noboru Yamagata for his valuable suggestions.

REFERENCES

Ben-Dor A., Friedman N., Yakhini Z. (2000). Scoring genes for relevance. Tech Report AGL-2000-13, Agilent Labs, Agilent Technologies.

Bittner M., Chen Y. (2001). Statistical methods: Identification of differentially expressed genes by weighted gene analysis.
Available at http://www.nejm.org/general/content/supplemental/hedenfalk/index.html.

Breiman L., Friedman J.H., Olshen R.A., Stone C.J. (1984). Classification and Regression Trees. Wadsworth.

Ge N., Huang F., Shaw P., Wu C.F.J. PIDEX: A statistical approach for screening differentially expressed genes using microarray analysis (unpublished manuscript).

Hedenfalk I., Duggan D., Chen Y., Radmacher M., Bittner M., Simon R., Meltzer P., Gusterson B., Esteller M., Kallioniemi O.P., Wilfond B., Borg A., Trent J. (2001). Gene-expression profiles in hereditary breast cancer. N Engl J Med 344(8): 539-548.

Ji Y., Kim K. Identification of gene sets from cDNA microarrays using classification trees (unpublished manuscript).

Kim H., Loh W.-Y. (2001). Classification trees with unbiased multiway splits. J. American Statistical Association 96:589-604.

Landis S.H., Murray T., Bolden S., Wingo P.A. (1999). Cancer Statistics, 1999. CA Cancer J Clin 49(1):8-31.

Loh W.-Y., and Shih, Y.-S. (1997). Split selection methods for classification trees. Statistica Sinica 7:815-840.

Radmacher M.D., Simon R. Statistical methods: Generating gene lists with permutation F and t Tests.
Available at http://www.nejm.org/general/content/supplemental/hedenfalk/index.html.

Reggia J.A., Tuhrim S. (1985). An overview of methods for computer assisted medical decision making, in *Computer-Assisted Medical Decision Making*, Vol. 1, J.A. Reggia & S.Tuhrim, eds. Springer-Verlag, New York.

Shyr Y. (2002). Analysis and interpretation of array data in human lung cancer using statistical class-prediction model. AACR meeting April, 2002; San Francisco, CA: Program/Proceeding Supplement, 41-42.

Tukey J.W. (1993). Tightening the clinical trial. Control Clin Trials 14(4): 266-285.

Tusher V.G., Tibshirani R., Chu G. (2001). Significance analysis of microarays applied to the ionizing radiation response. Proc Natl Acad Sci USA 98(9): 5116-5121.

Wolfinger R.D., Gibson G., Wolfinger E.D. (2001). Assessing gene significance from cDNA microarray expression data via mixed models. J Comput Biol 8(6): 625-37.

Chapter 11

CLASSIFICATION OF EXPRESSION PATTERNS USING ARTIFICIAL NEURAL NETWORKS

Markus Ringnér[1,2], Patrik Edén[2], Peter Johansson[2]

[1] *Cancer Genetics Branch, National Human Research Institute, National Institutes of Health, Bethesda, Maryland 20892, USA*

[2] *Complex Systems Division, Department of Theoretical Physics, Lund University, Lund, Sweden, e-mail:* {markus,patrik,peterjg}@thep.lu.se

1. INTRODUCTION

Artificial neural networks in the form of feed-forward networks (ANNs) have emerged as a practical technology for classification with applications in many fields. ANNs have in particular been used in applications for many biological systems (see (Almeida, 2002) for a review). For a general introduction to ANNs and their applications we refer the reader to the book by Bishop (Bishop, 1995). In this chapter we will show how ANNs can be used for classification of microarray experiments. To this aim, we will go through in detail a classification procedure shown to give good results and use the publicly available data set of small round blue-cell tumors (SRBCTs) (Khan et al., 2001) as an example[1]. The ANNs described in this chapter perform supervised learning, which means that the ANNs are calibrated to classify samples using a training set for which the desired target value of each sample is known and specified. The aim of this learning procedure is to find a mapping from input patterns to targets, in this case a mapping from gene expression patterns to classes or continuous values associated with samples. Unsupervised learning is another form of learning that does not require the specification of target data. In unsupervised learning the goal may instead be to discover clusters or other structures in the data. Unsupervised methods have been used extensively to analyze array data and

---

[1] The data is available at http://www.nhgri.nih.gov/DIR/Microarray/Supplement/.

are described in other chapters. The main reasons for choosing a supervised method are to obtain a classifier or predictor, and to extract the genes important for the classification. Here we will exemplify this by describing an ANN based classification of expression profiles of SRBCT samples into four distinct diagnostic categories: neuroblastoma (NB), rhabdomyosarcoma (RMS), Burkitt's lymphoma (BL) and Ewing's sarcoma (EWS). This data set consists of 63 training samples each belonging to one of the four categories and 25 test samples. There are many other supervised methods (discussed in other chapters) that have been used to classify array data, spanning from simple linear single gene discriminators to machine learning approaches similar to ANNs, in particular *support vector machines* (SVMs). A major advantage of using a machine learning approach such as ANNs is that one gains flexibility. Using an ANN framework, it is for example straightforward to modify the number of classes, or to construct both linear and non-linear classifiers. Sometimes this flexibility is gained at the expense of an intuitive understanding of how and why classification of a particular problem gives good results. We hope this chapter will provide the reader with an understanding of ANNs, such that some transparency is regained and the reader will feel confident in using a machine learning approach to analyze array data.

We begin with a discussion on how to reduce high-dimensional array data to make the search for good ANNs more effcient. This is followed by section 3 on ANNs. Section 3 is split into 5 subsections as follows. We describe how to design an ANN for classification, how to train the ANN to give small classification errors, how a *cross-validation* scheme can be used to obtain classifiers with good predictive ability, how *random permutation tests* can be used to assess the significance of classification results, and finally how one can extract genes important for the classification from ANNs. This chapter ends with a short section on implementation followed by a summary.

2.   DIMENSIONAL REDUCTION

For each sample in a typical study, the expression levels of several thousand genes are measured. Thus, each sample can be considered a point in "gene-space", where the dimensionality is very high. The number of considered samples, N, is usually of the order of 100, which is much smaller than the number of genes. As discussed below, an ANN using more inputs than available samples tend to become a poor predictor, and in microarray analyses it is therefore important to reduce the dimensionality before starting to train the ANN. Dimensional reduction and input selection in connection with supervised learning have attracted a lot of interest (Nguyen and Rocke, 2002).

The simplest way to reduce dimensionality is to select a few genes, expected to be relevant, and ignore the others. However, if the selection procedure involves tools less flexible than ANNs, the full potential of the ANN approach is lost. It is therefore preferable to combine the genes into a smaller set of components, and then chose among these for the ANN inputs.

For classification, we only need the *relative* positions of the samples in gene-space. This makes it possible to significantly reduce the dimensionality of microarray data without any loss of information relevant for classification. Consider the simple case of only two samples. We can then define one component as a *linear combination* of genes, corresponding to the line in gene-space going through the two sample points. This single component then fully specifies the distance in gene-space between the two samples. In the case of three samples, we can define as components two different linear combinations of genes, which together define a plane in gene-space which is going through the three data points. These two components then fully specify the relative location of the samples. This generalizes to N samples, whose relative locations in gene-space can be fully specified in an $N - 1$ dimensional subspace. Thus, with N samples, all information relevant for classification can be contained in $N - 1$ components, which is significantly less than the number of genes.

Reducing a large set of components (in our case, the genes) into a smaller set, where each new component is a linear combination of the original ones, is called a *linear projection*. In connection with ANNs, *principal component analysis*, PCA, is a suitable form of linear projection. PCA is described in detail in Chapter 5. In brief, it ranks the components according to the amount of *variance* along them, and maximizes the variance in the first components. Thus, the first component is along the direction which maximizes the variance of data points. The second component is chosen to maximize the variance, subject to the condition of orthogonality to the first component. Each new component must be orthogonal to all previous ones, and is pointing in the direction of maximal variance, subject to these constraints.

As an example, Figure 11.1 shows how much of the variance of the SRBCT gene expression data matrix is included in the different principal components. Using the 10 first principal components will in this case include more than 60% of the variance.

In our example, we used mean centered values for the PCA and did not perform any rescaling. In principle, however, any rescaling approach is possible, since the selection of principal components as ANN inputs can be done in a supervised manner.

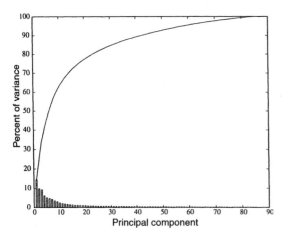

Figure 11.1. The percent of variance in the SRBCT gene expression data matrix (all 88 samples) contained by each principal component (bars). The cumulative contained variance is also shown (solid line). The 10 first principal components contain more than 60% of the variance in the data matrix for this example.

Our preference for PCA is based on the following arguments:

1. To allow for a biological interpretation of the ANN results, it is important that the connection between genes and ANN inputs can be easily reconstructed. Linear projections, e.g., PCA, fulfills this demand, in contrast to e.g., multi-dimensional scaling (Khan et al., 1998).

2. ANN analyses should involve cross-validation, described in Section 3.3. This implies that the ANN is trained several times, on different subsets of data, giving different parameter settings. An unsupervised dimensional reduction, e.g,. PCA, can use all available samples without introducing any bias. The result of this reduction can then be reused for every new training set. In contrast, a supervised reduction scheme can only rely on training samples to avoid bias, and must be redone as the training set changes.

3. In general, the $N - 1$ components containing all information are still too many for good ANN analyses. The ranking of principal components according to the variance gives information about inputs for which the separation of data points is robust against random fluctuations. Thus, PCA gives a hint on how to reduce the dimensionality further, without losing essential information.

4. The ANN inputs may need to be carefully selected, using a supervised evaluation of the components. The more correlated different input candidates are, the more difficult it is to select an optimal set. As principal components have no linear correlations, the PCA facilitates an efficient supervised selection of ANN inputs.

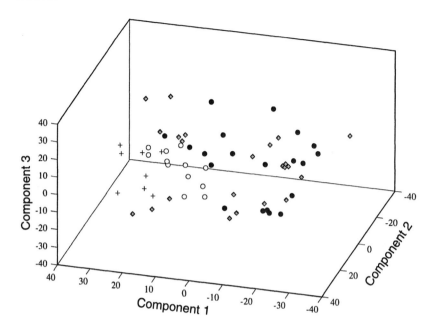

Figure 11.2. Projection of the 63 SRBCT training samples onto the 3 first principal components. The samples belong to four diagnostic categories, NB (circles), RMS (filled circles), BL (pluses) and EWS (diamonds). There is a tendency of separation between the categories. Of note, the first principal component essentially separates tumor samples (on the right) from cell lines (on the left).

In Figure 11.2, the 63 SRBCT training samples, projected onto the first three principal components, are shown. Along component one there is a clear separation, poorly related to the four classes. This separation distinguishes tissue samples from cell-lines. This illustrates that the major principal components do not need to be the most relevant for the classification of interest. Thus, it can be useful to select ANN inputs using a supervised method. In doing so, we are helped by point 4 above. Since a central idea in the ANN approach is not to presume linear solutions, it is reasonable to use ANNs also for supervised input ranking.

A simple method is to train a network using quite many inputs (maybe even all). This network will most likely be heavily overfitted and is of little interest for blind testing, but can be used to investigate how the network performance on the training set is affected as one input is excluded. Doing so for each input gives information for input selection. In the SRBCT example, there was no need to do a supervised input selection, as the classification was successful using the 10 first principal components.

3.   CLASSIFICATION USING ANNs

3.1   ANN Architecture

The simplest ANN-model is called a *perceptron*. It consists of an input layer and a single output (Figure 11.3). Associated with each input is a weight that decides how important that input is for the output. An input pattern can be fed into the perceptron and the responding output can be computed. The perceptron is trained by minimizing the error of this output. The perceptron is a linear classifier since the weights define a hyperplane that divides the input space into two parts.

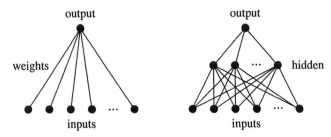

Figure 11.3. In a linear perceptron (left), the input data is fed into the input layer and triggers a response in the output layer. The weights are tuned such that the output ideally should correspond to the target value. In a multi-layer perceptron (right), a hidden layer is added in between the input and output layers.

In our example, we have four classes and a single perceptron does not suffice. Instead we use a system of four parallel perceptrons. Each perceptron is trained to separate one class from the rest, and as a classification the class with the largest output is chosen. It is recommended to first try a linear network. However, for more complicated problems a linear hyperplane is not good enough as a separator. Instead it is advantageous to have a nonlinear surface separating the classes. This can be achieved by using a *multi-layer perceptron*, in which several perceptrons are connected in a series (Figure 11.3).

Besides having an input and output layer, one also has one (or several) hidden layer(s) in between. The nodes in the hidden layer are computed from the inputs

$$h_j = f\left(\sum_i w_{ji}^{(1)} x_i\right),\qquad(11.1)$$

where x_i denotes the ith input and $w^{(1)}$ denotes the weights between the input and hidden layers.

The hidden nodes are used as input for the output (y) in the same manner

$$y = g\left(\sum_j w_j^{(2)} h_j\right) = g\left(\sum_j w_j^{(2)} f\left(\sum_i w_{ji}^{(1)} x_i\right)\right) \qquad (11.2)$$

where $w^{(2)}$ denotes the weights between the hidden and output layers,

$$g(x) = \frac{1}{1 + e^{-x}} \qquad (11.3)$$

is the logistic sigmoid activation function, and

$$f(x) = \tanh(x) = \frac{e^x - e^{-x}}{e^x + e^{-x}} \qquad (11.4)$$

is the "tanh" activation function. A way to view this is that the input space is mapped into a hidden space, where a linear separation is done as for the linear perceptron. This mapping is not fixed, but is also included in the training. This means that the number of parameters, i.e. the number of weights, is much larger. When not having a lot of training samples this might cause overfitting. How many parameters one should use varies from problem to problem, but one should avoid using more than the number of training samples. The ANN is probabilistic in the sense that the output may easily be interpreted as a probability. In other words, we are modelling the probability that, given a certain information (the inputs), a sample belongs to a certain class (Hampshire and Pearlmutter, 1990).

3.2 Training the ANN

Training, or calibrating, the network means finding the weights that give us the smallest possible classification error. There are several ways of measuring the error. A frequently used measure and the one we used in our example is the mean squared error (MSE)

$$E = \frac{1}{N} \sum_k^N \sum_l (y_{kl} - t_{kl})^2, \qquad (11.5)$$

where N is the number of training samples and y_{kl} and t_{kl} are the output and target for sample k and output node l, respectively. Since the outputs in classification are restricted (see Equations 11.2 and 11.3), the MSE is relatively insensitive to outliers, which typically results in a robust training and a good performance on a validation set. Additionally, the MSE is computationally inexpensive to use in the training.

There is a countless number of training algorithms. In each algorithm there are a few training parameters that must be tuned by the user in order to get good and efficient training. Here, we will briefly describe the parameters in the *gradient descent algorithm* used in our example.

Given a number of input patterns and corresponding targets, the classification error can be computed. The error will depend on the weights and in the training we are looking for its minimum. The idea of the gradient descent can be illustrated by a man wandering around in the Alps, looking for the lowest situated valley. In every step he walks in the steepest direction and hopefully he will end up in the lowest valley. Below, we describe four parameters: *epochs, step size, momentum coefficient* and *weight decay,* and how they can be tuned.

The number of steps, epochs, is set by the user. It can be tuned using a plot of how the classification error change during the calibration (see Figure 11.4).

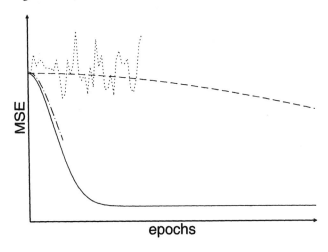

Figure 11.4 The MSE is plotted as a function of the number of training epochs. Using too many epochs (solid) or a too small step size (dashed) is time-consuming. Using too few epochs (dash-dotted) or a too large step size (dotted), the minimum is not reached.

Using too few epochs, the minimum is not reached, yielding a training plot in which the error is not flattening out but still decreasing in the end.

Using too many epochs is time consuming, since the network does not change once a minimum is reached.

The step size is normally proportional to the steepness, and the proportionality constant is given by the user. How large the step size should be depends on the typical scale of the error landscape. A too large step size results in a coarse-grained algorithm that never finds a minimum. A fluctuating classification error indicates a too large step size. Using a too small step is time consuming.

An illustration of how weights are updated, depending on step size, is shown in Figure 11.5. The corresponding development of the MSE is illustrated in Figure 11.4.

The gradient descent method can be improved by adding a momentum term. Each new step is then a sum of the step according to the pure gradient descent plus a contribution from the previous step. In general, this gives a faster learning and reduces the risk of getting stuck in a local minimum. How much of the last step that is taken into account is often defined by the user in a momentum coefficient, between 0 and 1, where 0 corresponds to pure gradient descent. Having a momentum coefficient of 1 should be avoided since each step then depends on all previous positions. The gradient descent method with momentum is illustrated in Figure 11.5.

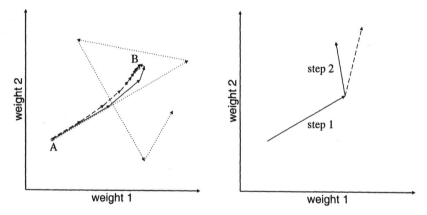

Figure 11.5. In the gradient descent (GD) algorithm the weights of an ANN are tuned to minimize the classification error. An ANN can during training be viewed as moving around in an error landscape defined by its weights. This is illustrated in two dimensions with an ANN that starts out in point A and has the lowest classification error in point B (left). With a suitable choice of the step size parameter, the minimum is reached (solid). It is time-consuming to reach the minimum with a small step size (dashed), whereas with a too large step size the algorithm fails to find the minimum (dotted). In GD with momentum (right), each new step (dashed step) is the sum of the step according to pure GD (step 2) plus a contribution from the previous step (step 1). In the figure a momentum coefficient of 1/3 was used.

The predictive ability of the network can be improved by adding a term that punishes large weights to the error measure. This so-called weight decay yields a smoother decision surface and helps avoiding over-fitting. However, too large weight decay results in too simple networks. Therefore, it is important to tune the size of this term in a cross-validation scheme.

3.3   Cross-Validation and Tuning of ANNs

In the case of array data, where the number of samples typically is much smaller than the number of measured genes, there is a large risk of overfitting. That is, among the many genes, we may always find those that perfectly classify the samples, but have poor predictive ability on additional samples. Here, we describe how a supervised learning process can be carefully monitored using a cross-validation scheme to avoid overfitting. Of note, overfitting is not specific to ANN classifiers but is a potential problem with all supervised methods.

To obtain a classifier with good predictive power, it is often fruitful to take the variability of the training samples into account. One appealing way to do this is to construct a set of classifiers, each trained on a different subset of samples, and use them in a committee such that predictions for test samples are given by the average output of the classifiers. Thus, another advantage with using a cross-validation scheme is that it results in a set of ANNs that can be used as a committee for predictions on independent test samples in a robust way. In a cross-validation scheme there is a competition between having large training sets, needed for constructing good committee members, and obtaining different training sets, to increase the spread of predictions of the committee members. The latter results in a decrease in the committee error (Krogh and Vedelsby, 1995).

In general, 3-fold cross-validation is appropriate and it is the choice in the SRBCT example. In 3-fold cross-validation, the samples are randomly split into three groups. Two groups are used to train an ANN, and the third group is used for validation. This is repeated three times using each of the three groups for validation such that every sample is in a validation set once. To obtain a large committee, the random separation into a training and a validation set can be redone many times so that a set of ANNs are calibrated. The calibration of each ANN is then monitored by plotting both the classification error of the training samples and the validation samples as a function of training epochs (see Figure 11.6). A decrease in the training and the validation error with increasing epochs demonstrates the ability of the ANN to classify the experiments.

There are no general values for the learning parameters of ANNs, instead they can be optimized by trial and error using a cross-validation scheme. Overfitting results in an increase of the error for the validation samples at the point where the models begin to learn features in the training set that are not present in the validation set. In our example, there was no sign of overfitting (Figure 11.6).

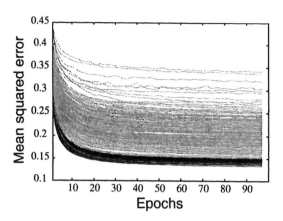

Figure 11.6 The mean squared error is plotted during the training iterations (epochs). A pair of lines, black (training) and gray (validation) represents one model (each corresponding to a random partitioning of the data into a training and validation set). Reproduced with permission (Khan et al., 2001). ©2001, Nature Publishing Group.

Overfitting can for example be avoided by *early stopping*, which means that one sets the maximal number of training iterations to be less than where overfitting begins or by tuning the weight decay. In addition, by monitoring the cross-validation performance one can also optimize the architecture of the ANN, for example the number of inputs to use. When tuning ANNs in this way, it is important to choose a cross-validation scheme that does not give a too small number of samples left in each validation set.

Even though cross-validation can be used to assess the quality of supervised classifiers, it is always important to evaluate the prediction performance using an independent test set that has not been used when the inputs or the parameters of the classifier were optimized. The importance of independent test sets should be appreciated for all supervised methods.

3.4   Random Permutation Tests

It is often stated that "black boxes" such as ANNs can learn to classify anything. Though this is not the case, it is instructive to evaluate if the classification results are significant. This can be accomplished using random permutation tests (Pesarin, 2001). In microarray analysis, random permutation tests have mostly been used to investigate the significance of genes with expression patterns that discriminate between disease categories of interest (Golub et al., 1999, Bittner et al., 2000). In our example, we randomly permute the target values for the samples and ANNs are trained to classify these randomly labeled samples. This random permutation of target values is performed many times to generate a distribution of the number of correctly classified samples that could be expected under the hypothesis of random gene expression. This distribution is shown for the 63 SRBCT samples using 3-fold cross-validation to classify the samples in Figure 11.7.

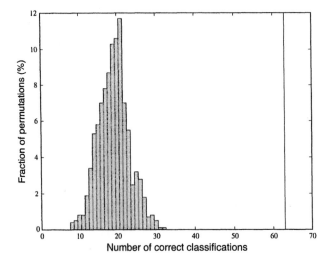

Figure 11.7. Validation results for randomly permuted sample labels (target values) using a committee of ANNs from a 3-fold cross-validation scheme. The number of correctly classified samples is histogrammed for the random permutations. Typically 20 samples are correctly classified for a random permutation, whereas all 63 samples are correct for the diagnostic categories (vertical line).

The classification of the diagnostic categories of interest resulted in all 63 samples being correctly classified in the validation, whereas a random classification typically resulted in only 20 correctly classified samples. This illustrates the significance of the classification results for the diagnostic categories of the SRBCT samples.

3.5   Finding the Important Genes

A common way to estimate the importance of an input to an ANN is to exclude it and see how much the mean squared error increases. This is a possible way to rank principal components, but the approach is ill suited for gene ranking. The large number of genes, many of which are correlated, implies that any individual gene can be omitted without any noticeable change of the ANN performance.

Instead, we may rank genes according to how much the ANN output is affected by a variation in the expression level of a gene, keeping all other gene expression levels and all ANN weights fixed. Since the ANN output is a continuous function of the inputs (cf. Section 3.1), this measure of the gene's importance is simply the *partial derivative* of the output, with respect to the gene input. To get the derivative, it is important that the dimensional reduction is mathematically simple, so the gene's influence on the ANN inputs can be easily calculated.

For well classified samples, the output is insensitive to all its inputs, as it stays close to 0 or 1 also for significant changes in the output function argument. Not to reduce the influence of well classified samples on the gene ranking, one may instead consider as sensitivity measure the derivative, with respect to the gene input, of the output function argument.

There are problems where neither single input exclusion, nor sensitivity measures as above, identifies the importance of an input (Sarle, 1998). Therefore, the gene list obtained with the sensitivity measure should be checked for consistency, by redoing the analysis using only a few top-ranked genes. If the performance is equally good as before or better, the sensitivity measure has identified important genes. If the performance is worse, the sensitivity measure may be misleading, but it could also be that too few top genes are selected, giving too little information to the ANN.

In Figure 11.8, the ANN performance of the SRBCT classification, as a function of included top genes, is shown. The good result when selecting the top 96 genes shows that these genes indeed are relevant for the problem. One can also see that selecting only the six top genes excludes too much information. The performance measure in this example is the average number of validation sample misclassifications of an ANN. One can also consider less restrictive measures, like the number of misclassifications by the combined ANN committee. If so, it may be that less than 96 top genes can be selected without any noticeable reduction in the performance. When more than 96 genes were selected the average performance of the committee members went down. This increase in the number of misclassifications is likely due to overfitting introduced by noise from genes with lower rank. However, the combined ANN committee still classified all the samples correctly when more than 96 genes were used.

Once genes are ranked, it is possible to further study gene expression differences between classes. For example, one can check how many top genes can be removed without significantly reducing the performance.

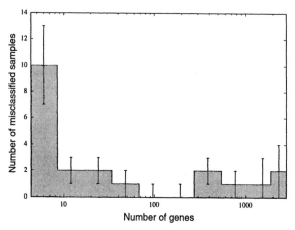

Figure 11.8. The number of misclassified samples for each ANN model is averaged over all models and plotted against increasing number of used genes. As can be seen, using the 96 highest ranked genes results in having cross-validation models that all on average give zero misclassifications for this example. Reproduced with permission (Khan et al., 2001). © 2001, Nature Publishing Group.

Alternatively, several analyses can be made, where the same number of genes have been selected, but from different regions of the ranked gene-list. This approach was used when classifying breast cancers according to estrogen receptor status (Gruvberger et al., 2001). Given the gene-list, 100 genes were selected according to rank in the ranges 1 to 100, 51 to 150, 101 to 200, etc. For each selection of 100 genes, the analysis was redone, and the ANN classification was shown to remain good using genes in the range 301 to 400, giving the conclusion that the gene expression patterns associated with estrogen receptor status are remarkably distinct.

4.　IMPLEMENTATION

The principal component analysis can be formulated in terms of a *singular value decomposition* (see Chapter 5 of this volume) of the expression data matrix. It is straightforward to implement the complete analysis procedure outlined in this chapter in MATLAB with the Neural Network application toolbox, both available from The MathWorks (Natick, Massachusetts).

5.　SUMMARY

We have presented an ANN-based method for classification of gene expression data. This method was successfully applied to the example of classifying SRBCTs into distinct diagnostic categories. The key components of this classification procedure are PCA for dimensional reduction and cross-validation to optimize the training of the classifiers. In addition, we described a way to rank the genes according to their importance for the classification. Random permutation tests were introduced to assess the significance of the classification results. There are other ANN methods that have been used to classify gene expression data (Selaru et al., 2002).

ACKNOWLEDGMENTS

The authors want to thank C. Peterson, P.S. Meltzer, J. Khan and S. Gruvberger for their encouragement and support. MR and PE were supported by postdoctoral fellowships from the Swedish Research Council.

REFERENCES

Almeida J.S. (2002). Predictive non-linear modeling of complex data by artificial neural networks. Curr Opin Biotechnol 13:72-6.

Bishop C.M. (1995). *Neural networks for pattern recognition*. Oxford: Oxford University Press.

Bittner M., Meltzer P., Chen Y., Jiang Y., Seftor E., Hendrix M., Radmacher M., Simon R., Yakhini Z., Ben-Dor A., Sampas N., Dougherty E., Wang E., Marincola F., Gooden C., Lueders J., Glatfelter A., Pollock P., Carpten J., Gillanders E., Leja D., Dietrich K., Beaudry C., Berens M., Alberts D., Sondak V., Hayward N., Trent J. (2000). Molecular classification of cutaneous malignant melanoma by gene expression profiling. Nature 406:536-40.

Golub T.R., Slonim D.K., Tamayo P., Huard C., Gaasenbeek M., Mesirov J.P., Coller H., Loh M.L., Downing J.R., Caligiuri M.A., Bloomfield C.D., Lander E.S. (1999). Molecular classification of cancer: Class discovery and class prediction by gene expression monitoring. Science 286:531-7.

Gruvberger S., Ringnér M., Chen Y., Panavally S., Saal L.H., Borg A., Fernoe M., Peterson C., Meltzer P.S. (2001). Estrogen receptor status in breast cancer is associated with remarkably distinct gene expression patterns. Cancer Res 61:5979-84.

Hampshire J.B., Pearlmutter B. (1990). Equivalence proofs for multi-layer perceptron classifiers and the Bayesian discriminant function. Proceedings of the 1990 connectionist models summer school. San Mateo, CA: Morgan Kaufman.

Khan J., Simon R., Bittner M., Chen Y., Leighton S.B., Pohida T., Smith P.D., Jiang Y., Gooden G.C., Trent J.M., Meltzer P.S. (1998). Gene expression profiling of alveolar rhabdomyosarcoma with cDNA microarrays. Cancer Res 58:5009-13.

Khan J., Wei J.S., Ringn´er M., Saal L.H., Ladanyi M., Westermann F., Berthold F., Schwab M., Atonescu C.R., Peterson C., Meltzer P.S. (2001). Classification and diagnostic prediction of cancers using gene expression profiling and artificial neural networks. Nat Med 7:673-79.

Krogh A., Vedelsby J. (1995). Neural network ensembles, cross validation and active learning. Advances in Neural Information Processing Systems, Volume 7. Cambridge, MA: MIT Press.

Nguyen D.V., Rocke D.M. (2002). Tumor classification by partial least squares using microarray gene expression data. Bioinformatics; 18:39-50.

Pesarin F. (2001). *Multivariate Permutation Tests: With Applications in Biostatistics.* Hoboken, NJ: John Wiley & Sons.

Sarle W.S. (1998). How to measure the importance of inputs? Technical Report, SAS Institute Inc, Cary, NC, USA. Available at ftp://ftp.sas.com/pub/neural/FAQ.html.

Selaru F.M., Xu Y., Yin J., Zou T., Liu T.C., Mori Y., Abraham J.M., Sato F., Wang S., Twigg C., Olaru A., Shustova V., Leytin A., Hytiroglou P., Shibata D., Harpaz N., Meltzer S.J. (2002). Artificial neural networks distinguish among subtypes of neoplastic colorectal lesions. Gastroenterology 122:606-13.

Chapter 12

GENE SELECTION AND SAMPLE CLASSIFICATION USING A GENETIC ALGORITHM AND *K*-NEAREST NEIGHBOR METHOD

Leping Li and Clarice R. Weinberg

Biostatistics Branch, National Institute of Environmental Health Sciences, Research Triangle Park, NC 27709, USA,
e-mail: {li3,weinberg}@niehs.nih.gov

1. INTRODUCTION

Advances in microarray technology have made it possible to study the global gene expression patterns of tens of thousands of genes in parallel (Brown and Botstein, 1999; Lipshutz et al., 1999). Such large scale expression profiling has been used to compare gene expressions in normal and transformed human cells in several tumors (Alon et al., 1999; Gloub et al., 1999; Alizadeh et al., 2000; Perou et al., 2000; Bhattacharjee et al., 2001; Ramaswamy et al., 2001; van't Veer et al., 2002) and cells under different conditions or environments (Ooi et al., 2001; Raghuraman et al., 2001; Wyrick and Young, 2002). The goals of these experiments are to identify differentially expressed genes, gene-gene interaction networks, and/or expression patterns that may be used to predict class membership for unknown samples. Among these applications, class prediction has recently received a great deal of attention. Supervised class prediction first identifies a set of discriminative genes that differentiate different categories of samples, e.g., tumor versus normal, or chemically exposed versus unexposed, using a learning set with known classification. The selected set of discriminative genes is subsequently used to predict the category of unknown samples. This method promises both refined diagnosis of disease subtypes, including markers for prognosis and better targeted treatment, and improved understanding of disease and toxicity processes at the cellular level.

1.1 Classification and Gene Selection Methods

Pattern recognition methods can be divided into two categories: *supervised* and *unsupervised*. A supervised method is a technique that one uses to develop a predictor or classification rule using a learning set of samples with known classification. The predictive strategy is subsequently validated by using it to classify unknown samples. Methods in this category include *neighborhood analysis* (Golub et al., 1999), *support vector machines* (SVM) (Ben-Dor et al., 2000; Furey et al., 2000; Ramaswamy et al., 2001), *k-nearest neighbors* (KNN) (Li et al., 2001a & 2001b), *recursive partitioning* (Zhang et al., 2001), *Tukey's compound covariate* (Hedenfalk et al., 2001), *linear discriminant analysis* (LDA) (Dudoit et al., 2002; Li and Xiong, 2002), and *nearest shrunken centroids* (Tibshirani et al., 2002). Unsupervised pattern recognition largely refers to clustering analysis for which class information is not known or not required. Unsupervised methods include *hierarchical clustering* (Eisen et al., 1998), *k-means clustering* (Tavazoie et al., 1999), and the *self-organizing map* (Toronen et al., 1999). Reviews of the classification methods can be found in Brazma and Vilo (2000), Dudoit et al. (2002) and Chapter 7 in this volume.

Usually, a small number of variables (genes) are used in the final classification. Reducing the number of variables is called *feature reduction* in pattern recognition (e.g., see Chapter 6). Feature reduction in microarray data is necessary, since not all genes are relevant to sample distinction. For certain methods such as LDA, feature reduction is a must. For other methods such as SVMs, ill-posed data (where the number of genes exceeds the number of samples) are more manageable (e.g., see Chapter 9).

The most commonly used methods for selecting discriminative genes are the standard two-sample *t*-test or its variants (Golub et al, 1999; Hedenfalk et al., 2001; Long et al., 2001; Tusher et al., 2001). Since typical microarray data consist of thousands of genes, a large number of *t*-tests are involved. Clearly, multiple testing is an issue as the number of chance findings, "false positives", can exceed the number of true positives. A common correction to individual p values is the Bonferroni correction. For a two-sided *t*-test, an adjusted significance level is $\alpha^* = \alpha / n$, where n is the number of genes, and α is the unadjusted significance level. When the sample size is small, as the case for most microarray data, the variances may be poorly estimated. One way to address this problem is to "increase" the sample size by using genes with similar expression profiles in variance estimation (Baldi and long 2001;

Tusher et al., 2001). Furthermore, *t*-test depends on strong parametric assumptions that may be violated and are difficult to verify with small sample size. To avoid the need for parametric assumptions, one may use permutation techniques (Dudoit et al., 2000; Tusher et al., 2001; Pan et al, 2002). Other methods for selecting differentially expressed genes include Wilcoxon rank sum test (Virtaneva et al., 2001). A comparative review of some of these methods can be found in (Pan, 2002).

Besides the *t*-test and its variants, one can use a classification method to select discriminative genes. For example, Li and Xiong (2002) used LDA in a stepwise fashion, sequentially building a subset of discriminative genes starting from a single gene. In SVM, Ramaswamy et al. (2001) started with all genes to construct a support vector and then recursively eliminated genes that provided negligible contribution to class separation (the smallest elements in a support vector w; see Chapter 9 of this volume). The GA/KNN approach (Li et al., 2001a & 2001b) utilizes KNN as the discriminating method for gene selection.

1.2   Why the *k*-Nearest Neighbors Method?

Many supervised classification methods perform well when applied to gene expression data (see, e.g., Dudoit et al., 2002). We chose KNN as the gene selection and classification method for the following reasons.

KNN is one of the simplest non-parametric pattern recognition methods. It has been shown to perform as well as or better than more complex methods in many applications (see, e.g., Vandeginste et al., 1998; Dudoit et al., 2002). Being a non-parametric method, it is free from statistical assumptions such as normality of the distribution of the genes. This feature is important, since the distributions of gene expression levels or ratios are not well characterized and the distributional shapes may vary with the quality of either arrays themselves or the sample preparation.

Like many other supervised methods, KNN is inherently multivariate, taking account of dependence in expression levels. It is known that the expression levels of some genes may be regulated coordinately and that the changes in expression of those genes may well be correlated. Genes that are jointly discriminative, but not individually discriminative, may be co-selected by KNN.

Perhaps most importantly, KNN defines the class boundaries implicitly rather than explicitly, accommodating or even identifying distinct subtypes within a class. This property is particularly desirable for studies of cancer where clinical groupings may represent collections of related but biologically distinct tumors. Heterogeneity within a single tumor type has been shown in many tumors including leukemia (Golub et al., 1999), lymphoma (Alizadeh et al., 2000), and breast cancer (Perou et al., 2000). When applied to a leukemia data set, the GA/KNN method selected a subset of genes that not only discriminated between acute lymphoblastic leukemia (ALL) and acute myeloid leukemia (AML) but also unmasked clinically meaningful subtypes within ALL (T-cell ALL versus B-cell ALL) – even though gene selection only used the ALL-AML dichotomy (Li et al., 2001a).

Finally, the resulting classification of KNN is qualitative and requires none of the hard-to-verify assumptions about the within-class variances or shift alternatives that are used in many other statistical methods. Typical microarray data contain many fewer samples than genes, and the variance-covariance matrix becomes singular (linear dependences exist between the rows/columns of the variance-covariance matrix), restricting attention to certain linear combinations of genes with non-zero eigenvalues. Moreover, methods that require variance-covariance estimation suffer in the face of outlying observations, disparate covariance structures, or heterogeneity within classes.

1.3   Why a Genetic Algorithm?

In KNN classification, samples are compared in multi-dimensional space. However, considering all possible subsets of genes from a large gene pool is not feasible. For instance, the number of ways to select 30 from 3,000 is approximately $6.7 \cdot 10^{71}$. Thus, an efficient sampling tool is needed. A natural choice would be a *genetic algorithm* (GA). A GA is a stochastic optimization method. First described by John Holland in the 70's (Holland, 1975), GAs mimic Darwinian natural selection (hence "genetic") in that selections and mutations are carried out to improve the "fitness" of the successive generations (Holland, 1975; Goldberg, 1989). It starts with a population of *chromosomes* (mathematical entities). Usually, the chromosomes are represented by a set of strings, either binary or non-binary, constituting the building blocks of the candidate solutions. The better the fitness of a chromosome, the larger its chance of being passed to the next generation. Mutation and crossover are carried out to introduce new chromosomes into the population (e.g., see Judson et al., 1997). Through evolution, a solution may evolve. After it was introduced, GA has been used in many optimization problems ranging from protein folding (Pedersen and Moult, 1996) to sequence alignment (Notredame et al., 1997). For reviews, see Forrest (1997) and Judson (1997). Although, it has been demonstrated that GAs are effective in searching high-dimensional space, they do not guarantee convergence to a global minimum, given the stochastic nature of the algorithm. Consequently, many independent runs of GAs are needed to ensure the convergence.

2. THE GA/KNN METHOD

2.1 Overall Methodology

The GA/KNN (Li et al., 2001a & 2001b) is a multivariate classification method that selects many subsets of genes that discriminate between different classes of samples using a learning set. It combines a search tool, GA, and a non-parametric classification method, KNN. Simply speaking, we employ the GA to choose a relatively small subset of genes for testing, with KNN as the evaluation tool. Details of the GA and KNN are given below.

For high dimensional microarray data with a paucity of samples, there may be many subsets of genes that can discriminate between different classes. Different genes with similar patterns of expression may be selected in different, but equally discriminative, subsets. Consequently, it is important to examine as many subsets of discriminative genes as possible. When a large number of such subsets has been obtained, the frequency with which genes are selected can be examined. The selection frequency should correlate with the relative predictive importance of genes for sample classification: the most frequently selected genes should be most discriminative whereas the least frequently selected genes should be less informative. The most frequently selected genes may be subsequently used to classify unknown samples in a test set.

2.2 KNN

Suppose that the number of genes under study is N and that $q \ll N$ is the number of genes in a much smaller subset. Let $G_m = (g_{1m}, g_{2m}, \ldots, g_{im}, \ldots, g_{qm})$, where g_{im} is the expression value (typically *log* transformed) of the ith gene in the mth sample; $m = 1, \ldots, M$. In the KNN method (e.g., Massart et al., 1988), one computes the distance between each sample, represented by its vector G_m, and each of the other samples (see, e.g., Table 12.1). For instance, one may employ the Euclidean distance. When values are missing, methods for missing value imputation can be found in Chapter 3. A sample is classified according to the class membership of its k nearest neighbors, as determined by the Euclidean distance in q-dimensional space. Small values of 3 or 5 for k have been alleged to provide good classification. In a classic KNN classification, an unknown sample is classified in the group to which the majority of the k objects belong. One may also apply a more stringent criterion to require all k nearest neighbors to agree in which case a sample would be considered unclassifiable if the *k* nearest neighbors do not all belong to the same class.

Figure 12.1 displays an example. The unknown, designated by **X**, is classified with the triangles, because its 3 nearest neighbors are all triangles.

Table 12.1. An example of two genes (g1 and g2) and 10 samples (S1-S10).

Sample	S1	S2	S3	S4	S5	S6	S7	S8	S9	S10
Class	△	△	△	△	△	○	○	○	○	○
g1	$g_{1,1}$	$g_{1,2}$	$g_{1,3}$	$g_{1,4}$	$g_{1,5}$	$g_{1,6}$	$g_{1,7}$	$g_{1,8}$	$g_{1,9}$	$g_{1,10}$
g2	$g_{2,1}$	$g_{2,2}$	$g_{2,3}$	$g_{2,4}$	$g_{2,5}$	$g_{2,6}$	$g_{2,7}$	$g_{2,8}$	$g_{2,9}$	$g_{2,10}$

Note that the data are well clustered, because each observation has a class that agrees with the class of its 3 nearest neighbors.

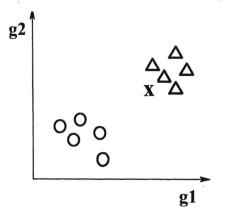

Figure 12.1. KNN classification. For clarity, only two dimensions are shown ($q = 2$), that is, each sample is represented by a vector of two genes (*g1* and *g2*). Triangles and circles represent two distinct classes. A 3-NN classification would assign the unknown sample **X** to the class of triangle.

2.3   A Genetic Algorithm

2.3.1   Chromosomes

In GAs, each "chromosome" (a mathematical entity, not the biological chromosome) consists of *q* distinct genes randomly selected from the gene "pool" (all genes studied in the experiment). Thus, a chromosome can be viewed as a string containing *q* gene index labels. An example is shown in Figure 12.2. In the example, genes 1, 12, 23, 33, and so on, are selected. The set of *q* genes in the chromosome constitutes a candidate solution to the gene selection problem, as the goal of each run of the GA is to identify a set of *q* discriminative genes. Typically, $q = 20$, 30 or 40 should work well for most microarray data sets. A set of such "chromosomes" (e.g., 100) constitutes a *population* or *niche*. We work with 10 such niches in parallel.

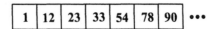

Figure 12.2. An example of a chromosome in GA.

2.3.2 Fitness

The *fitness* of each chromosome is subsequently evaluated by its ability to correctly classify samples using KNN. For each chromosome (a set of q selected genes), we compute the pair-wise Euclidean distances between the samples in the q-dimensional space. The class membership of a sample is then declared by its k-nearest neighbors. If the actual class membership of the sample matches its KNN-declared class, a score of one is assigned to that sample; otherwise, a score of zero is assigned. Summing these scores across all samples provides a fitness measure for the chromosome. A perfect score would correspond to the number of samples in the training set.

2.3.3 Selection and Mutation

Once the fitness score of each chromosome in a niche is determined, the fittest chromosomes, one from each niche, are combined and used to replace the corresponding number of least fit chromosomes (the lowest scoring chromosomes) in *each* niche. This enrichment strategy allows the single best chromosome found in each niche to be shared with all the other niches. For a typical run with 10 niches, each of which consists of 100 chromosomes, the 10 least fit chromosomes in each niche are replaced by the 10 best chromosomes, one from each niche.

Next, the chromosomes in each niche are ranked, with the best chromosome assigned a rank of 1. The single best chromosome in a niche is passed deterministically to the next generation for that niche *without* subsequent mutation. This guarantees that the best chromosome at each generation is preserved. The remaining chromosomes in the niche are chosen by sampling all chromosomes including the best chromosome in the niche with probability proportional to the chromosome's fitness. This is the so-called "*roulette-wheel selection*" in which the high scoring chromosomes are given high probability of being selected whereas the low scoring chromosomes are given low, but non-zero, probability of being passed to the next generation. Including less fit chromosomes may prevent the search from being trapped at a local minimum. Chromosomes selected based on this sampling strategy are next subject to mutation.

Once a chromosome is selected for mutation, between 1 and 5 of its genes are randomly selected for mutation. The number of mutations (from 1 to 5) is assigned randomly, with probabilities, 0.53125, 0.25, 0.125, 0.0625, and 0.03125 ($1/2^r$, where r is 1 to 5; 0.03125 is added to the probability

$r = 1$, so that the total probability is equal to 1.0), respectively. In this way, a single replacement is given the highest probability while simultaneous multiple replacements have lower probability. This strategy prevents the search from behaving as a random walk as it would if many new genes were introduced at each generation. Once the number of genes to be replaced in the chromosome has been determined, these replacement genes are randomly selected and replaced randomly from the genes not already in the chromosome. An example of a single point mutation is shown in Figure 12.3.

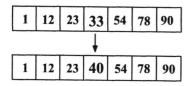

Figure 12.3. A single point mutation. For simplicity, only 7 genes are shown. Upon a single point mutation, gene 33 is replaced by gene 40.

2.3.4   Termination Criterion

Niches are allowed to evolve by repeating the above steps until at least one of the chromosomes achieves a targeted fitness criterion. A targeted fitness criterion is considered to be reached when most of the samples (e.g., 90% of them) have been correctly classified. Because we do not require perfect classification, gene selection may be less sensitive to outliers or occasional misclassified samples in the data. A less stringent criterion is also computationally faster.

Intuitively, the more distinct classes, the more difficult it will be to find a subset of discriminative genes. For toxicogenomics data or tumor data, multiple classes are not uncommon. For those datasets, the above 90% requirement may be too stringent. For instance, Ramaswamy et al. (2001) did gene expression profiling on 218 tumor samples, covering 14 tumor types, and 90 normal tissue samples using oligonucleotide arrays. When we applied the GA/KNN method to the training set (144 samples and 14 classes), requiring 90% of the 144 samples to be correctly classified was not possible. For such circumstances, one should start with a test run to see how the fitness score evolves from generation to generation. One might choose a fitness score based on what can be achieved in 20 to 40 generations as the targeted fitness value, to balance the computation speed and discrimination power. It should be pointed out that gene selection is relatively insensitive to this choice of the targeted fitness criterion. The other cases where a less stringent criterion may be needed are time-course and dose-response microarray data, where there are again multiple, potentially similar classes.

We refer to a chromosome that achieves this targeted fitness score as a *near-optimal chromosome*. When a near-optimal chromosome evolves in any niche, that chromosome is retrieved and added to a list; then the entire niche is re-initialized. Because typical microarray data consist of a large number of genes and a small number of samples, for a given data set there may exist many different subsets of genes (near-optimal chromosomes) that can discriminate the classes of samples very well. Hence, the GA/KNN procedure must be repeated through many evolutionary runs, until many such near-optimal chromosomes (e.g., 10,000) are obtained. Once a large number of near-optimal chromosomes have been obtained, genes can be ranked according to how often they were selected into these near-optimal chromosomes. The most frequently selected genes should be more relevant to sample distinction whereas the least frequently selected genes should be less informative.

It may not be practically possible or necessary to obtain a very large number of near-optimal chromosomes. However, one should check to see if one has sampled enough of the GA solution space for results to stabilize. To do that, one may divide the near-optimal solutions into two groups of equal size and compare their frequency distributions and ranks for the top genes. A tight diagonal line indicates that the ranks for the top genes are nearly reproducible, suggesting that enough near-optimal solutions have been obtained to achieve stability (Figure 12.4).

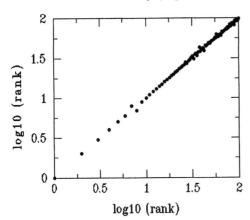

Figure 12.4. An example of plot of the log_{10}-transformed ranks of the 100 top-ranked genes from two independent runs of the GA/KNN procedure. The genes were ranked according to frequency of occurrence in the 500,000 near-optimal chromosomes, with the most frequent gene assigned rank 1 (0 after transformation). Similar result was obtained using fewer near-optimal chromosomes (e.g., 10,000).

2.4   Statistical Analysis of the Near-Optimal Chromosomes

The next step is to develop a predictive algorithm to apply to the test set, by selecting a certain number of top-ranked genes and using those genes, with the KNN method, on the test set samples. A simple way to choose the number of discriminative genes is to take the top 50. Although fewer genes (e.g., 10) may be preferred in classification, for microarray data, a few more genes might be useful. More genes might provide more insight for the underlying biology. With more genes, the classification should be less sensitive to the quality of data, since the current microarray technology is not fully quantitative. Alternatively, one may choose the number of top-ranked genes that give optimal classification for the training set (Li et al., 2001b). It may also be helpful to plot the Z score of the top-ranked genes (Figure 12.5). Let $Z = (S_i - E(S_i)) / \sigma$, where S_i is the number of times gene i was selected, $E(S_i)$, is the expected number of times for gene i being selected, σ is the square root of the variance. Let A = number of near-optimal chromosomes obtained (not necessarily distinct), and $P_i = q$ / number of genes on the microarray, the probability of gene i being selected (if random). Then,

$$E(S_i) = P_i \times A, \text{ and } \sigma = \sqrt{P_i \cdot (1 - P_i) \cdot A}.$$

A sharp decrease in Z score may suggest that only a few of the top-ranked genes should be chosen as the discriminative genes.

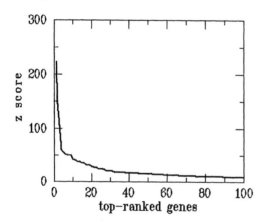

Figure 12.5. A plot of Z scores for 100 top-ranked genes for the breast cancer data set (Hedenfalk et al., 2001). The Z scores decrease quickly for the first 5 to 10 genes. The decrease is much slower after 30 genes. In this case, it seems reasonable to choose 20 to 30 top-ranked genes as the most discriminative genes.

2.5   Comparison between Near-Optimal Chromosomes and the Top-Ranked Genes

As pointed out earlier, for high-dimensional microarray data with a paucity of samples, many subsets of genes that can discriminate between different classes of samples may exist. Different genes with similar patterns of expression may be selected in different, but equally discriminative subsets, especially when a qualitative classification method, such as KNN, is used. The overlap between q top-ranked genes and each of the near-optimal chromosomes (q genes in length) can be low. For instance, for the breast cancer data set (Hedenfalk et al., 2001), we obtained 500,000 near-optimal chromosomes that can distinguish between *BRCA1* and *BRCA2* tumors. Among the 500,000 near-optimal chromosomes, only 13% of them had 6 or more genes listed among the 30 top-ranked genes. Moreover, classifications in a leave-one-out cross-validation procedure (e.g., Chapter 7) using the individual near-optimal chromosomes revealed bad performance (data not shown). On the other hand, empirically, we found that substantially larger separation between *BRCA1* and *BRCA2* samples was achieved with the 30 top-ranked genes than with any individual near-optimal chromosome (data not shown). These results suggest that the top-ranked genes do much better than any of the individual near-optimal chromosomes for sample classification.

Although ranking the genes by selecting those individual genes that occur most frequently in near-optimal chromosomes may seem to sacrifice correlation structure, this selection process appears to retain aspects of multivariate structure important for class discrimination. Heuristically, when a subset of genes can discriminate among classes jointly, but not singly, that subset of genes should tend to appear together in near-optimal chromosomes and, consequently, each gene in the jointly discriminative subset may tend to have high frequency of occurrence.

2.6   Computation Cost

The GA/KNN method is computationally intensive, as it searches for many near-optimal solutions (chromosomes). For a typical run, as many as 10,000 near-optimal solutions may be needed. For a small data set with 10 samples in each of two categories, obtaining that many near-optimal solutions can be achieved in a few hours or less. However, for a large data set with multiple classes (e.g., the MIT's 14 categories tumor data set) (Ramaswamy et al., 2001), it may take a few days to complete the GA/KNN on a Linux machine with reasonable speed.

2.7   Availability

The GA/KNN method will be available on the Web site: http://dir.niehs.nih.gov/microarray/datamining/ for downloading in September 2002.

3.   CONCLUDING REMARKS

In summary, the GA/KNN method is non-parametric, multivariate, and able to accommodate (and potentially detect) the presence of heterogeneous subtypes within classes. As the quantitative aspects of the microarray technology improve and computational methods that mine the resulting large data sets are developed further, the technology will have a great impact on biology, toxicology, and medicine.

ACKNOWLEDGEMENTS

We thank David Umbach and Shyamal Peddada for insightful discussions and careful reading of the manuscript. LL also thanks Lee Pedersen and Thomas Darden for advice and support.

REFERENCES

Alizadeh A.A., Eisen M.B., Davis R.E., Ma C., Lossos I.S., Rosenwald A., Boldrick J.C., Sabet H., Tran T., Yu X., Powell J.I., Yang L., Marti G.E., Moore T., Hudson J., Jr, Lu L., Lewis D.B., Tibshirani R., Sherlock G., Chan W.C., Greiner T.C., Weisenburger D.D., Armitage J.O., Warnke R., Staudt L.M. et al (2000). Distinct types of diffuse large B-cell lymphoma identified by gene expression profiling. Nature 403:503-11.

Alon U., Barkai N., Notterman D.A., Gish K., Ybarra S., Mack D., Levine A.J. (1999). Broad patterns of gene expression revealed by clustering analysis of tumor and normal colon tissues probed by oligonucleotide arrays. Proc Natl Acad Sci USA 96:6745-50.

Baldi P., Long A.D. (2001). A Bayesian framework for the analysis of microarray expression data: regularized t -test and statistical inferences of gene changes. Bioinformatics 17:509-19.

Ben-Dor A., Bruhn L., Friedman N., Nachman I., Schummer M., Yakhini Z. (2000). Tissue classification with gene expression profiles. J Comput Biol 2000; 7:559-83.

Bhattacharjee A., Richards W.G., Staunton J., Li C., Monti S., Vasa P., Ladd C., Beheshti J., Bueno R., Gillette M., Loda M., Weber G., Mark E.J., Lander E.S., Wong W., Johnson B.E., Golub T.R., Sugarbaker D.J., Meyerson M. (2001). Classification of human lung carcinomas by mRNA expression profiling reveals distinct adenocarcinoma subclasses. Proc Natl Acad Sci USA 98:13790-5.

Brazma A., Vilo J. (2000). Gene expression data analysis. FEBS Lett 480:17-24.

Brown P.O., Botstein D. (1999). Exploring the new world of the genome with DNA microarrays. Nat Genet 21(1 Suppl):33-7.

Dudoit S., Yang Y.H., Callow M.J., Speed T. (2000). Statistical methods for identifying differentially expressed genes in replicated cDNA microarray experiments. Technical Report, Number 578, Department of Statistics, University of California, Berkeley, California.

Dudoit S., Fridlyand J., Speed T.P. (2002). Comparison of discrimination methods for the classification of tumors using gene expression data. J Am Stat Assoc 97:77-87.

Eisen M.B., Spellman P.T., Brown P.O., Botstein D. (1998). Cluster analysis and display of genome-wide expression patterns. Proc Natl Acad Sci USA 95:14863-8.

Forrest S. (1993). Genetic algorithms: principles of natural selection applied to computation. Science 261:872-8.

Furey T.S., Cristianini N., Duffy N., Bednarski D.W., Schummer M., Haussler D. (2000). Support vector machine classification and validation of cancer tissue samples using microarray expression data. Bioinformatics 16:906-14.

Goldberg D.E. (1989) *Genetic algorithms in search, optimization, and machine learning.* Massachusetts: Addison-Wesley, 1989.

Golub T.R., Slonim D.K., Tamayo P., Huard C., Gaasenbeek M., Mesirov J.P., Coller H., Loh M.L., Downing J.R., Caligiuri M.A., Bloomfield C.D., Lander E.S. (1999). Molecular classification of cancer: class discovery and class prediction by gene expression monitoring. Science 286:531-7.

Hedenfalk I., Duggan D., Chen Y., Radmacher M., Bittner M., Simon R., Meltzer P., Gusterson B., Esteller M., Kallioniemi O.P., Wilfond B., Borg A., Trent J. (2001). Gene-expression profiles in hereditary breast cancer. N Engl J Med 344:539-48.

Holland J.H. (1975). *Adaptation in Natural and Artificial Systems.*, Ann Arbor: University of Michigan Press.

Judson R. (1997). Genetic algorthms and their use in chemistry. In *Reviews in computational chemistry*, Kenny B. Lipowitz and Donald B. Boyd, eds. New York: VCH publishers, Vol. 10.

Li L., Darden T.A., Weinberg C.R., Levine A.J., Pedersen L.G. (2001). Gene assessment and sample classification for gene expression data using a genetic algorithm/k-nearest neighbor method. Comb Chem High Throughput Screen 4:727-39.

Li L., Weinberg C.R., Darden T.A., Pedersen L.G. (2001). Gene selection for sample classification based on gene expression data: study of sensitivity to choice of parameters of the GA/KNN method. Bioinformatics 17:1131-42.

Li W., Xiong M. (2002). Tclass: tumor classification system based on gene expression profile. Bioinformatics 18:325-326.

Lipshutz R.J., Fodor S.P., Gingeras T.R., Lockhart D.J. (1999). High density synthetic oligonucleotide arrays. Nat Genet 21(1 Suppl):20-4.

Long A.D., Mangalam H.J., Chan B.Y., Tolleri L., Hatfield G.W., Baldi P. (2001). Improved statistical inference from DNA microarray data using analysis of variance and a Bayesian statistical framework. Analysis of global gene expression in Escherichia coli K12. J Biol Chem 276:19937-44.

Massart, D.L., Vandeginste B.G.M., Deming S.N., Michotte Y., Kaufman, L. (1988). *Chemometrics: a textbook (Data Handling in Science and Technology, vol 2)*, Elsevier Science B.V: New York.

Notredame C., O'Brien E.A., Higgins D.G. (1997). RAGA: RNA sequence alignment by genetic algorithm. Nucleic Acids Res 25:4570-80.

Ooi S.L., Shoemaker D.D., Boeke J.D. (2001). A DNA microarray-based genetic screen for nonhomologous end-joining mutants in Saccharomyces cerevisiae. Science 294:2552-6.

Pan W. (2002). A comparative review of statistical methods for discovering differentially expressed genes in replicated microarray experiments. Bioinformatics 2002; 18:546-54.

Pedersen J.T., Moult J. (1996). Genetic algorithms for protein structure prediction. Curr Opin Struct Biol 6:227-31.

Perou C.M., Sørlie T, Eisen M.B., van de Rijn M., Jeffrey S.S., Rees C.A., Pollack J.R., Ross D.T., Johnsen H., Aksien L.A., Fluge O., Pergamenschikov A., Williams C., Zhu S.X., Lonning P.E., Borresen-Dale A.L., Brown P.O., Botstein D. (2000). Molecular portraits of human breast tumours. Nature 406: 747-52.

Raghuraman M.K., Winzeler E.A., Collingwood D., Hunt S., Wodicka L., Conway A., Lockhart D.J., Davis R.W., Brewer B.J., Fangman W.L. (2001). Replication dynamics of the yeast genome. Science 294:115-21.

Ramaswamy S., Tamayo P., Rifkin R., Mukherjee S., Yeang C.H., Angelo M., Ladd C., Reich M., Latulippe E., Mesirov J.P., Poggio T., Gerald W., Loda M., Lander E.S., Golub T.R. (2001). Multiclass cancer diagnosis using tumor gene expression signatures. Proc Natl Acad Sci USA 98:15149-54.

Tavazoie S., Hughes J.D., Campbell M.J., Cho R.J., Church G.M. (1999). Systematic determination of genetic network architecture. Nat Genet 22:281-5.

Tibshirani R., Hastie T., Narasimhan B., Chu G. (2002). Diagnosis of multiple cancer types by shrunken centroids of gene expression. Proc Natl Acad Sci USA 99:6567-72.

Toronen P., Kolehmainen M., Wong G., Castren E. (1999). Analysis of gene expression data using self-organizing maps. FEBS Lett 451:142-6.

Tusher V.G., Tibshirani R., Chu G. (2001). Significance analysis of microarrays applied to the ionizing radiation response. Proc Natl Acad Sci USA 98:5116-21.

Vandeginste B.G.M., Massart D.L., Buydens L.M.C., De Jong S., Lewi P.J., Smeyers-Verbeke J. (1998). *Handbook of Chemometrics and Qualimetrics. Vol 20B*. The Netherlands: Elsevier Science.

van 't Veer L.J., Dai H., van de Vijver M.J., He Y.D., Hart A.A., Mao M., Peterse H.L., van der Kooy K., Marton M.J., Witteveen A.T., Schreiber G.J., Kerkhoven R.M., Roberts C., Linsley P.S., Bernards R., Friend S.H. (2002). Gene expression profiling predicts clinical outcome of breast cancer. Nature 415:530-6.

Virtaneva K., Wright F.A., Tanner S.M., Yuan B., Lemon W.J., Caligiuri M.A., Bloomfield C.D., de La Chapelle A., Krahe R. (2001). Expression profiling reveals fundamental biological differences in acute myeloid leukemia with isolated trisomy 8 and normal cytogenetics. Proc Natl Acad Sci USA 98:1124-9.

Wyrick J.J., Young R.A. (2002). Deciphering gene expression regulatory networks. Curr Opin Genet Dev 12:130-6.

Zhang H., Yu C.Y., Singer B., Xiong M. (2001). Recursive partitioning for tumor classification with gene expression microarray data. Proc Natl Acad Sci USA 98:6730-5.

Chapter 13

CLUSTERING GENOMIC EXPRESSION DATA: DESIGN AND EVALUATION PRINCIPLES

Francisco Azuaje and Nadia Bolshakova

University of Dublin, Trinity College, Department of Computer Science, Dublin 2, Ireland,
e-mail: {Francisco.Azuaje, Nadia.Bolshakova}@cs.tcd.ie

1. INTRODUCTION: CLUSTERING AND GENOMIC EXPRESSION ANALYSIS

The analysis of expression data is based on the idea that genes that are involved in a particular pathway, or respond to a common environmental stimulus, should be co-regulated and therefore should exhibit similar patterns of expression. Thus, a fundamental task is to identify groups of genes or samples showing similar expression patterns.

Clustering may be defined as a process that aims to find partitions or groups of similar objects. It can be seen as an unsupervised recognition procedure whose products are known as *clusters*. In a genomic expression application, a cluster may consist of a number of samples (or genes) whose expression patterns are more similar than those belonging to other clusters. Figure 13.1 depicts a situation, in which two types of genes, each one associated with a different biological function, are clustered based on their expression profiles. The clusters are represented by circles, and the genes that are linked to each cluster are depicted randomly within the correspondent circle.

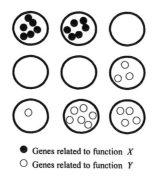

● Genes related to function X
○ Genes related to function Y

Figure 13.1. Clustering of genes according to their expression patterns.

Clustering has become a fundamental approach to analyzing genomic expression data. It can support the identification of existing underlying relationships among a set of variables such as biological conditions or perturbations. Clustering may represent a basic tool not only for the classification of known categories, but also (and perhaps most importantly) for the discovery of relevant classes. The description and interpretation of its outcomes may also allow the detection of associations between samples or variables, the generation of rules for decision-making support and the evaluation of experimental models. In the expression domain it has provided the basis for novel clinical diagnostic and prognostic studies (Bittner et al., 2000), and other applications using different model organisms (Ideker et al., 2001).

Several clustering methods have been proposed for expression analysis, and many other options will surely be applied in the future. Moreover, post-genome scientists deal with highly complex and diverse biological problem domains. Therefore, it would not be reasonable to expect the existence of universal clustering solutions. This chapter provides an overview of the major types of clustering problems and techniques for genomic expression data. It focuses on crucial design and analytical aspects of the clustering process. We hope that this chapter will guide our readers to address questions such as: Which clustering strategy should I use? How many clusters should it find? Is this a good partition? Is there a better partition?

Section 2 introduces important concepts for the effective application of clustering techniques. It overviews some of the major types of clustering algorithms for genomic expression data: their advantages, limitations and applications. It provides the reader with some important criteria for the selection of clustering methods. Section 3 approaches the systematic evaluation of clustering results based on their relevance and validity (both computational and biological). Two evaluation models will be presented:

Cluster validity strategies based on the *Dunn's index*, and the *silhouette method*. As a way of illustration these methods are implemented using two expression data sets, which were obtained from different clinical diagnostic studies. The results demonstrate that such validity frameworks may represent a useful tool to support biomedical knowledge discovery. Section 4 concludes with a discussion of the results, future work and recommendations.

2.   DESIGN PRINCIPLES FOR CLUSTERING STUDIES

Typical clustering algorithms are based on the optimisation of a partition quality measure. Generally these measures are related to the following factors: a) the heterogeneity of the clusters, also known as the cluster cohesion or compactness; and b) their separation from the rest of the data, also known as cluster isolation. Thus, a basic clustering approach may aim to search for a partition that a) minimize intra-cluster distances, and b) maximize inter-cluster distances.

There are several types of metrics to assess the distance or similarity, between samples and between clusters (Everitt, 1993). A clustering algorithm commonly requires the data to be described by a matrix of values, x_{ij} $(i = 1,..., m)$ $(j = 1,..., n)$, where x_{ij} refers to the value of the jth feature associated with the ith sample. In an expression data application x_{ij} may represent, for instance, the expression value of gene i during a perturbation j.

Other techniques require a matrix of *pairwise* values, p_{ij} $(i, j = 1,..., m)$, where p_{ij} represents the similarity (or dissimilarity) value between the ith and j^{th} objects to be clustered. In an expression data application p_{ij} may represent, for instance, the similarity or dissimilarity between the ith and j^{th} genes under a biological condition.

Some basic measures for heterogeneity or compactness assessment are the *sum of squares, L_1 measures, intra-cluster diameter metrics* and the *sum of distances* (Everitt, 1993). Isolation may be measured by, for example, calculating the minimum distance between clusters, or the sum of dissimilarities between samples in a particular cluster and samples belonging to other clusters. The reader is referred to (Hansen and Jaumard, 1997) for a more detailed description on heterogeneity and isolation measures for clustering processes.

The second part of this section will introduce relevant clustering systems for expression data applications. This overview addresses three major types of clustering systems: a) hierarchical clustering, b) techniques based on iterative relocation, and c) adaptive solutions and other advances.

2.1 Key Clustering Approaches For Expression Data

2.1.1 Hierarchical Clustering

Hierarchical clustering is perhaps the best-known clustering method for expression data analyses. Chapter 14 discusses its implementation and applications in more detail. The main objective of this technique is to produce a tree like structure in which the nodes represent subsets of an expression data set. Thus, expression samples are joined to form groups, which are further joined until a single hierarchical tree (also known as *dendrogram*) is produced. There are different versions of hierarchical clustering, which depend, for example, on the metric used to assess the separation between clusters.

Several studies on the molecular classification of cancers and biological modelling have been based on this type of algorithms. Pioneering studies include an investigation by Eisen et al. (1998), which found that hierarchical clustering might be used to group genes of known similar function in *Saccharomyces cerevisiae*. Dhanasekaran et al. (2001) illustrates how dendrograms can reveal the variation in gene expression pattern between distinct pools of normal prostate samples. Perou et al. (2000) measured the variation in the expression of 1,753 genes in 84 experimental breast cancer samples "before and after" chemotherapy. This study shows how these patterns provide a distinctive molecular portrait of each tumour. Moreover, the tumours could be classified into subtypes based on the differences of their gene expression patterns.

2.1.2 Models based on Iterative Relocation

This type of clustering algorithms involves a number of "learning" steps to search for an optimal partition of samples. Such processes may require: a) the specification of an initial partition of objects into a number of classes; b) the specification of a number of clustering parameters to implement the search process and assess the adequacy of its outcomes; c) a set of procedures to transform the structure or composition of a partition; and d) a repetitive sequence of such transformation procedures.

Some techniques included in this category are the *k-means* or *c-means* algorithms, and the *Kohonen Self-organizing Map* (SOM). The *k*-means method categorizes samples into a fixed number (k) of clusters, but it requires a priori knowledge on the number of clusters representing the expression data under study. SOMs have been applied to analyze expression profiles in several biomedical and *systems biology* studies (Quackenbush, 2001). This is a clustering approach based on hypothetical neural structures called feature maps, which are adapted by the effect of the input expression samples to be classified. Thus, users may use SOMs to find and visualize clusters of similar expression patterns. The SOM-based model was one of the first machine learning techniques used to illustrate the molecular classification of cancer. Golub and colleagues (1999) reported a model to discover the distinction between acute myeloid leukemia and acute lymphoblastic leukemia. To illustrate the value of SOMs Tamayo and coworkers applied it to hematopoietic differentiation data (Tamayo et al., 1999). In this research SOMs organized samples into biologically relevant clusters that suggest, for example, genes involved in differentiation therapy used in the treatment of leukemia. Ideker and colleagues (2001) used SOMs to support an integrated approach to building and refining a cellular pathway model. Based on this method they identified a number of mRNAs responding to key perturbations of the yeast galactose-utilization pathway. Chapter 15 illustrates the application of SOMs in expression data.

2.1.3   Adaptive Systems And Other Advances

Some of these clustering solutions, unlike the methods introduced in Section 2.1.2, may not require the specification of an initial partition or knowledge on the underlying class structure. That is the case of some adaptations of the original SOM, such as Growing Cell Structures (GCS), which has been applied for the discovery of relevant expression patterns in biomedical studies (Azuaje, 2001a). Chapter 15 introduces the design and application of GCS-based clustering models.

Recent advances for expression data analysis include *Biclustering*, which consists of a one-step process to find direct correlations between a subset of features (genes or perturbations) and a subset of samples (genes or tissues) (Cheng and Church, 2000). From a biological perspective this is a useful approach because it allows the simultaneous clustering of genes and conditions, as well as the representation of multiple-cluster membership.

Other contributions have demonstrated how a supervised neural network can be used to perform automatic clustering or discovery of classes. A model based on a supervised neural network called *Simplified Fuzzy ARTMAP* (Kasuba, 1993) has been used to recognize relevant expression patterns for the classification of lymphomas (Azuaje, 2001b). From a user's point of view this type of models also offers a number of computational advantages. For example, the user only needs to specify a single clustering parameter, and the clustering process can be executed with a single processing iteration.

2.2   Basic Criteria For The Selection Of Clustering Techniques

Even when one would not expect the development of universal clustering solutions for genomic expression data, it is important to understand fundamental factors that may influence the choice and performance of the most appropriate technique. This section provides readers with basic criteria to select clustering techniques. These guidelines address questions such as: Which clustering algorithm should I use? Should I apply an alternative solution? How can results be improved by using different methods? This discussion does not intend to offer a formal framework for the selection of clustering algorithms, but to highlight important dimensions that may have to be taken into account for improving the quality of clustering-based studies.

Choosing "the best" algorithm for a particular problem may represent a challenging task. There are multiple clustering techniques that can be used to analyze expression data. Advantages and limitations may depend on factors such as the statistical nature of the data, pre-processing procedures, number of features etc. Moreover, it is not uncommon to observe inconsistent results when different clustering methods are tested on a particular data set. In order to make an appropriate choice is important to have a good understanding of:

(1) the problem domain under study, and
(2) the clustering options available.

Knowledge on the underlying biological problem may allow a scientist to choose a tool that satisfies certain requirements, such as the capacity to detect overlapping classes. Knowledge on the mathematical properties or processing dynamics of a clustering technique may significantly support the selection process. How does this algorithm represent similarity (or dissimilarity)?, how much relevance does it assign to cluster heterogeneity?, how does it implement the process of measuring cluster isolation?. Answers to these questions may indicate crucial directions for the selection of an adequate clustering algorithm.

Empirical studies have defined several *mathematical criteria of acceptability* (Fisher and Van Ness, 1971). For example, there may be clustering algorithms that are capable of guaranteeing the generation of partitions whose cluster structures do not intersect. Such algorithms may be called *convex admissible*. There are algorithms capable of generating partition results that are insensitive to the duplication of data samples. These techniques may be called *point proportion admissible*. Other clustering algorithms may be known as *monotone admissible* or *noise-tolerant* if their clustering outcomes are not affected by monotone transformations on the data. It has been demonstrated, for instance, that both single-linkage and complete-linkage hierarchical clustering should be characterized as non-convex admissible, point proportion admissible and monotone admissible. The reader is referred to Fisher and Van Ness (1971) for a review on these and other mathematical criteria of acceptability.

Several algorithms indirectly assume that the cluster structure of the data under consideration exhibits particular characteristics. For instance, the k-means algorithm assumes that the shape of the clusters is spherical; and single-linkage hierarchical clustering assumes that the clusters are well separated. Unfortunately, this type of knowledge may not always be available in an expression data study. In this situation a solution may be to test a number of techniques on related data sets, which have previously been classified (a reference data set). Thus, a user may choose a clustering method if it produced consistent categorization results in relation to such reference data set.

Specific user requirements may also influence a selection decision. For example, a scientist may be interested in observing direct relationships between classes and subclasses in a data partition. In this case, a hierarchical clustering approach may represent a basic solution. But in some studies hierarchical clustering results could be difficult to visualize because of the number of samples and features involved. Thus, for instance, a SOM may be considered to guide an exploratory analysis of the data.

In general, the application of two or more clustering techniques may provide the basis for the synthesis of accurate and reliable results. A scientist may be more confident about the clustering experiments if very similar results are obtained by using different techniques. This approach may also include the implementation of voting strategies, consensus classifications, clustering fusion techniques and statistical measures of consistency (Everitt, 1993).

3.   CLUSTER VALIDITY AND EVALUATION FRAMEWORKS FOR EXPRESSION DATA

Several clustering techniques have been applied to the analysis of expression data, but fewer approaches to the evaluation and validation of clustering results have been studied.

Once a clustering algorithm has been selected and applied, scientists may deal with questions such as: Which is the best data partition?, which clusters should we consider for further analysis? What is the right number of clusters?

Answering those questions may represent a complex and time-consuming task. However, it has been shown that a robust strategy may consist of estimating the correct number of clusters based on validity indices (Azuaje, 2002a).

Such indices evaluate a measure, $Q(U)$, of quality of a partition, U, into c clusters. Thus, the main goal is to identify the partition of c clusters for which $Q(U)$ is optimal.

Two such cluster validity approaches are introduced and tested on expression data sets: The *Dunn's based indices* (Bezdek and Pal, 1998) and the *silhouette* method (Rousseeuw, 1987).

3.1   Assessing Cluster Quality With Dunn's Validity Indices

This index aims at identifying sets of clusters that are compact and well separated. For any partition $U \leftrightarrow X: X_1 \cup ... X_i \cup ... X_c$, where X_i represents the ith cluster of such partition, the Dunn's validation index, V, is defined as:

$$V(U) = \min_{1 \leq i \leq c}\left\{ \min_{\substack{1 \leq j \leq c \\ j \neq i}}\left[\frac{\delta(X_i, X_j)}{\max\limits_{1 \leq k \leq c}\{\Delta(X_k)\}} \right]\right\} \tag{13.1}$$

$\delta(X_i, X_j)$ defines the distance between clusters X_i and X_j (intercluster distance); $\Delta(X_k)$ represents the intracluster distance of cluster X_k; and c is the number of clusters of partition U. The main goal of this measure is to maximize intercluster distances whilst minimizing intracluster distances. Thus, large values of V correspond to good clusters. Therefore, the number of clusters that maximises V is taken as the optimal number of clusters, c (Bezdek and Pal, 1998).

In this study, eighteen validity indices based on Equation 13.1 were compared. These indices consist of different combinations of intercluster and intracluster distance techniques. Six intercluster distances, δ_i, $1 \leq i \leq 6$; and 3 intracluster distances, Δ_j, $1 \leq j \leq 3$ were implemented. Thus, for example, V_{13}, represents a validity index based on an intercluster distance, δ_1, and an intracluster distance Δ_3. The mathematical definitions of these intercluster and intracluster distances are described in Tables 13.1 and 13.2, respectively.

Table 13.1. Intercluster distances used to implement the Dunn's index. S and T are clusters from partition U; $d(x, y)$ defines the distance between any two samples, x and y, belonging to S and T respectively; $|S|$ and $|T|$ provide the number of samples included in clusters S and T, respectively.

$$\delta_1(S,T) = \min_{x \in S, y \in T}\{d(x,y)\}$$

$$\delta_4(S,T) = d(vs, vt)$$

$$vs = \frac{1}{|S|}\sum_{x \in S} x$$

$$vt = \frac{1}{|T|}\sum_{y \in T} y$$

$$\delta_2(S,T) = \max_{x \in S, y \in T}\{d(x,y)\}$$

$$\delta_5(S,T) = \frac{1}{|S|+|T|}\left(\sum_{x \in S} d(x,vt) + \sum_{y \in T} d(y,vs)\right)$$

$$\delta_3(S,T) = \frac{1}{|S\|T|}\sum_{\substack{x \in S \\ y \in T}} d(x,y)$$

$$\delta_6(S,T) = \max\{\delta(S,T), \delta(T,S)\}$$

$$\delta(S,T) = \max_{x \in S}\left\{\min_{y \in T}\{d(x,y)\}\right\}$$

$$\delta(T,S) = \max_{y \in T}\left\{\min_{x \in S}\{d(x,y)\}\right\}$$

Table 13.2. Intracluster distances used to implement the Dunn's index. S is a cluster from partition U; $d(x, y)$ defines the distance between any two samples, x and y, belonging to S; $|S|$ represents the number of samples included in cluster S.

$$\Delta_1(S) = \max_{x,y \in S}\{d(x,y)\}$$

$$\Delta_2(S) = \frac{1}{|S| \cdot (|S|-1)}\sum_{\substack{x,y \in S \\ x \neq y}} d(x,y)$$

$$\Delta_3(S) = 2\left(\frac{\sum_{x \in S} d(x,\bar{v})}{|S|}\right)$$

$$\bar{v} = \frac{1}{|S|}\sum_{x \in S} x$$

As a way of illustration, this validation process is tested on expression data from a study on the molecular classification of lymphomas. Clustering is performed using the SOM algorithm. The expression levels from a number of genes with suspected roles in processes relevant in diffuse large B-cell lymphoma (DLBCL) were used as the features for the automatic clustering of a number of B-cell samples. The data consisted of 63 cases (45 DLBCL and 18 normal) described by the expression levels of 23 genes. These data were obtained from an investigation published by Alizadeh and colleagues

(2000), who identified subgroups of DLBCL based on the analysis of the patterns generated by a specialized cDNA microarray technique. A key goal of this study was to distinguish two categories of DLBCL: Germinal Centre B-like DLBCL (GC B-like DLBCL) (22 samples) and Activated B-like DLBCL (23 samples) (Alizadehn et al., 2000). The full data and experimental methods are available on the Web site of Alizadeh et al. (http://llmpp.nih.gov/lymphoma).

Table 13.3 shows the values of the 18 validity indices and the average index at each number of clusters, c, for $c = 2$ to $c = 6$. The shaded entries correspond to the highest values of the indices, and $d(x, y)$ was calculated using the Euclidean distance. Fifteen of the indices indicated the correct value $c = 2$ while the remaining favour $c = 5$.

An examination of these partitions confirms that the case $c = 2$ represents the most appropriate prediction from a biomedical point of view. This partition accurately allows the identification of the two DLBCL subtypes: GC B-like and activated B-like. Table 13.4 describes the clusters obtained using the optimal value $c = 2$. Cluster 1 may be referred to as the cluster representing activated B-like DLBCL, while Cluster 2 recognizes the subclass GC B-like DLBCL.

A more robust way to predict the optimal value for c may consist of: a) implementing a voting procedure, or b) calculating the average index value for each cluster configuration. Table 13.3 indicates that based on such criteria the best partition consist of two clusters.

Table 13.3. Predicting the correct number of clusters: Validity indices for expression clusters originating from B-cells. The entries represent the Dunn's values using 3 types of intracluster measures and 6 types of intercluster measures. Bold-faced entries represent the optimal number of clusters, c, predicted by each index.

Validity index	$c = 2$	$c = 3$	$c = 4$	$c = 5$	$c = 6$
V_{11}	0.29	0.29	0.29	**0.31**	0.26
V_{21}	**1.46**	0.98	0.77	0.86	0.69
V_{31}	**0.72**	0.60	0.53	0.54	0.50
V_{41}	**0.50**	0.37	0.30	0.30	0.27
V_{51}	**0.62**	0.50	0.45	0.44	0.41
V_{61}	**0.83**	0.71	0.58	0.62	0.52
V_{12}	0.51	0.51	0.51	**0.52**	0.45
V_{22}	**2.57**	1.76	1.36	1.47	1.20
V_{32}	**1.27**	1.08	0.94	0.93	0.87
V_{42}	**0.88**	0.66	0.54	0.51	0.47
V_{52}	**1.09**	0.90	0.79	0.76	0.71

V_{62}	**1.47**	1.27	1.02	1.05	0.91
V_{13}	0.37	0.37	0.37	**0.38**	0.34
V_{23}	**1.86**	1.28	0.99	1.08	0.90
V_{33}	**0.92**	0.79	0.69	0.68	0.65
V_{43}	**0.64**	0.48	0.39	0.37	0.35
V_{53}	**0.79**	0.66	0.58	0.56	0.54
V_{63}	**1.06**	0.93	0.75	0.77	0.68
Average	**0.99**	0.79	0.66	0.68	0.60

Table 13.4. A relevant partition for a study on lymphoma data.

Cluster	Description .
1 (Activated B-like DLBCL)	23 samples belonging to subtype Activated B-like DLBCL, 1 sample belonging to subtype GC B-like DLBCL 9 Normal samples
2 (GC B-like DLBCL)	21 samples belonging to subtype GC B-like DLBCL 9 Normal samples

Table 13.5. Validity indices for expression clusters originating from a study on DLBCL. The entries represent the average Dunn's values based on the distances shown in Tables 13.1 and 13.2, and using three measures for $d(x, y)$. Bold-faced entries represent the optimal number of clusters, c, predicted by each method. *E.dist.*: Euclidean distance; *M.dist.*: Manhattan distance; *C.dist.*: Chebychev distance.

Index based on	$c = 2$	$c = 3$	$c = 4$	$c = 5$	$c = 6$
E.dist.	**0.99**	0.79	0.66	0.68	0.60
M.dist.	**1.57**	1.21	1.02	1.04	0.92
C.dist.	**0.97**	0.79	0.70	0.69	0.63

The results shown in Table 13.3 were obtained when $d(x, y)$ was calculated using the well-known Euclidean distance (Tables 13.1 and 13.2). However there are several ways to define $d(x, y)$ such as the *Manhattan* and *Chebychev* metrics (Everitt, 1993). Therefore, an important problem is to know how the choice of $d(x, y)$ may influence the prediction process. Table 13.5 summarizes the effects of three measures, $d(x, y)$, on the calculation of the Dunn's cluster validity indices. This analysis suggests that the estimation of the optimal partition is not sensitive to the type of metric, $d(x, y)$, implemented.

3.2   Assessing Cluster Validity With Silhouettes

For a given cluster, X_j $(j = 1,..., c)$, this method assigns to each sample of X_j a quality measure, $s(i)$ $(i = 1,..., m)$, known as the *silhouette width*. The silhouette width is a confidence indicator on the membership of the ith sample in cluster X_j.

The silhouette width for the ith sample in cluster X_j is defined as:

$$s(i) = \frac{b(i) - a(i)}{\max\{a(i), b(i)\}} \tag{13.2}$$

where $a(i)$ is the average distance between the ith sample and all of samples included in X_j, 'max' is the maximum operator, and $b(i)$ is implemented as:

$$b(i) = \min_{X_k \neq X_j} (d(i, X_k)) \tag{13.3}$$

where $d(i, X_k)$ is the average distance between the ith sample and all of the samples clustered in X_k; and 'min' represents the minimum value of $d(i, X_k)$ $(k = 1,..., c; k \neq j)$. It is easily seen from Equation 13.2 that $-1 \leq s(i) \leq 1$.

When a $s(i)$ is close to 1, one may infer that the ith sample has been "well-clustered", i.e. it was assigned to an appropriate cluster. When a $s(i)$ is close to zero, it suggests that the ith sample could also be assigned to the nearest neighbouring cluster, i.e. such a sample lies equally far away from both clusters. If $s(i)$ is close to -1, one may argue that such a sample has been "misclassified".

Thus, for a given cluster, X_j $(j = 1,..., c)$, it is possible to calculate a cluster silhouette S_j, which characterizes the heterogeneity and isolation properties of such a cluster:

$$S_j = \frac{1}{m} \sum_{i=1}^{m} s(i) \tag{13.4}$$

It has been shown that for any partition $U \leftrightarrow X: X_1 \cup ... X_i \cup ... X_c$, a *global silhouette value*, GS_u, can be used as an effective validity index for U (Rousseeuw, 1987).

$$GS_u = \frac{1}{c} \sum_{j=1}^{c} S_j \tag{13.5}$$

Furthermore, it has been demonstrated that Equation 13.5 can be applied to estimate the most appropriate number of clusters for U. In this case the partition with the maximum S_u is taken as the optimal partition.

By way of example, this technique is tested on expression data originating from a study on the molecular classification of leukemias (Golub et al., 1999). Clustering is again performed using SOM. The analyzed data consisted of 38 bone marrow samples: 27 acute lymphoblastic leukemia (ALL) and 11 acute myeloid leukemia (AML), whose original descriptions and experimental protocols can be found on the *MIT Whitehead Institute* Web site (http://www.genome.wi.mit.edu/MPR).

Table 13.6. Silhouette values for expression clusters originating from leukemia samples. The entries represent the global silhouette values, GS_u, for each partition, and the silhouette values, S, for each cluster defining a partition. Bold-faced entries highlight the optimal number of clusters, c, predicted by this method.

c	GS_u	S_1	S_2	S_3	S_4	S_5	S_6
2	**0.43**	**0.17**	**0.57**	-	-	-	-
3	0.14	0.11	0.35	0.11	-	-	-
4	**0.25**	**0.15**	**0.31**	**0.31**	**0.26**	-	-
5	0.19	0.07	0.45	0.23	0.23	0.21	-
6	0.23	0.28	0.23	0.28	0.42	0.14	0.14

Table 13.6 shows the global silhouette values, GS_u, for each partition, and the silhouette values, S, at each number of clusters, c, for $c = 2$ to $c = 6$. The shaded entries correspond to the optimal values for this validation method. It predicts $c = 2$ as the best clustering configuration. Table 13.7 describes the clusters obtained using $c = 2$, which adequately distinguish ALL from AML samples.

Table 13.7. An optimal partition of leukemia samples which distinguishes ALL from AML samples.

Cluster	Description
1 (AML class)	11 AML samples 2 ALL samples
2 (ALL class)	25 ALL samples

Table 13.6 suggests that the partition consisting of 4 clusters may also be considered as a useful partition, because it generates the second highest GSu. An examination of such a partition confirms that it represents relevant information relating to the detection of the ALL subclasses, B-cell and T-

cell, as demonstrated by Golub and colleagues (1999). The composition of this alternative partition is described in Table 13.8.

Table 13.8. Predicting appropriate partitions in a leukemia study: distinction of subtypes of ALL samples.

Cluster	Description
1 (AML class)	10 AML samples
2 (Unlabeled class)	2 B-ALL samples 1 T-ALL samples 1 AML sample
3 (T-ALL subclass)	7 T-ALL samples 2 B-ALL samples
4 (B-ALL subclass)	15 B-ALL samples

The results shown in Table 13.6 were obtained using the well-known Euclidean distance. Alternative measures include, for example, the *Manhattan* and the *Chebychev* metrics. Table 13.9 summarizes the effects of three distance measures on the calculation of the highest global silhouette values, GS_u. These results indicate that the estimation of the optimal partition is not sensitive to the type of distance metric chosen to implement Equation 13.2.

Table 13.9. Prediction of the optimal partition based on silhouettes and different distance metrics for leukaemia data. The entries represent the global silhouette values, GS_u, for each partition. Bold-faced entries highlight the optimal number of clusters, *c*, predicted by each method. *E.dist.*: Euclidean distance; *M.dist.*: Manhattan distance; *C.dist.*: Chebychev distance.

GS_u based on	$c = 2$	$c = 3$	$c = 4$	$c = 5$	$c = 6$
E.dist.	**0.43**	0.14	0.25	0.19	0.23
M.dist.	**0.43**	0.14	0.25	0.19	0.23
C.dist.	**0.43**	0.14	0.25	0.19	0.23

4.   CONCLUSIONS

This chapter has introduced key aspects of clustering systems for genomic expression data. An overview of the major types of clustering approaches, problems and design criteria was presented. It addressed the evaluation of clustering results and the prediction of optimal partitions. This problem, which has not traditionally received adequate attention from the expression research community, is crucial for the implementation of advanced clustering-based studies. A cluster evaluation framework may have a major impact on the generation of relevant and valid results. This paper shows how it may also support or guide biomedical knowledge discovery tasks. The clustering and validation techniques presented in this chapter may be applied to expression data of higher sample and feature set dimensionality.

A general approach to developing clustering applications may consist of the comparison, synthesis and validation of results obtained from different algorithms. For instance, in the case of hierarchical clustering there are tools that can support the combination of results into *consensus trees* (Bremer, 1990). However, additional methods will be required to automatically compare different partitions based on validation indices and/or graphical representations.

Other problems that deserve further research are the development of clustering techniques based on the direct correlation between subsets of samples and features, multiple-membership clustering, and context-oriented visual tools for clustering support (Azuaje, 2002b). Furthermore, there is the need to improve, adapt and expand the use of statistical techniques to assess uncertainty and significance in genomic expression experiments.

ACKNOWLEDGEMENTS

This contribution was partly supported by the *Enterprise Ireland Research Innovation Fund* 2001.

REFERENCES

Alizadeh A.A., Eisen M.B., Davis R.E., Ma C., Lossos I.S., Rosenwald A., Boldrick J.C., Sabet H., Tran T., Yu X., Powell J.I., Yang L., Marti G.E., Moore T., Hudson J., Lu L., Lewis D.B., Tibshirani R., Sherlock G., Chan W.C., Greiner T.C., Weisenburger D.D., Armitage J.O., Warnke R., Levy R., Wilson W., Grever M.R., Bird J.C., Botstein D., Brown P.O., Staudt M. (2000). Distinct types of diffuse large B-cell lymphoma identified by gene expression profiling. Nature 403:503-511.

Azuaje F. (2001a). An unsupervised neural network approach to discovering gene expression patterns in B-cell lymphoma. Online Journal of Bioinformatics 1:23-41.

Azuaje F. (2001b). A computational neural approach to support the discovery of gene function and classes of cancer. IEEE Transactions on Biomedical Engineering 48:332-339.

Azuaje F. (2002a) A cluster validity framework for genome expression data. Bioinformatics 18:319-320.

Azuaje F. (2002b). In silico approaches to microarray-based disease classification and gene function discovery. Annals of Medicine 34.

Bezdek J.C., Pal N.R. (1998). Some new indexes of cluster validity. IEEE Transactions on Systems, Man and Cybernetics, Part B 28:301-315.

Bittner M., Meltzer P., Chen Y., Jiang Y., Seftor E., Hendrix M., Radmacher M., Simon R., Yakhini Z., Ben-Dor A., Sampas N., Dougherty E., Wang E., Marincola F., Gooden C., Lueders J., Glatfelter A., Pollock P., Carpten J., Gillanders E., Leja D., Dietrich K., Beaudry C., Berens M., Alberts D., Sondak V., Hayward N., Trent J. (2000). Molecular classification of cutaneous malignant melanoma by gene expression profiling. Nature 406:536-540.

Bremer K. (1990). Combinable component consensus. Cladistics 6:69-372.

Cheng Y., Church G.M. (2000). Biclustering of expresssion data. Proceedings of ISMB 8th International Conference on Intelligent Systems for Molecular Biology; 2000 August 19 - 23; La Jolla. California.

Dhanasekaran S.M., Barrete T., Ghosh ., Shah R., Varambally S., Kurachi K., Pienta K., Rubin M., Chinnaiyan A. (2001). Delineation of prognostic biomarkers in prostate cancer. Nature 412:822-826.

Eisen M.B., Spellman P., Brown P.O., Botstein D. (1998). Cluster analysis and display of genome-wide expression patterns. Proc. Natl. Acad. Sci. USA 95:14863-14868.

Everitt B. (1993) *Cluster Analysis*. London: Edward Arnold.

Fisher L., Van Ness J.W. (1971). Admissible clustering procedures. Biometrika 58:91-104.

Golub T.R., Slonim D.K., Tamayo P., Huard C., Gassenbeck M., Mesirov J.P., Coller H., Loh M.L., Downing J.R., Caligiuri M.A., Bloomfield C.D., Lander E.S. (1999). Molecular classification of cancer: class discovery and class prediction by gene expression monitoring. Science 286:531-537.

Hansen P., Jaumard B. (1997). Cluster analysis and mathematical programming. Mathematical Programming 79:191-215.

Ideker T., Thorsson V., Ranish J.A., Christmas R., Buhler J., Eng J.K., Bumgarner R., Goodlett D.R., Aebersol R., Hood L. (2001). Integrated genomic and proteomic analyses of a systematically perturbated metabolic network. Science 292:929-933.

Kasuba T. (1993). Simplified fuzzy ARTMAP. AI Expert 8:19-25.

Perou C.M., Sorlie T., Eisen M.B., Van de Rijn M., Jeffrey S.S., Rees C.A., Pollack J.R., Ross D.T., Johnsen H., Aksien L.A., Fluge O., Pergamenschikov A., Williams C., Zhu S.X., Lonning P.E., Borresen-Dale A.L., Brown P.O., Botstein D. (2000). Molecular portraits of human breast tumours. Nature 406:747-752.

Quackenbush J. (2001). Computational analysis of microarray data. Nature Reviews Genetics 2:418-427.

Rousseeuw P.J. (1987). Silhouettes: a graphical aid to the interpretation and validation of cluster analysis. Journal of Computational and Applied Mathematics 20:53-65.

Tamayo P., Slonim D., Mesirov J., Zhu Q., Kitareewan S., Dmitrovsky E., Lander E.S., Golub R. (1999). Interpreting patterns of gene expression with self-organizing maps: methods and application to hematopoietic differentiation. Proc. Natl. Acad. Sci. USA 96:2907-291.

Chapter 14

CLUSTERING OR AUTOMATIC CLASS DISCOVERY: HIERARCHICAL METHODS

Derek C. Stanford, Douglas B. Clarkson, Antje Hoering
Insightful Corporation, 1700 Westlake Avenue North, Seattle, WA, 98109, USA
e-mail: {stanford, clarkson, hoering}@insightful.com

1. INTRODUCTION

Given a set of data, a hierarchical clustering algorithm attempts to find naturally occurring groups or clusters in the data. It is an exploratory technique that can give valuable insight into underlying relationships not otherwise easily displayed or found in multidimensional data. Microarray data sets present several challenges for hierarchical clustering: the need to scale the algorithms to a very large number of genes, selection of an appropriate clustering criterion, the choice of an appropriate distance measure (a *metric*), the need for methods which deal with missing values, screening out unimportant genes, and selection of the number of clusters, to name a few. This chapter discusses standard methods for *hierarchical agglomerative clustering*, including *single linkage*, *average linkage*, *complete linkage*, and *model-based clustering*; it also presents *adaptive single linkage clustering*, which is a new clustering method designed to meet the challenges of microarray data.

There are two general forms of hierarchical clustering. In the agglomerative method (bottom-up approach), each data point initially forms a cluster, and the two "closest" clusters are merged in each step. The "closest" clusters are defined by a *clustering criterion* (see Section 3). The divisive method (top-down approach) starts with one large cluster that contains all data points and splits off a cluster at each step. This chapter concentrates on the agglomerative method, the most commonly used method in practice.

Applications of hierarchical clustering in microarray data are diverse. One application is to group together genes with similar regulation. The underlying idea is that genes with similar expression levels might code for the same protein or proteins that exhibit similar functions on a molecular level. This clustering information helps researchers to better understand biological processes. For example, suppose that a time series of microarrays is observed and the main interest is to find clusters of genes forming genetic networks or regulatory circuits. The number of networks is not usually known and may be quite large; hierarchical clustering is particularly useful here because it allows for any number of gene clusters. Applications in this area range from searching for genetic networks in the yeast cell cycle, to the development of the central nervous system in the rat, or neurogenesis in the Drosophila (Erb et al., 1999).

Among other uses, hierarchical clustering of microarray data has been used to find cancer diagnostics (Ramaswamy et al., 2001), to investigate cancer tumorigenesis mechanisms (Welcsh et al., 2002), and to identify cancer subtypes (Golub et al., 1999). In the case of B-cell lymphoma, for example, hierarchical cluster analysis has been used to discover a new subtype which laboratory work was unable to detect (Alizadeh et al., 2000).

In addition to identifying groups of related genes, hierarchical clustering methods also offer tools to screen out uninteresting genes (see Section 3.3 for discussion and Section 4.2 for an example). Removal of uninteresting genes is especially important in microarray data, where a very large number of genes may be observed.

Section 2 discusses underlying issues and challenging aspects of the hierarchical clustering of microarray data such as scalability, metrics, and missing data. Section 3 presents several hierarchical clustering methods, including adaptive single linkage clustering, which is a new method designed to provide adaptive cluster detection while maintaining scalability. Section 4 provides examples using both simulated and real data. A brief discussion, including mention of existing hierarchical clustering software, is given in Section 5.

2.   SCALABILITY, METRICS, AND MISSING DATA

Scalability issues must be addressed both for a large number of data points (N) and for a large number of dimensions (P). Microarray data sets routinely make use of thousands of genes analyzed across hundreds or thousands of samples. *Nonparametric* hierarchical clustering methods usually make use of a distance (or similarity) matrix, which contains all inter-point distances for the data. With N data points, the distance matrix has N^2 entries. When N is large (e.g., $N > 10,000$) the size of the distance matrix becomes prohibitive in terms of both memory usage and speed; a nonparametric algorithm can fail if it requires an explicit computation of the distance matrix, as is the case with most standard hierarchical clustering software. *Model-based* clustering methods use a probability density function to model the location, size, and shape of each cluster. A potentially serious problem with model-based clustering, especially in gene expression data, is that the number of parameters can be quite large unless extensive constraints are placed on the model density. For example, a multivariate normal (or Gaussian) density is often used to model each cluster; with no constraints, this requires P parameters for the mean vector for each cluster and $P(P + 1)/2$ parameters for the covariance matrix for each cluster. Estimation of this many parameters is impractical when P is large (e.g., $P > 50$).

The choice of a metric appropriate for use on a particular microarray data set is inherently problem specific, depending on both the goals of the analysis and the properties of the data. The metric chosen can have a significant impact on both clustering results and computational speed. Hierarchical clustering software usually provides a selection of metrics, and may also allow users to define their own metrics. Aside from choosing a metric based on obvious data characteristics (e.g., discrete or continuous), researchers must consider their own perceptions of what it means to say that two observations are similar.

The use of different metrics can lead to quite different clustering results, as the following simple example illustrates. Suppose we have four genes (numbered from one to four) and two experiments. In the first experiment, only the first two genes are expressed, while in the second experiment only the even numbered genes are expressed. Then a metric giving positive weight to only the first experiment results in two clusters (genes 1 and 2, and genes 3 and 4), while a metric giving positive weight to the second experiment results in two very different clusters (genes 2 and 4 and genes 1 and 3). The metric must also account for scale – if the expression levels in experiment 1 have larger magnitude than in experiment 2, then a simple Euclidean metric will yield the experiment 1 clustering results.

Popular metrics for clustering microarray data are Euclidean distance and metrics based upon correlation measures. One correlation-based metric is computed as one minus the correlation coefficient. This metric yields a distance measure between zero and 2, with a correlation of one yielding a distance of zero, and a correlation of minus one yielding the largest possible distance of two. A variation of this is the use of one minus the absolute correlation. In this case, a correlation of either one or minus one yields a distance of zero, while a correlation of zero yields the largest possible distance of one. This allows clustering of genes responding to a stimulus in the same or opposite ways. For binary (0 or 1) or categorical data, metrics based upon probabilities are common, e.g., the probability, over all experiments, that both genes are expressed above some threshold.

In addition to the metric, one must also consider the genes and experiments used in the analysis. Inclusion of variables that do not differentiate (i.e., which are purely noise) can lead to poor clustering results. Removal of these prior to clustering can significantly improve clustering performance.

A significant difficulty with microarray data sets is the high rate of missing values. This reduces accuracy, and it also creates problems for computational efficiency. This problem is sometimes resolved by simply deleting the rows and columns with missing data. This cannot only yield biased estimates, it can also eliminate nearly all of the data. A common alternative is to define a metric that accepts missing values, usually by averaging or up-weighting of the observed data. For example, missing values can be handled by the pairwise exclusion of missing observations in distance calculations or, when mean values are required, by using all available data. Weighting schemes that account for the missing dimensions may also be desirable. Results obtained from metrics defined in this way can be misleading since there is no guarantee that the observed data behaves in the same manner as the missing data – it is easy to find simple examples showing that the metric chosen for dealing with missing data can have a large impact on the estimated distance. Often, imputation of missing data (see Chapter 3) is more appropriate. Imputation methodology, and missing data in general, is an open research topic with no easy solution, especially when the location of the missing data may be causally related to the unknown clusters.

3.   HIERARCHICAL CLUSTERING METHODS

The basic algorithm in hierarchical agglomerative clustering is to begin with each data point as a separate cluster and then iteratively merge the two "closest" clusters until only a single cluster remains. Here "close" is defined by the *clustering criterion*, which defines how to determine the distance between two clusters. In nonparametric clustering, this criterion consists of two parts: the distance measure or metric, which specifies how to compute the distance between two points; and the *linkage*, which specifies how to combine these distances to obtain the between-cluster distance. In model-based clustering, the clustering criterion is based on the *likelihood* of the data given the model (see Section 3.1).

With N data points, this approach of iteratively merging the two closest clusters provides a nested sequence of clustering results, with one result for each number of clusters from N to 1. Hierarchical clustering does not seek to globally optimize a criterion; instead, it proceeds in a stepwise fashion in which a merge performed at one step cannot be undone at a later step.

The clustering results are typically displayed in a *dendrogram* showing the cluster structure; for example, see Figure 14.1. Clusters or nodes forming lower on the dendrogram are closer together, while upper nodes represent merges of clusters that are farther apart. Since each data point begins as a single cluster, the *leaves* (terminal nodes at the bottom of the dendrogram) each represent one data point, while interior nodes represent clusters of more than one data point. The top node of the dendrogram denotes the entire data set as a single cluster. The *y*-axis is usually the *merge height*, the distance between two clusters when they are merged. Some methods do not have an explicit height associated with each merge; for example, model-based clustering chooses each merge by seeking to maximize a clustering criterion based on the likelihood. In these cases, a dendrogram can still be constructed, but the *y*-axis may represent the value of a clustering criterion or simply the order of clustering. The *x*-axis is arbitrary; the sequence of data points along the *x*-axis is generally chosen to avoid crossing lines in the display of the dendrogram.

3.1   Clustering Criteria and Linkage

Three common nonparametric approaches to hierarchical clustering are *single linkage*, *complete linkage*, and *average linkage*; a comprehensive review is given by (Gordon, 1999). In single linkage clustering, the distance between any two clusters of points is defined as the smallest distance between any point in the first cluster and any point in the second cluster. The single linkage approach is related to the minimum spanning tree (MST) of the data set (Gower and Ross, 1969). This leads to a significant advantage of single linkage clustering: efficient algorithms can be used to obtain the single linkage clustering result without allocating the order N^2 memory units usually required by other nonparametric hierarchical clustering algorithms.

Complete linkage defines the inter-cluster distance as the largest distance between any point in the first cluster and any point in the second cluster. Average linkage is often perceived as a compromise between single and complete linkage because it uses the average of all pair-wise distances between points in the first cluster and points in the second cluster. This is also called *group average linkage* (Sokal and Michener, 1958). *Weighted average linkage* (Sokal and Sneath, 1963) is defined in terms of its updating method: when a new cluster is created by merging two smaller clusters, the distance from the new cluster to any other cluster is computed as the average of the distances of the two smaller clusters. Thus, the two smaller clusters receive equal weight in the distance calculation, as opposed to average linkage, which accords equal weight to each data point.

Ward's method (Ward, 1963) examines the sum of squared distances from each point to the centroid or mean of its cluster, and merges the two clusters yielding the smallest increase in this sum of squares criterion. This is equivalent to modeling the clusters as multivariate normal densities with different means and a single hyper-spherical covariance matrix.

Ward's method is a special case of model-based clustering (Banfield and Raftery, 1993). Model-based clustering is based on an assumption of a within-cluster probability density as a model for the data. If the model chosen is incorrect or inappropriate, erroneous results will be obtained. Though models can be based on any density, the most common choice is the multivariate normal; this density is parametrized by a *mean vector* and a *covariance matrix* for each cluster. The mean vector determines the location of the cluster, while the covariance matrix specifies its shape and size. At each step, two clusters are chosen for a merge by maximizing the *likelihood* of the data given the model. The likelihood is the value of the probability density model evaluated using the observed data (Arnold, 1990). The likelihoods of all data points are combined into an overall likelihood; various approaches exist for this, such as a mixture likelihood or a classification likelihood. (Stanford, 1999) gives two theorems linking optimal choice of the form of the overall likelihood to the goals of the clustering procedure.

When choosing a hierarchical clustering method, some consideration should be given to the type of clusters expected. Complete linkage algorithms tend to yield compact clusters similar to the multivariate normal point clouds modeled in Ward's method, while single linkage clusters can be "stringy" or elongated, adapting well to any pattern of closely spaced points. In model-based clustering, the chosen density will have a strong impact on the resulting cluster shapes. For example, if the covariance matrix is constrained to be the same for all clusters, then clusters with the same size and shape will most likely be observed. If the covariance matrices are constrained to be multiples of the identity matrix, then hyper-spherical clusters (as in Ward's method) will be found.

3.2 Choosing the Final Clusters

The traditional approach for determining the final set of clusters is to specify the number of clusters desired and then cut the dendrogram at the height, which yields this number. This procedure only works well if the merges near the top of the dendrogram have large children, i.e. when the final agglomeration steps involve large subsets of the data. Single linkage clustering only exhibits this structure if the clusters are all well separated; otherwise, they tend to show a *chaining effect*, in which many of the upper dendrogram nodes are merely merges of distant points with a main group.

For example, if the data consist of two large clusters near each other and a single distant data point, then the two cluster result from single linkage clustering will give the single distant point as one cluster and everything else as the other cluster. Dendrograms based on other criteria, such as average and complete linkage, are not as prone to the chaining effect, but they also generally have nodes with large children near the top even when only one cluster is present in the data. For model-based clustering, the number of clusters can be assessed by examining the likelihood, though this requires severe assumptions about the data.

The chaining effect in a single linkage dendrogram contains important information: it indicates that the clusters are not well separated. Adaptive single linkage clustering utilizes the information in the single linkage dendrogram, but it uses a better method for determining the final clusters.

3.3   Adaptive Single Linkage Clustering

Adaptive single linkage (ASL) clustering (McKinney, 1995) begins with a single linkage dendrogram but extracts clusters in a bottom-up rather than a top-down manner. Generally, desirable clusters represent *modal* regions, regions of the data with higher point densities than the surrounding regions. To find these modal regions, each node is analyzed to determine its *runt* value (Hartigan and Mohanty, 1992), where the runt value of a node is defined as the size of its smaller child. Large runt values provide evidence against unimodality because they indicate the merge of two large subgroups. Clusters are found by selecting nodes with runt values larger than a specified threshold. The threshold provides a lower bound on the cluster size, and also determines the number of clusters found. For microarray analysis, small threshold values (e.g., 5) can be used to identify small, highly similar clusters; this might be suited to finding potential regulatory pathways for further analysis. Larger threshold values (e.g., 30) are more appropriate for finding larger groups of genes, such as groups involved in large-scale cellular activities or responses to experimental conditions.

Adaptive single linkage clustering presents at least two advantages over traditional methods. First, as noted above, fast and memory-efficient algorithms exist for computing the single linkage dendrogram – all that is required is that the data fit into memory. Second, because nodes with size less than the threshold can be regarded as "noise" or "fluff", the method can be used to automatically eliminate a large number of genes (or experiments) from further consideration. This use as a screening tool is illustrated in the example in Section 4.2. Further details on the algorithms underlying adaptive single linkage clustering can be found in (Glenny et al., 1995).

4. EXAMPLES

We provide two examples of cluster analysis. The first is a small simulation comparing average linkage with adaptive single linkage. The second uses a real microarray data set to demonstrate several analyses.

4.1 Simulated Data

We begin by presenting a simulation that demonstrates the utility of adaptive single linkage clustering; this example uses two-dimensional data to allow visual inspection of the clustering results. Our data consist of two spherical Gaussian clusters located at $[-1,-1]$ and $[1,1]$ with 100 points each, and 100 points of Poisson background noise over the rectangle from $[-5,-5]$ to $[5,5]$. We use a Euclidean metric and compare results from adaptive single linkage clustering and average linkage clustering. We examine a range of average linkage results with up to 7 clusters, as well as the adaptive single linkage result with 2 clusters (the unclassified noise points in adaptive single linkage might be considered to be a third cluster).

The average linkage dendrogram (Figure 14.1) shows a confusing amount of structure; the dendrogram suggests that it might be reasonable to have several clusters. We must drill down to 7 clusters (Figure 14.2) before the large central group finally splits into a reasonable approximation of the original two Gaussian clusters.

In contrast, the adaptive single linkage approach detects clusters through an analysis of the cluster merges rather than cutting the dendrogram from the top. The dendrogram (Figure 14.3) shows two main clusters surrounded by many outlying points. These two clusters provide a close approximation of the true location of the two underlying Gaussian clusters (Figure 14.4). This example illustrates several points: different clustering methods can lead to very different results, the method for choosing the clusters has a significant impact on the interpretation of results and the ease of cluster detection, and adaptive single linkage clustering can provide reasonable results even for touching clusters with background noise (regarded as a difficult case for traditional single linkage).

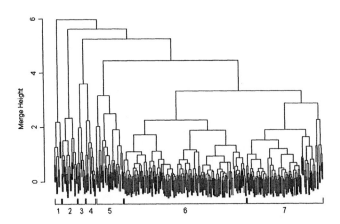

Figure 14.1. Average linkage dendrogram for the simulated data set, with labels indicating the seven cluster result.

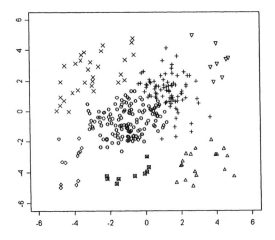

Figure 14.2. Average linkage clustering result for the simulated data set, with plot symbols indicating the seven cluster result.

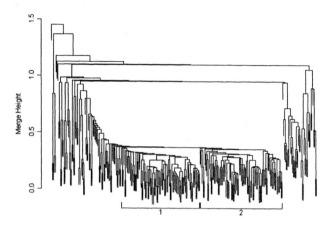

Figure 14.3. Adaptive single linkage (ASL) dendrogram for the simulated data set, with labels indicating the two cluster result.

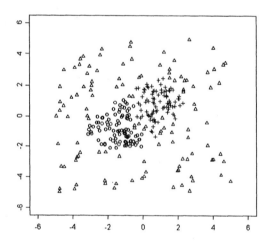

Figure 14.4. Adaptive single linkage (ASL) clustering result for the simulated data set, with plot symbols indicating the two cluster

4.2 Lymphoma Data

We analyzed a lymphoma data set from (Alizadeh et al., 2000) to illustrate the results of adaptive single linkage clustering on real microarray data. This data set consists of 4,026 genes across 96 tissue samples, and we used a correlation metric with pairwise exclusion of missing observations. We performed clustering first on the genes and then on the tissues; we also show an example using an imputation method for the missing values.

A full single linkage dendrogram for the gene clustering is shown in Figure 14.5. It is difficult to make out structure in this plot because of the large size of the data set; other clustering methods lead to similarly dense dendrograms. Adaptive single linkage clustering allows us to focus only on the points in locally high-density regions of the data. Clusters with runt statistics of size 10 or larger are shown in Figure 14.6; this application of adaptive single linkage clustering screens out 87% of the data points and leaves 26 clusters for further consideration.

We clustered the tissue samples using adaptive single linkage clustering and compared our results to the known tissue classes. The known tissue classes are shown by numbered labels on the single linkage dendrogram in Figure 14.7. These can be compared to the same dendrogram in Figure 14.8, which shows the clustering obtained with a runt threshold of 4. Adaptive single linkage clustering automatically finds clusters corresponding to all of the known tissue types except for the two-point clusters (labeled as 4 and 6 on Figure 14.7). It also suggests that there may be evidence of two subgroups within class 1 of Figure 14.7.

Figure 14.9 displays the known tissue classes on a single linkage dendrogram computed following imputation for all 19,667 missing values in the data. The imputation process was based on a Gaussian model (Schafer, 1997), using software provided in S-PLUS (Schimert et al., 2001). Markov chain Monte Carlo was used to compute Bayesian estimates of the unknown random variables, including the imputed missing values. See Chapter 3 for alternative imputation methods. Figure 14.9 is extremely similar to the result in Figure 14.7. Note that the ordering of the data on the x-axis is arbitrary; the ordering is chosen for simplicity of display. Switching the two children of a node from right to left has no impact on the structure of the dendrogram. For example, a large part of group 1 is displayed on the left side in Figure 14.7, while it is on the right side in Figure 14.9. This results from one dendrogram node for which the two sub-trees have been flipped; there are several other such switches apparent between these two figures. In the actual clustering result, there are only a few small changes, such as the two points in group 6, which appear to blend into group 1 more in Figure 14.9. We must keep in mind that our missing data methods, whether implicit in the metric or explicit via imputation, can have an impact on our analysis results when there are many missing values.

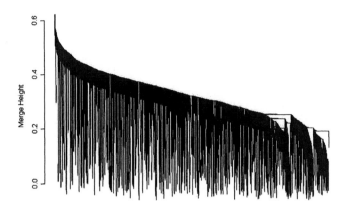

Figure 14.5. Single linkage dendrogram of the lymphoma data set.

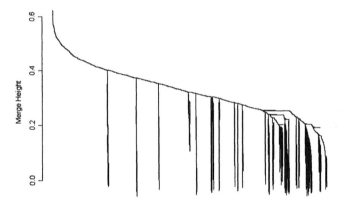

Figure 14.6. Adaptive single linkage dendrogram of the lymphoma data set showing clusters of size ten or larger.

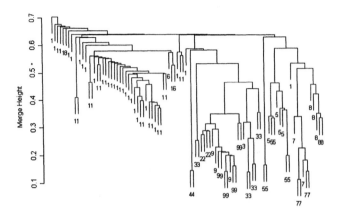

Figure 14.7. Adaptive single linkage dendrogram of the lymphoma data set with known tissue classes labeled.

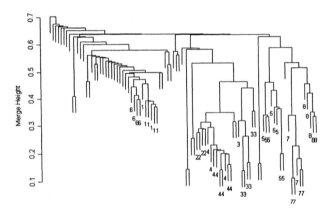

Figure 14.8. Adaptive single linkage dendrogram of lymphoma data with clusters of size four or larger labeled (unclassified points are unlabeled).

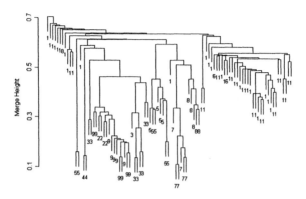

Figure 14.9. Adaptive single linkage dendrogram of the lymphoma data following imputation for missing values, with known tissue classes labeled.

5. DISCUSSION AND SOFTWARE

Hierarchical clustering is an effective exploratory tool for the analysis of microarray data sets or other high throughput screening data; the resulting dendrogram simultaneously presents 1 to N clusters, yielding insight into relationships not otherwise easily found in multidimensional data. Model-based methods offer good cluster detection and a statistical model which can help in determining the number of clusters, but they require modeling assumptions which can be severe. Nonparametric methods avoid modeling assumptions and adapt to any cluster shape, which allows their use when density assumptions are not appropriate or not verifiable. Only a few methods, such as adaptive single linkage, can be scaled to handle the large data sizes which are becoming common in microarray data sets.

There are many software packages, both commercial and free, that perform hierarchical clustering. The most widely used statistical packages, such as S-PLUS, SAS, and SPSS, provide general hierarchical clustering tools that can be used to display and manipulate dendrograms. The adaptive single linkage clustering software used here is new; a preliminary version, suitable for use in S-PLUS, can be obtained by contacting the authors. See Chapter 19 for a survey of microarray analysis tools.

REFERENCES

Alizadeh A.A., Eisen M.B., Davis R.E., Ma C., Lossos I.S., Rosenwald A., Boldrick J.C., Sabet H., Tran T., Yu X., Powell J.I., Yang L., Marti G.E., Moore T., Hudson J., Lu L., Lewis D.B., Tibshirani R., Sherlock G., Chan W.C., Greiner T.C., Weisenberger D.B.,

Armitage J.O., Warnke R., Levy R., Wilson W., Grever M.R., Byrd J.C., Botstein D., Brown P.O., Staudt L.M. (2000). Distinct types of diffuse large B-cell lymphoma identified by gene expression profiling. Nature 403:503-11.

Arnold S.F. (1990). *Mathematical Statistics*. Englewood Cliffs: Prentice-Hall.

Banfield J.D., Raftery A.E. (1993). Model-Based Gaussian and Non-Gaussian Clustering. Biometrics 49:803-21.

Erb R.S., Michael G.S. (1998). Cluster Analysis of Large Scale Gene Expression Data. Computing Science and Statistics 30:303-8.

Glenny R.W., Polissar N.L., McKinney S., Robertson H.T. (1995). Temporal heterogeneity of regional pulmonary perfusion is spatially clustered. J Appl Physiol 79(3):986-1001.

Golub T.R., Slonim D.K., Tamayo P., Huard C., Gaasenbeek M., Mesirov J.P., Coller H., Loh M.L., Downing J.R., Caligiuri M.A., Bloomfield C.D., Lander E.S. (1999). Molecular Classification of Cancer: Class Discovery and Class Prediction by Gene Expression Monitoring. Science 286:531-7.

Gordon A.D. (1999). *Classification, 2nd Ed*. New York: Chapman & Hall.

Gower J.C., Ross G.J.S. (1969). Minimum Spanning Trees and Single Linkage Cluster Analysis. Applied Statistics 18(1):54-64.

Hartigan J.A., Mohanty S. (1992). The Runt Test for Multimodality. Journal of Classification 9:63-70.

McKinney S. (1995). *Autopaint: A Toolkit for Visualizing Data in Four or More Dimensions*. PhD Thesis, University of Washington Biostatistics Department.

Prim R. (1957). Shortest Connection Networks and Some Generalizations. Bell Systems Technical Journal 1389-1401.

Ramaswamy S., Tamayo P., Rifkin R., Mukherjee S., Yeang C.H., Angelo M., Ladd C., Reich M., Latulippe E., Mesirov J.P., Poggio T., Gerald W., Loda M., Lander E.S., Golub T.R. (2001). Multiclass cancer diagnosis using tumor gene expression signatures. Proc. Nat. Acad. Sc. USA 98(26):15149-54.

Sokal R.R., Michener C.D. (1958). A Statistical Method for Evaluating Systematic Relationships. University of Kansas Science Bulletin 38:1409-38.

Schafer J.L. (1997). *Analysis of Incomplete Multivariate Data*, London:Chapman & Hall.

Schimert J., Schafer J.L., Hesterberg T., Fraley C., Clarkson D.B. (2001). *Analyzing Data with Missing Values in S-PLUS*, Seattle:Insightful.

Sokal R.R., Sneath P.H.A. (1963). *Principles of Numerical Taxonomy*, San Francisco:Freeman.

Stanford D.C. (1999). *Fast Automatic Unsupervised Image Segmentation and Curve Detection in Spatial Point Patterns*. PhD Thesis, University of Washington Statistics Department.

Ward J. (1963). Hierarchical groupings to optimize an objective function. Journal of the American Statistical Association 58:234-44.

Welcsh P.L., Lee M.K., Gonzalez-Hernandez R.M., Black D.J., Mahadevappa M., Swisher E.M., Warrington J.A., King M.C. (2002). BRCA1 transcriptionally regulates genes involved in breast tumorigenesis. Proc. Nat. Acad. Sc. USA 99(11):7560-5.

Zahn C.T. (1971). Graph-Theoretical Methods for Detecting and Describing Gestalt Structures. IEEE Transactions on Computers C-20:68-86.

Chapter 15

DISCOVERING GENOMIC EXPRESSION PATTERNS WITH SELF-ORGANIZING NEURAL NETWORKS

Francisco Azuaje

University of Dublin, Trinity College, Department of Computer Science, Dublin 2, Ireland, e-mail: Francisco.Azuaje@cs.tcd.ie

1.   INTRODUCTION

Self-organizing neural networks represent a family of useful clustering-based classification methods in several application domains. One such technique is *the Kohonen Self-Organizing Feature Map* (SOM) (Kohonen, 2001), which has become one of the most successful approaches to analysing genomic expression data. This model is relatively easy to implement and evaluate, computationally inexpensive and scalable. In addition, it exhibits significant advantages in comparison to other options. For instance, unlike hierarchical clustering it facilitates an automatic detection and inspection of clusters. Unlike Bayesian-based clustering it does not require prior hypotheses or knowledge about the data under consideration. Compared to the *k*-means clustering algorithm, the SOM exemplifies a robust and structured classification process.

Self-organizing neural networks are based on the principle of transforming a set of *p-variate* observations into a spatial representation of smaller dimensionality, which may allow a more effective visualization of correlations in the original data. Murtagh and Hernández-Pajares (1995), among many others, have discussed the connections between SOMs and alternative data analysis techniques. Before its introduction to the area of functional genomics, SOMs had been extensively applied in different biomedical decision support tasks, including coronary heart risk assessment (Azuaje et al., 1998), electrocardiogram-based diagnostic studies

(Papadimitriou et al., 2001) and tissue characterization in cancer studies (Schmitz et al., 1999).

Scientists may use SOMs to detect clusters of similar expression patterns. The SOM-based model was one of the first machine learning techniques implemented for the molecular classification of cancer. Golub and colleagues (1999) reported a model to discover the distinction between acute myeloid leukemia and acute lymphoblastic leukemia. The application of SOMs was part of a systematic expression monitoring method based on DNA microarrays. They were able to illustrate not only a classification process to distinguish known categories of leukemia samples, but also a class discovery process to identify unknown relevant subtypes. The authors suggested that it would be possible to achieve a sub-classification of higher resolution with a larger sample collection. Moreover, this classification technique may provide the basis for the prediction of clinical outcomes, such as drug response or survival. This research is a good example of how a SOM-based classifier together with other statistical tools may support a complex knowledge discovery function.

Another relevant study consisted of the application of SOMs to organize thousands of genes into biologically relevant clusters using hematopoietic differentiation data (Tamayo et al., 1999). This classification system indicated, for example, genes involved in differentiation therapy used in the treatment of leukemia. It discussed some of the key attributes that make the SOM an adequate clustering technique for expression data. It shows how SOMs can primarily be used to perform exploratory data analysis and facilitate visualisation-based interpretations. The authors developed *Genecluster*, which is a computer package to perform SOM-based classification of genomic expression data. It has assisted, for instance, the generation of interpretations relating to the yeast cell cycle, macrophage differentiation in HL-60 cells and hematopoietic differentiation across different cell lines (Tamayo et al., 1999).

Ideker and colleagues (2001) also used SOMs in an integrated approach to refining a cellular pathway model. Based on this method they identified a number of mRNAs responding to key perturbations of the yeast galactose-utilization pathway.

The remainder of this chapter addresses two important questions on self-organizing neural networks applications for expression data: a) How do these systems work? and b) How can we use them to support genomic expression research? It focuses on the application of SOMs in different expression data analysis problems. Advantages and limitations will be discussed. Moreover, an alternative solution based on the principle of adaptive self-organization will be introduced. This chapter will end with an overview of current challenges and opportunities.

2. SOMS AND MICROARRAY DATA ANALYSIS

The SOM is based on hypothetical neural structures called feature maps, which are configured and adapted by the effect of sensory signals or data observations (Kohonen, 2001). Their processing components, known as *neurones, prototypes* or *cells*, are spatially correlated after completing a learning or training process, such that those prototypes at nearby points on the resulting structure are more similar than those widely separated. Each prototype is associated with a weight vector m_i. Thus, SOMs can be used to perform clustering functions (Murtagh and Hernández-Pajares, 1995). Figure 15.1 shows a typical SOM.

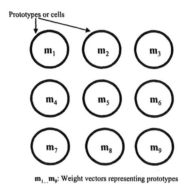

Prototypes or cells

$m_{1...}m_9$: Weight vectors representing prototypes

Figure 15.1. A typical SOM.

2.1 The SOM Clustering Algorithm

The SOM learning algorithm transforms any p-dimensional space into an ordered two-dimensional coordinate system. Also one may say that the SOM algorithm implements a "nonlinear projection" of the probability density function, $p(x)$, of the input data vector x onto a two-dimensional space (Kohonen, 2001).

Given a number of samples, N, each one represented by a number of features, p, a *Kohonen map* (Kohonen, 2001) consists of a grid of k prototypes, $m_j \in \mathfrak{R}^p$ (vector defined by p elements) (Figure 15.1). The main goal is then to define associations between each sample or observation and the prototypes represented on the map. The number of prototypes, k, and other learning parameters need to be defined by the user. Before starting the learning process the prototypes m_j are randomly initialized. Each of the k prototypes, m_j, may also be encoded with respect to an integer coordinate pair $r_j \in Q_1 \otimes Q_2$. Where $Q_1 = \{1,...,q_1\}$, $Q_2 = \{1,...,q_2\}$ and $k = q_1 \times q_2$. Figure 15.2 illustrates a SOM consisting of 9 prototypes, which are used to categorise a number of samples. The SOM learning process is summarised as follows.

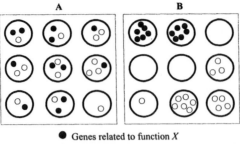

Figure 15.2. A SOM network before (panel A) and after (panel B) performing a learning process, based on a hypothetical data set of expression profiles linked to two classes of genes. The right panel indicates that the algorithm has successfully separated the classes under consideration.

Each observation, x_i, is processed one at a time. The first step in each *learning cycle* is to find the closest prototype m_j to x_i using, for example, the Euclidean distance in $\mathfrak{R}^p$. Then for all neighbours m_k of m_j, the idea is to make m_k closer to x_i, based on the following formula:

$$m_{k\text{-}new} = m_k + \alpha \times (x_i - m_k), \text{ where } m_k \in N_j \qquad (15.1)$$

where $m_{k\text{-}new}$ represents the new value for m_k, α is called the learning rate, and N_j represents the neighbourhood of m_j, which always includes m_j.

The main purpose of Equation 15.1 is not only to move the SOM prototypes closer to the data, but also to develop a smooth spatial relationship between the prototypes. This process is summarized in Figure 15.3.

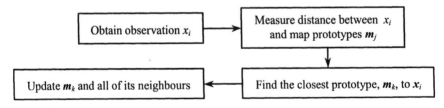

Figure 15.3. The SOM learning algorithm: a single learning cycle.

The neighbours of m_j are defined to be all m_k, such that the distance between r_j and r_k is *small*. Commonly this is calculated using the Euclidean distance, and *small* is defined by a threshold value, *Th*. The selection of the size of N_j is crucial to achieve a proper clustering process. For example, if the neighbourhood is too small at the beginning of the learning process, the

SOM will not be ordered globally. Thus, one can initiate it with a fairly wide N_j and let its size (threshold *Th*) decrease linearly during the learning process.

The performance of the SOM learning algorithm strongly depends on the selection of the learning rate, α. Typically α is linearly decreased from 1 to 0 over a few thousand learning cycles. For more information on the design principles of the SOM, the reader is referred to (Kohonen, 2001).

A SOM can also be seen as a constrained version of the k-means clustering algorithm. If we define *Th* small enough such that each neigbourhood contains only one prototype, then the spatial interrelation between the prototypes is not achieved. In that case it is possible to demonstrate that the SOM algorithm is a version of the *k*-means clustering method, which stabilizes at one of the local minima found by the *k*-means (Hasti et al., 2001).

Figure 15.2 illustrates a hypothetical situation, in which two types of genes, each one associated with a different biological function, are clustered based on their expression profiles. Panel A of Figure 15.2 shows a SOM at the very beginning of the learning process, while panel B portrays the clusters formed after completing a learning process. The prototypes are represented by circles, and the genes that are linked to each prototype are depicted randomly within the correspondent circle. One may, for example, run the algorithm during 2,600 learning cycles through this data set of 26 genes (100 cycles for each gene), and let *Th* and α decrease linearly over the 2,600 iterations. This example depicts a case in which a SOM network has successfully detected a class structure in a data set, which may allow one to differentiate its samples in terms of their patterns of functional similarity.

Once a SOM has been properly trained, one can use it to classify an unknown observation, which can also be referred to as a *testing* sample. In this situation the prediction process consists of identifying the closest SOM prototype to the sample under consideration, and use that prototype as its class or cluster predictor.

The following sub-section illustrates the application of the SOM to a genomic expression classification problem.

2.2   Illustrating Its Application

By way of example, this technique is first tested on expression data from a study on the molecular classification of leukemias. The data analysed consisted of 38 bone marrow samples: 27 acute lymphoblastic leukemia (ALL) and 11 acute myeloid leukemia (AML) samples. Each sample is described by the expression levels of 50 genes with suspected roles in this disease. These data were obtained from a study published by Golub and coworkers (1999). The original data descriptions and experimental protocols can be found at the *MIT Whitehead Institute* Web site (http://www.genome.wi.mit.edu/MPR).

The data were normalised such that the mean and variance of the genes are set to 0 and 1 respectively, which is the traditional pre-processing method used in expression analysis. The SOM networks were trained with 3800 learning cycles. The initial value of the learning parameter α was equal to 0.1 in all of the clustering experiments. Both values for α and *Th* were linearly decreased during the learning processes.

Figure 15.4 displays the clustering results based on a SOM network, which is defined by two prototypes: A and B. All of the AML samples were grouped by prototype A. The samples belonging to the class ALL were assigned to prototype B, except two of them that were located in the first prototype. This configuration indicates that the cluster defined by the prototype A is representative of the class AML, and the cluster defined by the prototype B is associated with the class ALL. Therefore, one may argue that this learning process was able to distinguish between the classes ALL and AML based on the expression values of 50 genes (Golub et al. 1999).

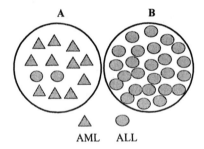

Figure 15.4. Expression data clustering using the SOM: two clusters of AML and ALL samples.

This type of clustering technique may also be used to predict the existence of subclasses or discover unknown categories. Figure 15.5 displays the clustering results based on 4 prototypes, which were used to categorise the same leukemia data set.

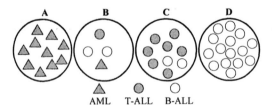

Figure 15.5. Expression data clustering using the SOM: Four clusters of AML and ALL samples. Clusters C and D are associated with two subtypes of ALL samples.

These results again suggest that it is possible to distinguish AML from ALL samples. AML samples are encoded by prototype A, except one that was included in cluster B. Clusters B, C and D include the samples belonging to the ALL class. A previous systematic study of these data demonstrated that the ALL samples may indeed be classified into two subtypes: T-ALL and B-ALL (Golub et al., 1999). The SOM clustering results depicted in Figure 15.5 offers a useful insight into the existence of those subclasses. Based on the composition of the clusters obtained in Figure 15.5, one may point out, for example, that cluster C can be labelled as the T-ALL cluster, while cluster D identifies the samples belonging to B-ALL.

A second example deals with the molecular classification of diffuse large B-cell lymphoma (DLBCL) samples. The data consisted of 63 cases (45 DLBCL and 18 normal) described by the expression levels of 23 genes with suspected roles in processes relevant to DLBCL (Alizadeh et al., 2000). These data were obtained from a study published by Alizadeh and colleagues (2000), who identified subgroups of DLBCL based on the systematic analysis of the patterns generated by a specialized cDNA microarray technique. The full data and experimental methods are available on the Web site of their research group (http://llmpp.nih.gov/lymphoma).

In this case, a SOM network was trained with 12,600 learning cycles, and the other learning parameters were defined as above. The data were normalised such that the mean and variance of the genes are set to 0 and 1 respectively. Figure 15.6 shows the clustering results based on two prototypes A and B.

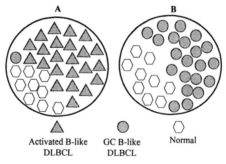

Figure 15.6. Expression data clustering using the SOM: Distinguishing subtypes of DLBCL.

Because both clusters A and B include Normal samples, this clustering configuration does not clearly distinguish Normal from DLBCL samples. The reader is referred to (Alizadeh et al., 2000) for a discussion on the relationships between the Normal and DLBCL samples in terms of their expression patterns, which are indicative of different stages of B-cell differentiation. Nevertheless, these clustering results represent relevant information to recognise the two subtypes of DLBCL reported by Alizadeh et al. (2000): Activated B-like DLBCL and germinal centre B-like

DLBCL (GC B-like DLBCL). In this case, cluster A can be labelled as the cluster representing Activated B-like DLBCL samples, and cluster B may be used to identify GC B-like DLBCL.

This section has dealt with the implementation and application of SOM networks for the analysis of expression data. The following section introduces some modifications to the original SOM, which may be useful to facilitate a knowledge discovery task based on this type of data.

3. SELF-ADAPTIVE AND INCREMENTAL LEARNING NEURAL NETWORKS FOR MICROARRAY DATA ANALYSIS

A number of research efforts have addressed some of the pattern processing and visualisation limitations exhibited by the original SOM. It has been shown how these limitations have negatively influenced several data mining, visualisation and cluster analysis applications (Alahakoon et al., 2000). A SOM system requires the user to predetermine the network structure and the number of prototypes. This trial-and-error task may represent a time-consuming and complex problem. Another important limitation is the lack of tools for the automatic detection of cluster boundaries. Different approaches have been proposed to improve the original SOM algorithm. Investigations have suggested the application of *self-adaptive and incremental learning neural networks* (SANN), instead of static topology networks in order to improve several data classification applications (Nour and Madey, 1996), (Fritzke, 1994).

Some of these approaches aim to determine the prototype composition, shape and size of the self-organizing structure during the learning process. These learning techniques are well adapted to application domains, such as expression analysis, which are characterised by incomplete data and knowledge.

Recent advances include a neural network model known as *Double Self-Organizing Map* (Su and Chang, 2001), which has been suggested for data projection and reduction applications. The *Fast Self-Organizing Feature Map* algorithm (Su and Chang, 2000) aims to automatically reduce the number of learning cycles needed to achieve a proper clustering process. Other authors have proposed to combine the SOM approach and advanced supervised learning techniques. One example is the *Supervised Network Self-Organizing Map* (sNet-SOM) (Papadimitriou et al., 2001). In this case a variant of SOM provides a global approximation of a data partition, while a supervised learning algorithm is used to refine clustering results in areas categorised as ambiguous or more critical for discovery purposes. Other models designed to implement automatic map generation and cluster boundary detection include the *Growing Cell Structure Network* (GCS) (Fritzke, 1994), the *Incremental Grid Growing Neural Network* (IGG) (Blackmore, 1995) and the *Growing Self-Organizing Map* (GSOM) (Alahakoon et al., 2000). The following subsection illustrates the application of one of these techniques to the problem of recognising relevant genomic expression patterns.

3.1   A GCS-Based Approach To Clustering Expression Data

GCS is an adapted version of the SOM, which has been applied to improve a number of pattern recognition and decision support systems (Azuaje et al., 1999), (Azuaje et al., 2000). One type of GCS can be described as a two-dimensional space, where its prototypes are inter-connected and organised in the form of triangles. An initial topology for the GCS is organised as one two-dimensional triangle (Figure 15.7.a). The connections between cells reflect their separation distance on the prototype space. Like in the original SOM, each cell is represented by a weight vector m_i, which is of the same dimension as the input data. At the beginning of the learning process the weight vectors are assigned random values. The learning process comprises the processing of input vectors and the adaptation of weight vectors, m_i. But unlike the SOM there is no need to define prototype neighbourhoods. Moreover, the learning rate, α, is substituted by two constant values, ε_w and ε_n, which represent the learning rates for the closest prototype to a sample (winning cell) and its neighbours respectively. The value of these learning rates ranges between 0 and 1.

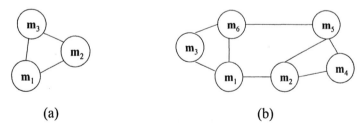

(a)   (b)

Figure 15.7. Growing Cell Structures. (a) An initial topology of GCS. (b) A GCS topology after a number of learning cycles.

GCS also performs an adaptation of the overall structure by inserting new cells into those regions that represent large portions of the input data (Fritzke, 1994). Also, in some cases, when one is interested in more accuracy or when the probability density of the input space consists of several separate regions, a better modelling can be obtained by removing those cells that do not contribute to the input data classification. This adaptation process is performed after a number of learning cycles. Figure 15.7.b depicts a typical GCS after performing a number of learning cycles. The reader is referred to (Fritzke, 1994) for a complete description of this algorithm. Section 4 discusses some of the advantages and limitations of this type of models.

In order to exemplify some of the differences between the SOM and the GCS clustering models the hypothetical classification problem described in Section 2.1 is retaken. Panel A of Figure 15.8 depicts the results that one may have obtained using a standard SOM, whose shape and size were defined by the user. Panel B of the same figure portrays the type of results that one may expect from a GCS clustering model. In this situation the insertion and deletion of cells allowed the categorisation of the two types of genes into two separated regions of cells. Thus, one major advantage is the automatic detection of cluster boundaries. Moreover, the distance between cells may be used as a measure of similarity between groups of genes.

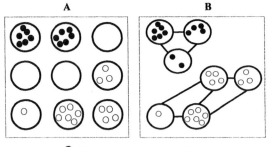

Figure 15.8. Comparing SOM-based (panel A) and GCS-based (panel B) clustering, using the hypothetical classification example introduced in Section 2.1

● Genes related to function X
○ Genes related to function Y

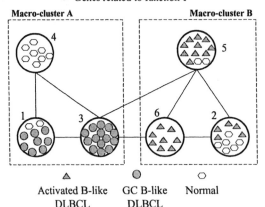

Figure 15.9. Expression data clustering using GCS: Distinguishing subtypes of DLBCL

Figure 15.9 shows the clusters obtained using a GCS and the DLBCL expression data presented in Section 2.2. The GCS network was trained with

2500 input presentation epochs (2,500 × 63 learning cycles), inserting a new cell every 500 epochs and deleting irrelevant cells every 1,000 epochs. The learning parameters, ε_w and ε_n, were equal to 0.095 and 0.010 respectively. For a complete description of this and other experiments the reader is referred to (Azuaje, 2001).

The resulting GCS consists of 6 cells or clusters containing the normal and DLBCL samples. The cell connections shown in Figure 15.9 do not reflect the weight vector distances. It shows that each cell corresponds to a representative cluster of the normal and DLBCL classes. For instance, cells 4 and 6 categorise only normal and DLBCL samples respectively. The majority of the samples recognised by Cells 1 and 5 belong to the class DLBCL. Cell 3 recognizes samples belonging only to the category DLBCL. Cells 1 and 3 comprise all of the GC B-like DLBCL subjects. Cells 2, 5 and 6 represent the clusters encoding the Activated B-like DLBCL subjects. Thus, this GCS network consists of two regions or *macro-clusters*, A and B, which identify the GC B-like and the Activated B-like DLBCL subjects respectively. Unlike the results obtained from the SOM-based clustering, the GCS was also able to separate normal from DLCBL samples (Cell 4). Further descriptions and experimental procedures can be implemented to validate the statistical (Azuaje, 2001) and biomedical significance of these results.

4. DISCUSSION

This chapter has introduced the application of self-organizing neural networks for the analysis of genomic expression data. Several studies have suggested the SOM model as a basic approach to expression clustering (Section 2). Some of its advantages were illustrated and alternative solutions based on advanced principles of network self-organization were overviewed. It has been indicated that the application of SANN (Section 3) may support a deeper comprehension of a pattern discovery problem. This chapter has illustrated how a SANN model called GCS may be implemented to specify interesting molecular patterns or confirm known functional categories.

SANN systems, such as GCS, offer several advantages in relation to the SOM and other expression data classification techniques. In contrast to the SOM, SANN structures are determined automatically from the expression data. Most of these models do not require the definition of time-dependence of decay schedule parameters. SANN's ability to insert and delete cells allows a more accurate estimation of probability densities of the input data. Its capacity to interrupt a learning process or to continue a previously interrupted one, permits the implementation of incremental clustering systems. SANN have demonstrated its strength to process both small and high dimensionality data in several application domains (Alahakoon et al., 2000), (Azuaje et al., 2000), (Papadimitriou et al., 2001). Some SANN may be implemented in either unsupervised or supervised learning modes (Fritzke, 1994), (Papadimitriou et al., 2001). However, there are important limitations that need to be addressed. For example, in the GCS model there is not a standard way to define a priori the number of learning cycles and the exact number of cells required to properly develop a network. Some models, such the GSOM (Alahakoon et al., 2000), partially address this problem by introducing *spread factors* to measure and control the expansion of a network. In a number of applications it has been shown that techniques, like GCS and IGG, may be more susceptible to variations in the initial parameter settings than the SOM clustering model (Blackmore, 1995), (Köhle and Merkl, 1996).

There are additional problems that merit further research in order to contribute to the advance of clustering-based genomic expression studies. Among them: The implementation of hierarchical clustering using SANN, faster clustering algorithms, specialised techniques for the processing of time-dependent or statistically-dependent data, and methods to automatically measure the contribution of a variable to the clustering results.

It is crucial to develop frameworks to assist scientists during the design and evaluation of clustering applications. Some of such guidelines and methods were examined in Chapter 13. Evaluation techniques may support not only the validation of clusters obtained from SOM, SANN or any other procedures, but also they may enable an effective and inexpensive mechanism for the automatic description of relevant clusters.

REFERENCES

Alahakoon D., Halgamuge S.K., Srinivasan B. (2000). Dynamic self-organizing maps with controlled growth for knowledge discovery. IEEE Transactions on Neural Networks 11: 601-614.

Alizadeh A.A., Eisen M.B., Davis R.E., Ma C., Lossos I.S., Rosenwald A., Boldrick J.C., Sabet H., Tran T., Yu X., Powell J.I., Yang L., Marti G.E., Moore T., Hudson J., Lu L., Lewis D.B., Tibshirani R., Sherlock G., Chan W.C., Greiner T.C., Weisenburger D.D., Armitage J.O., Warnke R., Levy R., Wilson W., Grever M.R., Bird J.C., Botstein D., Brown P.O., Staudt M. (2000). Distinct types of diffuse large B-cell lymphoma identified by gene expression profiling. Nature 403:503-511.

Azuaje F., Dubitzky W., Lopes P., Black N., Adamson K., Wu X., White J. (1998). Discovery of incomplete knowledge in electrocardiographic data. Proceedings of the Third International Conference on Neural Networks and Expert Systems in Medicine and Healthcare; 1998 September 2-4; Pisa. World Scientific: Singapore.

Azuaje F., Dubitzky W., Black N., Adamson K. (1999). Improving clinical decision support through case-based fusion. IEEE Transactions on Biomedical Engineering; 46: 1181-1185.

Azuaje F., Dubitzky W., Black N., Adamson K. (2000). Discovering relevance knowledge in data: a growing cell structure approach. IEEE Transactions on Systems, Man and Cybernetics, Part B: Cybernetics 30: 448-460.

Azuaje F. (2001). An unsupervised neural network approach to discovering gene expression patterns in B-cell lymphoma. Online Journal of Bioinformatics 1: 23-41.

Blackmore J. (1995). Visualizing high-dimensional structure with the incremental grid growing neural network. M.S. thesis, University of Texas at Austin.

Fritzke B. (1994). Growing cell structure--a self-organizing network for unsupervised and supervised learning. Neural Networks 7: 1441-1460.

Golub T.R., Slonim D.K., Tamayo P., Huard C., Gassenbeck M., Mesirov J.P., Coller H., Loh M.L., Downing J.R., Caligiuri M.A., Bloomfield C.D., Lander E.S. (1999). Molecular classification of cancer: class discovery and class prediction by gene expression monitoring. Science 286:531-537.

Hasti T., Tibshirani R., Friedman J. (2001). *The Elements of Statistical Learning*. NY: Springer.

Ideker T., Thorsson V., Ranish J.A., Christmas R., Buhler J., Eng J.K., Bumgarner R., Goodlett D.R., Aebersol R., Hood L. (2001). Integrated genomic and proteomic analyses of a systematically perturbated metabolic network. Science 292:929-933.

Köhle M., Merkl D. (1996). Visualizing similarities in high dimensional input spaces with a growing and splitting neural network. Proceedings of the International Conference of Artificial Neural Networks (ICANN'96), pp. 581-586.

Kohonen T. (2001). *Self-Organizing Maps*. Berlin: Springer.

Murtagh F., Hernández-Pajares M. (1995). The Kohonen self-organizing map method: an assessment. Journal of Classification 12:165-190.

Nour M.A., Madey G.R. (1996). Heuristic and optimization approaches to extending the Kohonen self organizing algorithm. European Journal of Operational research 93: 428-448.

Papadimitriou S., Mavroudi S., Vladutu L., Bezerianos A. (2001). Ischemia detection with a self-organizing map supplemented by supervised learning. IEEE Transactions on Neural Networks 12: 503-515.

Schmitz G., Ermert H., Senge T. (1999). Tissue-characterization of the prostate using radio frequency ultrasonic signals. IEEE Transactions on Ultrasonics, Ferroelectrics and Frequency Control 46: 126-138.

Su M.C, Chang H.T. (2000). Fast self-organizing feature map algorithm. IEEE Transactions on Neural Networks 11: 721-733.

Su M.C, Chang H.T. (2001). A new model of self-organizing neural networks and its application in data projection. IEEE Transactions on Neural Networks 12:153-158.

Tamayo P., Slonim D., Mesirov J., Zhu Q., Kitareewan S., Dmitrovsky E., Lander E.S., Golub R. (1999). Intepretating patterns of gene expression with self-organizing maps: methods and application to hematopoietic differentiation. Proc. Natl. Acad. Sci. USA 96:2907-2912.

Chapter 16

CLUSTERING OR AUTOMATIC CLASS DISCOVERY: NON-HIERARCHICAL, NON-SOM

Clustering Algorithms and Assessment of Clustering Results

Ka Yee Yeung

Department of Microbiology, University of Washington, Seattle, WA 98195, USA,
e-mail: kayee@cs.washington.edu

1. INTRODUCTION

DNA microarrays offer a global view on the levels of activity of many genes simultaneously. In a typical gene expression data set, the number of genes is usually such larger than the number of experiments. Even a simple organism like yeast has approximately six thousand genes. It is estimated that humans have approximately thirty thousand to forty thousand genes (Lander et al., 2001).

The goal of cluster analysis is to assign objects to clusters such that objects in the same cluster are more similar to each other while objects in different clusters are as dissimilar as possible. Clustering is a very well-studied problem, and there are many algorithms for cluster analysis in the literature. Please refer to (Anderberg, 1973), (Jain and Dubes, 1988), (Kaufman and Rousseeuw, 1990), (Hartigan, 1975) and (Everitt, 1993) for a review of the clustering literature. Because of the large number of genes and the complexity of biological networks, clustering is a useful exploratory technique for analysis of gene expression data.

In this chapter, we will examine a few clustering algorithms that have been applied to gene expression data, including *Cluster Affinity Search Technique* (CAST) (Ben-Dor and Yakhini, 1999), (Ben-Dor et al., 1999), *k*-means (MacQueen, 1965), (Tavazoie et al., 1999), *Partitioning Around Medoids* (PAM) (Kaufman and Rousseeuw, 1990), and *model-based clustering* (Fraley and Raftery, 1998), (Yeung et al., 2001a).

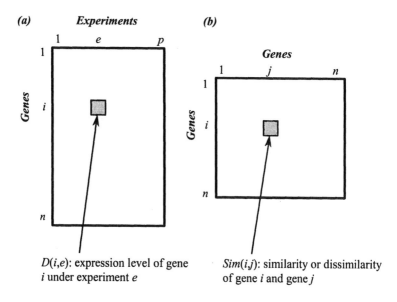

Figure 16.1. (a) A data matrix. (b) A similarity matrix.

2. BACKGROUND AND NOTATIONS

A data set containing objects to be clustered is usually represented in one of two formats: the data matrix or the similarity (or dissimilarity) matrix. In a data matrix, the rows usually represent objects, and the columns usually represent features or attributes of the objects. Suppose there are n objects and p attributes. We assume the rows represent genes and the columns represent experiments, such that entry (i, e) in the data matrix D represents the expression level of gene i under experiment e, where $1 \leq i \leq n$ and $1 \leq e \leq p$ (see Figure 16.1a). The ith row in the data matrix D (where $1 \leq i \leq n$), D_i, represents the expression vector of gene i across all p experiments. In clustering genes, the objects to be clustered are the genes. The similarity (or dissimilarity) matrix contains the pairwise similarity (or dissimilarity) of genes. Specifically, entry (i,j) in the similarity (or dissimilarity) matrix Sim represents the similarity (or dissimilarity) of gene i and gene j, where $1 \leq i, j \leq n$ (see Figure 16.1b). The similarity (or dissimilarity) of gene i and gene j can be computed using the expression vectors of gene i and gene j from the data matrix. Hence, the similarity (or dissimilarity) matrix Sim can be computed from the data matrix D. For the rest of the chapter, the objects to be clustered are the genes in a given gene expression data set unless otherwise stated.

3.   SIMILARITY METRICS

The measure used to compute similarity or dissimilarity between a pair of objects is called a *similarity metric*. Many different similarity metrics have been used in clustering gene expression data, among which the two most popular similarity metrics are *correlation coefficient* and *Euclidean distance*. Correlation coefficient is a similarity measure (a high correlation coefficient implies high similarity) while Euclidean distance is a dissimilarity measure (a high Euclidean distance implies low similarity).

The correlation coefficient between a pair of genes i and j ($1 \le i, j \le n$) is defined as

$$\sum_{e=1}^{p} \frac{(D(i,e) - \mu_i)(D(j,e) - \mu_j)}{(\| D_i \| \| D_j \|)} \tag{16.1}$$

where $\mu_i = \sum_{e=1}^{p} D(i,e)/p$ is the average expression level of gene i over all p experiments, and $\| D_i \| = \sqrt{\sum_{e=1}^{p} (D(i,e) - \mu_i)^2}$ is the norm of the expression vector D_i with the mean subtracted. Correlation coefficients range from -1 to 1. Two genes with correlation coefficient equal to 1 are perfectly correlated, i.e. their expression levels change in the same direction across the experiments. On the other hand, a correlation coefficient of -1 means that two genes are anti-correlated, i.e. their expression levels change in opposite directions. Geometrically, correlation coefficients capture the patterns of expression levels of two genes. For example, two genes with different average expression levels but with expression levels peaking at the same experiments have a high correlation coefficient.

The Euclidean distance between a pair of genes i and j ($1 \le i, j \le n$) is defined as

$$\sqrt{\sum_{e=1}^{p} (D(i,e) - D(j,e))^2} \tag{16.2}$$

A high Euclidean distance between a pair of genes indicates low similarity between the genes. Unlike correlation coefficients, Euclidean distances measure both the direction and amplitude difference in expression levels. For example, two genes peaking at the same experiments but with different average expression levels may lead to a large Euclidean distance, especially if the difference in average expression levels is high.

4.   CLUSTERING ALGORITHMS

There is a rich literature in clustering algorithms, and there are many different classifications of clustering algorithms. One classification is *model-based* versus *heuristic-based* clustering algorithms. The objects to be clustered are assumed to be generated from an underlying probability framework in the model-based clustering approach. In the heuristic-based approach, an underlying probability framework is not assumed. The inputs to a heuristic-based clustering algorithm usually include the similarity matrix and the number of clusters. CAST and PAM are examples of the heuristic-based approach. The *k*-means algorithm was originally proposed as a heuristic-based clustering algorithm. However, it was shown to be closely related to the model-based approach (Celeux and Govaert, 1992).

4.1   CAST

The *Cluster Affinity Search Technique* (CAST)} (Ben-Dor and Yakhini, 1999), (Ben-Dor et al., 1999) is a graph-theoretic algorithm developed to cluster gene expression data. In graph-theoretic clustering algorithms, the objects to be clustered (genes in this case) are represented as nodes, and pairwise similarities of genes are represented as weighted edges in a graph. The inputs to CAST include the similarity matrix *Sim*, and a threshold parameter *t* (which is a real number between 0 and 1), which indirectly controls the number of clusters.

4.1.1   Algorithm Outline

CAST is an iterative algorithm in which clusters are constructed one at a time. The current cluster under construction is called C_{open}. The affinity of a gene g, $a(g)$, is defined as the sum of similarity values between g and all the genes in C_{open}, i.e. $a(g) = \sum_{x \in C_{open}} Sim(g,x)$. A gene g is said to have high affinity if $a(g) \geq t \mid C_{open} \mid$. Otherwise, g is said to have low affinity. Note that the affinity of a gene depends on the genes that are already in C_{open}. When a new cluster C_{open} is started, the initial affinity is zero because C_{open} is empty. A gene not yet assigned to any clusters and having the maximum average similarity to all unassigned genes is chosen to be the first gene in C_{open}. The algorithm alternates between adding high affinity genes to C_{open}, and removing low affinity genes from C_{open}. C_{open} is closed when no more genes can be added to or removed from it. Once a cluster is closed, a new C_{open} is formed. The algorithm iterates until all the genes have been assigned to clusters and the current C_{open} is closed. After the CAST algorithm converges (assuming it does), there is an additional iterative step, in which all clusters are considered at the same time, and genes are moved to the cluster with the highest average similarity. For details of CAST, please refer to (Ben-Dor and Yakhini, 1999).

4.1.2 Algorithm Properties

Correlation coefficient is usually used as the similarity metric for CAST. From our experience, the iterative step in CAST may not converge if Euclidean distance is used as the similarity metric.

In contrast to the hierarchical clustering approach in which objects are successively merged into clusters, objects can be added to or removed from the current open cluster through the iterative steps. CAST tends to produce relatively high quality clusters, compared to the hierarchical approach (Yeung et al., 2001b).

4.2 *K*-means

K-means is another popular clustering algorithm in gene expression analysis. For example, Tavazoie et al. (1999) applied *k*-means to cluster the yeast cell cycle data (Cho et al., 1998).

4.2.1 Algorithm Outline

K-means (MacQueen, 1965) is a classic iterative clustering algorithm, in which the number of clusters, *k*, together with the similarity matrix are inputs to the algorithm. In the *k*-means clustering algorithm, clusters are represented by *centroids*, which are cluster centers. The goal of *k*-means is to minimize the sum of distances from each object to its corresponding centroid. In each iteration, each gene is assigned to the centroid (and hence cluster) with the minimum distance (or equivalently maximum similarity). After the gene re-assignment, new centroids of the *k* clusters are computed. The steps of assigning genes to centroids and computing new centroids are repeated until no genes are moved between clusters (and centroids are not changed). *K*-means was shown to converge for any metric (Selim and Ismail, 1984).

4.2.2 Effect of Initialization

Initialization plays an important role in the *k*-means algorithm. In the random initialization approach, the *k* initial centroids consist of *k* randomly chosen genes. An alternative approach is to use clusters from another clustering algorithm as initial clusters, for example, from hierarchical average-link. The advantage of the second approach is that the algorithm becomes deterministic (the algorithm always yields the same clusters). (Yeung et al., 2001b) showed that the iterative *k*-means step after the hierarchical step tends to improve cluster quality.

4.2.3 Algorithm Properties

Clusters obtained from the k-means algorithm tend to be equal-sized and spherical in shape. This is because the k-means algorithm is closely related to the equal volume spherical model in the model-based clustering approach (Celeux and Govaert, 1992).

4.2.4 Implementation

K-means is implemented in many statistical software packages, including the commercial software Splus (Everitt, 1994), and the GNU free software R (Ihaka and Gentleman, 1996). It is also available from other clustering packages tailored toward gene expression analysis, such as XCLUSTER from Gavin Sherlock, which is available at http://genome-www.stanford.edu/~sherlock/cluster.html.

4.3 PAM

Partitioning around Medoids (PAM) (Kaufman and Rousseeuw, 1990) searches for a representative object for each cluster from the data set. These representative objects are called *medoids*. The clusters are obtained by assigning each data point to the nearest medoid. The objective is to minimize the total dissimilarity of objects to their nearest medoid. This is very similar to the objective of k-means, in which the total dissimilarity of objects to their centroids is minimized. However, unlike centroids, medoids do not represent the mean vector of data points in clusters.

4.3.1 Algorithm Outline

The inputs to PAM include the similarity or dissimilarity matrix and the number of clusters k. The algorithm of PAM consists of two stages. In the first BUILD stage, an initial clustering is obtained by successive selection of representative objects until k objects are found. In the second SWAP stage, all pairs of objects (i, h), for which object i is in the current set of medoids and object h is not, are considered. The effect on the object function is studied if object h is chosen as a medoid instead of object i.

4.3.2 Algorithm Properties

PAM can be considered as a robust version of k-means since medoids are less affected by outliers. Similar to k-means, PAM also tends to produce spherical clusters (Kaufman and Rousseeuw, 1990).

4.3.3   Implementation

PAM is implemented in statistical packages such as Splus and R.

5.   ASSESSMENT OF CLUSTER QUALITY

We have discussed three different heuristic-based clustering algorithms to analyze gene expression data. Different clustering algorithms can potentially generate different clusters on the same data set. However, no clustering method has emerged as the method of choice in the gene expression community. A biologist with a gene expression data set is faced with the problem of choosing an appropriate clustering algorithm for his or her data set. Hence, assessing and comparing the quality of clustering results is crucial.

Jain and Dubes (1988) classified cluster validation procedures into two main categories: *external* and *internal criterion analysis*. External criterion analysis validates a clustering result by comparing to a given "gold standard", which is another partition of the objects. Internal criterion analysis uses information from within the given data set to represent the goodness of fit between the input data set and the clustering results.

5.1   External Validation

In external validation, a clustering result with a high degree of agreement to the "gold standard" is considered to contain high quality clusters. The gold standard must be obtained by an independent process based on information other than the given data. This approach has the strong benefit of providing an independent, hopefully unbiased assessment of cluster quality. On the other hand, external criterion analysis has the strong disadvantage that an external gold standard is rarely available.

Both clustering results and the external criteria can be considered as partitions of objects into groups. There are many statistical measures that assess the agreement between two partitions, for example, the adjusted Rand index (Hubert and Arabie, 1985). The adjusted Rand index is used to assess cluster quality in (Yeung and Ruzzo, 2001) and (Yeung et al., 2001a).

5.2   Internal Validation

Internal criterion analysis does not require any independent external criteria. Instead, it assesses the goodness of fit between the input data set and the clustering results. We will briefly describe three internal validation approaches.

5.2.1 Homogeneity and Separation

Since objects in the same cluster are expected to be more similar to each other than objects in different groups and objects in different clusters are expected to be dissimilar, *homogeneity* of objects in the same cluster and *separation* between different clusters are intuitive measures of cluster quality (Shamir and Sharan, 2001). Homogeneity is defined as the average similarity between objects and their cluster centers, while separation is defined as the weighted average similarity between cluster centers. A high homogeneity indicates that objects in clusters are similar to each other. A low separation means that different clusters are not well-separated.

5.2.2 Silhouette

Silhouettes can be used to evaluate the quality of a clustering result. Silhouettes are defined for each object (gene in our context) and are based on the ratio between the distances of an object to its own cluster and to its neighbor cluster (Rousseeuw, 1987). A high silhouette value indicates that an object lies well within its assigned cluster, while a low silhouette value means that the object should be assigned to another cluster. Silhouettes can also be used to visually display clustering results. The objects in each cluster can be displayed in decreasing order of the silhouette values such that a cluster with many objects with high silhouette values is a pronounced cluster. Silhouettes are implemented in Splus and R. In order to summarize the silhouette values in a data set with k clusters, the *average silhouette width*, is defined to be the average silhouette value over all the objects in the data. The average silhouette width can be used as an internal validation measure to compare the quality of clustering results.

5.2.3 Figure of Merit

Yeung et al. (2001b) proposed the *figure of merit* (FOM) approach to compare the quality of clustering results. The idea is to apply a clustering algorithm to all but one experiment in a given data set, and use the left-out experiment to assess the predictive power of the clustering algorithm.

Intuitively, a clustering result has possible biological significance if genes in the same cluster tend to have similar expression levels in additional experiments that were not used to form the clusters. We estimate this predictive power by removing one experiment from the data set, clustering genes based on the remaining data, and then measuring the within-cluster similarity of expression values in the left-out experiment. The figure of merit is a scalar quantity, which is an estimate of the predictive power of a clustering algorithm.

6.   MODEL-BASED APPROACH

Clustering algorithms based on probability models offer a principled alternative to heuristic algorithms. The issues of selecting a "good" clustering method and determining the "correct" number of clusters are reduced to model selection problems in the probability framework. This provides a great advantage over heuristic clustering algorithms, for which there is no rigorous method to determine the number of clusters or the best clustering method. (Yeung et al., 2001a) applied the model-based approach to various gene expression and synthetic data, and showed that the model-based approach tends to produce higher cluster quality than the heuristic-based algorithms.

6.1   The Model-based Framework

In *model-based clustering*, the data is assumed to be generated from a finite mixture of underlying probability distributions.[1] In other words, we assume the data consists of different groups (or components), and each group (or component) is generated from a known probability distribution. Based on this assumption, the goal of model-based clustering algorithms is to recover clusters that correspond to the components in the data.

There are many possible probability distributions underlying each group (or component). In this chapter, we assume a *Gaussian mixture model* in which each component is generated by the multivariate normal distribution (also known as the multivariate Gaussian distribution).[2] Gaussian mixture models have been shown to be a powerful tool for clustering in many applications, for example, (Banfield and Raftery, 1993), (Celeux and Govaert, 1993), (McLachlan and Basford, 1988), (MacLachlan and Peel, 2000).

The multivariate normal distribution is parameterized by the mean vector μ and covariance matrix Σ. When the objects to be clustered are the genes, the mean vector μ is of dimension p (which is the number of experiments). The mean vector of a component is equal to the average expression level of all the genes in that component. Hence, the mean vector represents the location where the component is centered at. The covariance matrix Σ is a p by p matrix such that $\Sigma(i, j)$ represents the covariance of experiment i and

---

[1] A probability distribution is a mathematical function which describes the probability of possible events.

[2] A multivariate normal distribution is a generalization of the normal distribution to more than one variable.

experiment j. The diagonal entries in the covariance matrix are the variances of the p experiments.[3]

Let G be the number of components in the data. In the Gaussian mixture assumption, each component k (where $k = 1,...,G$) is generated by the multivariate normal distribution with parameters μ_k (mean vector) and Σ_k (covariance matrix). The number of components, G, is assumed to be known. The goal is to estimate the parameters μ_k and Σ_k from the data (where $k = 1,...,G$), and find clusters corresponding to these parameter estimates.

In order to make estimation of the parameters easier, (Banfield and Raftery, 1993) proposed a general framework to decompose the covariance matrix

$$\Sigma_k = \lambda_k D_k A_k D_k^T, \qquad (16.3)$$

where D_k is an orthogonal matrix, A_k is a diagonal matrix, and λ_k is a scalar. The matrix D_k determines the orientation of the component, A_k determines its shape, and λ_k determines its volume. Hence, the covariance matrix Σ_k controls the shape, volume and orientation of each component.

Allowing some but not all of the parameters in Equation 16.3 to vary results in a set of models within this general framework. In particular, constraining $D_k A_k D_k^T$ to be the identity matrix I corresponds to Gaussian mixtures in which each component is spherical.

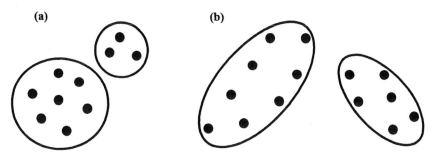

Figure 16.2. Fictitious examples illustrating (a) the unequal volume spherical model in which clusters are spherical but may have different volumes, and (b) the unconstrained model in which clusters may have different volume, orientation, and shape.

For example, the *equal volume spherical* model, which is parameterized by $\Sigma_k = \lambda I$, represents the most constrained model under this framework,

---

[3] The variance of an experiment is the average of the squared deviation of the experiment from its mean, while the covariance of two experiments measures their tendency to vary together.

with the smallest number of parameters.[4] The classical k-means clustering algorithm has been shown to be closely related to this model (Celeux and Govaert, 1992). However, there are circumstances in which this model may not be appropriate. For example, if some groups of genes are much more tightly co-regulated than others, a model in which the spherical components are allowed to have different volumes may be more appropriate. The unequal volume spherical model (see Figure 16.2a), $\Sigma_k = \lambda_k I$, allows the spherical components to have different volumes by allowing a different λ_k for each component k. We have also observed considerable correlation between experiments in time-series experiments, coupled with unequal variances. An elliptical model may better fit the data in these cases, for example, the unconstrained model (see Figure 16.2b) allows all of D_k, A_k and λ_k to vary between components. The unconstrained model has the advantage that it is the most general model, but has the disadvantage that the maximum number of parameters need to be estimated, requiring relatively more data points in each component. There is a range of elliptical models with other constraints, and hence requiring fewer parameters.

6.2   Algorithm Outline

Assuming the number of clusters, G, is fixed, the model parameters are estimated by the expectation maximization (EM) algorithm. In the EM algorithm, the expectation (E) steps and maximization (M) steps alternate. In the E-step, the probability of each observation belonging to each cluster is estimated conditionally on the current parameter estimates. In the M-step, the model parameters are estimated given the current group membership probabilities. When the EM algorithm converges, each observation is assigned to the group with the maximum conditional probability. The EM algorithm can be initialized with model-based hierarchical clustering (Dasgupta and Raftery, 1998), (Fraley and Raftery, 1998), in which a maximum-likelihood pair of clusters is chosen for merging in each step.

6.3   Model Selection

Each combination of a different specification of the covariance matrices and a different number of clusters corresponds to a separate probability model. Hence, the probabilistic framework of model-based clustering allows the issues of choosing the best clustering algorithm and the correct number of clusters to be reduced simultaneously to a model selection problem. This is important because there is a trade-off between probability model, and number of clusters. For example, if one uses a complex model, a small

---

[4] Only the parameter λ needs to be estimated to specify the covariance matrix for the equal model spherical model.

number of clusters may suffice, whereas if one uses a simple model, one may need a larger number of clusters to fit the data adequately.

Let D be the observed data, and M_k be a model with parameter θ_k. The *Bayesian Information Criterion* (BIC) (Schwarz, 1978) is an approximation to the probability that data D is observed given that the underlying model is M_k, $p(D \mid M)$.

$$2\log p(D \mid M_k) \approx 2\log p(D \mid \widehat{\theta}_k, M_k) - v_k \log(n) = BIC_k \qquad (16.4)$$

where v_k is the number of parameters to be estimated in model M_k, and $\widehat{\theta}_k$ is the maximum likelihood estimate for parameter θ_k.. Intuitively, the first term in Equation 16.4, which is the maximized mixture likelihood for the model, rewards a model that fits the data well, and the second term discourages overfitting by penalizing models with more free parameters. A large BIC score indicates strong evidence for the corresponding model. Hence, the BIC score can be used to compare different models.

6.4   Implementation

Typically, different models of the model-based clustering algorithm are applied to a data set over a range of numbers of clusters. The BIC scores for the clustering results are computed for each of the models. The model and the number of clusters with the maximum BIC score are usually chosen for the data. These model-based clustering and model selection algorithms (including various spherical and elliptical models) are implemented in MCLUST (Fraley and Raftery, 1998). MCLUST is written in Fortran with interfaces to Splus and R. It is publicly available at http://www.stat.washington.edu/fraley/mclust.

7.   A CASE STUDY

We applied some of the methods described in this chapter to the yeast cell cycle data (Cho et al., 1998), which showed the fluctuation of expression levels of approximately 6,000 genes over two cell cycles (17 time points). We used a subset of this data, which consists of 384 genes whose expression levels peak at different time points corresponding to the five phases of cell cycle (Cho et al., 1998). We expect clustering results to approximate this five-class partition. Hence, the five phases of cell cycle form the external criterion of this data set.

Before any clustering algorithm is applied, the data is pre-processed by standardization, i.e. the expression vectors are standardized to have mean 0 and standard deviation 1 (by subtracting the mean of each row in the data, and then dividing by the standard deviation of the row). Data pre-processing techniques are discussed in detail in Chapter 2.

We applied CAST, PAM, hierarchical average-link and the model-based approach to the standardized yeast cell cycle data to obtain 2 to 16 clusters. The clustering results are evaluated by comparing to the external criterion of the 5 phases of cell cycle, and the adjusted Rand indices are computed. The results are illustrated in Figure 16.3. A high-adjusted Rand index means high agreement to the 5-phase external criterion. The results from three different models from the model-based approach are shown in Figure 16.3: the equal volume spherical model (denoted by EI), the unequal volume spherical model (denoted by VI), and the unconstrained model (denoted by VVV). The equal volume spherical model (EI) and CAST achieved the highest adjusted Rand indices at 5 clusters. Figure 16.4 shows a silhouette plot of the 5 clusters produced using PAM.

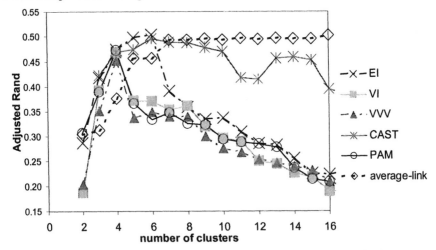

Figure 16.3. Adjusted Rand indices for the standardized yeast cell cycle data.

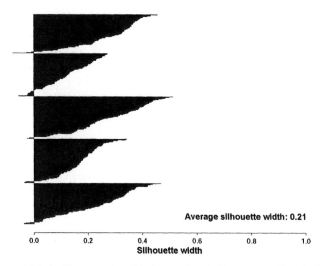

Figure 16.4. A silhouette plot of 5 clusters from PAM on the cell cycle data.

Three of the five clusters show higher silhouette values than the other two, and hence, they are relatively more pronounced clusters. In each cluster, there are a few genes with very low silhouette values, and they represent outliers in the clusters.

REFERENCES

Anderberg, M.R. (1973). Cluster analysis for applications. Academic Press.

Baneld, J.D. and Raftery, A.E. (1993). Model-based Gaussian and non-Gaussian clustering. Biometrics, 49:803-821.

Ben-Dor, A., Shamir, R., and Yakhini, Z. (1999). Clustering gene expression patterns. Journal of Computational Biology, 6:281-297.

Ben-Dor, A. and Yakhini, Z. (1999). Clustering gene expression patterns. In RECOMB99: Proceedings of the Third Annual International Conference on Computational Molecular Biology, pages 33-42, Lyon, France.

Celeux, G. and Govaert, G. (1992). A classification EM algorithm for clustering and two stochastic versions. Computational Statistics and Data Analysis, 14:315-332.

Celeux, G. and Govaert, G. (1993). Comparison of the mixture and the classification maximum likelihood in cluster analysis. Journal of Statistical Computation and Simulation, 47:127-146.

Cho, R.J., Campbell, M.J., Winzeler, E.A., Steinmetz, L., Conway, A., Wodicka, L., Wolfsberg, T.G., Gabrielian, A. E., Landsman, D., Lockhart, D. J., and Davis, R. W. (1998). A genome-wide transcriptional analysis of the mitotic cell cycle. Molecular Cell, 2:65-73.

Dasgupta, A. and Raftery, A.E. (1998). Detecting features in spatial point processes with clutter via model-based clustering. Journal of the American Statistical Association, 93:294-302.

Everitt, B. (1994). *A handbook of statistical analyses using S-plus*. Chap man and Hall, London.

Everitt, B.S. (1993). *Clustering Analysis*. John Wiley and Sons.

Fraley, C. and Raftery, A.E. (1998). How many clusters? Which clustering method? - Answers via model-based cluster analysis. The Computer Journal, 41:578-588.

Hartigan, J.A. (1975). *Clustering Algorithms*. John Wiley and Sons.

Hubert, L. and Arabie, P. (1985). Comparing partitions. Journal of Classification, 2:193-218.

Ihaka, R. and Gentleman, R. (1996). R: A language for data analysis and graphics. Journal of Computational and Graphical Statistics, 5(3):299-314.

Jain, A.K. and Dubes, R.C. (1988). Algorithms for Clustering Data. Prentice Hall, Englewood Cliffs, NJ.

Kaufman, L. and Rousseeuw, P.J. (1990). *Finding Groups in Data: An Introduction to Cluster Analysis*. John Wiley & Sons, New York.

Lander, E.S. et al. (2001). Initial sequencing and analysis of the human genome. Nature, 409(6822):860-921. International Human Genome Sequencing Consortium.

MacQueen, J. (1965). Some methods for classification and analysis of multivariate observations. In Proceedings of the 5th Berkeley Symposium on Mathematical Statistics and Probability, pages 281-297. McLachlan, G. J. and Basford, K. E. (1988). Mixture models: inference and applications to clustering. Marcel Dekker New York.

McLachlan, G.J. and Peel, D. (2000). *Finite Mixture Models*. New York: Wiley.

Rousseeuw, P. J. (1987). Silhouettes: a graphical aid to the interpretation and validation of cluster analysis. Journal of Computational and Applied Mathematics, 20:53-65.

Schwarz, G. (1978). Estimating the dimension of a model. Annals of Statistics, 6:461-464.

Selim, S.Z. and Ismail, M.A. (1984). *K*-means type algorithms: a generalized convergence theorem and characterization of local optimality. IEEE Transactions on Pattern Analysis and Machine Intelligence, PAMI-6(1):81-86.

Shamir, R. and Sharan, R. (2001). Algorithmic approaches to clustering gene expression data. In Current Topics in Computational Biology. MIT Press.

Tavazoie, S., Huges, J.D., Campbell, M.J., Cho, R.J., and Church, G.M. (1999). Systematic determination of genetic network architecture. Nature Genetics, 22:281-285.

Yeung, K.Y., Fraley, C., Murua, A., Raftery, A.E., and Ruzzo, W.L. (2001a). Model-based clustering and data transformations for gene expression data. Bioinformatics, 17:977-987.

Yeung, K.Y., Haynor, D.R., and Ruzzo, W.L. (2001b). Validating clustering for gene expression data. Bioinformatics, 17(4):309-318.

Yeung, K.Y. and Ruzzo, W.L. (2001). Principal component analysis for clustering gene expression data. Bioinformatics, 17:763-774.

Chapter 17

CORRELATION AND ASSOCIATION ANALYSIS

Simon M. Lin and Kimberly F. Johnson

Duke Bioinformatics Shared Resource, Box 3958, Duke University Medical Center, Durham, NC 27710,USA,
email: {Lin00025, Johns001}@mc.duke.edu

1.   INTRODUCTION

Establishing an association between variables is always of interest in the life sciences. For example, is increased blood pressure associated with the expression level of the angiotensin gene? Does the PKA gene down-regulate the Raf1 gene?

To answer these questions, we usually have a collection of objects (samples) and the measurements on two or more attributes (variables) of each object. Table 17.1 illustrates a collection of n pairs of observations (x_i, y_i).

Table 17.1. Input data format for studying the relationship between two variables.

Objects	Variable X	Variable Y
Object 1	x_1	y_1
Object 2	x_2	y_2
...	...	...
Object i	x_i	y_i
...	...	...
Object n	x_n	y_n

In this chapter, we first define types of variables, the nature of which determines the appropriate type of analysis. Next we discuss how to statistically measure the strength between two variables and test their significance. Then we introduce machine learning algorithms to find association rules. Finally, after discussing the association vs. causality inference, we conclude with a discussion of microarray applications.

2.   TYPES OF VARIABLES

Different statistical procedures are developed for different types of variables. Here we define three types of variables. *Nominal variables* are orderless non-numerical categories, such as sex and marital status. *Ordinal variables* are ordered categories; sometimes they are also called *rank variables*. Different from nominal and ordinal variables, *metric variables* have numerical values that can be mathematically added or multiplied. Examples of different types of variables are in Table 17.2.

Table 17.2. Examples of different types of variables.

Type of Variable	Examples
Nominal	smoking history (yes / no), eye color (green / black / brown)
Ordinal	aggressiveness of the tumor (+ / ++ / +++),
	birth weight (low / medium / high)
Metric	blood pressure, gene expression intensity

Sometimes it is convenient to convert the metric variables into ordinal variables. For example, rather than using the exact expression values of each gene, we discretize them into high, medium, and low values (Berrar et al., 2002). Although this conversion will lose some information from the original data, it makes the computation efficient and tractable, (Chang et al., 2002) or allows the use of algorithms for ordinal data (see Section 3.4; Chang et al., 2002).

3.   MEASUREMENT AND TEST OF CORRELATION

In many situations it is often of interest to measure the degree of association between two attributes (variables) when we have a collection of objects (samples). Sample correlation coefficients provide such a numerical measurement. For example, *Pearson's product-moment correlation coefficient* ranges from -1 to $+1$. The magnitude of this correlation coefficient indicates the strength of the linear relationship, while its sign indicates the direction. More specifically, -1 indicates perfect negative linear correlation; 0 indicates no correlation; and $+1$ indicates perfect positive linear correlation. According to the type of the variables, there are different formulas for calculating correlation coefficients. In the following sections, we will first discuss the most basic *Pearson's product moment correlation coefficient* for a metric variable; then, rank-based correlation coefficients including *Spearman's rho* and *Kendall's tau*; and finally, *Pearson's contingency coefficient* for nominal variables. Conceptually, there is a difference between the sample statistics vs. the true population parameters (Sheskin, 2000); for example, the sample correlation coefficient *r* vs. the population correlation coefficient ρ. In the following discussion we focus on the summary statistics of the samples.

3.1   Pearson's Product-Moment Correlation Coefficient

For metric variables, the *Pearson's product-moment correlation coefficient* (or Pearson's rho) is calculated according to the following formula:

$$r_p = \frac{\sum_{i=1}^{n}(x_i - \bar{x})(y_i - \bar{y})}{\sqrt{\sum_{i=1}^{n}(x_i - \bar{x})^2}\sqrt{\sum_{i=1}^{n}(y_i - \bar{y})^2}} \tag{17.1}$$

where $\bar{x}$ and $\bar{y}$ are the average of variable x and y, respectively. This formula measures the strength of a *linear relationship* between variable x and y. Since r_p is the most commonly used correlation coefficient, most of the time it is referred to simply as r.

3.2   Spearman's Rank-Order Correlation Coefficient

For ordinal variables, *Spearman's rank-order correlation coefficient* (or, Spearman's rho), is given by

$$r_s = 1 - 6\sum_{i=1}^{n}\frac{d_i^2}{n^3 - n} \tag{17.2}$$

where n is the total number of objects, and d_i is the difference between the variable pair of rankings associated with the ith object. Actually, r_s is simplified from Pearson's product-moment correlation coefficient r_p when the variable values are substituted with ranks.

Spearman's rho measures the degree of *monotonic relationship* between two variables. A relationship between two variables x and y is monotonic if, as x increases, y increases (monotonic increasing) or as x decreases, y decreases (monotonic decreasing).

3.3   Kendall's Tau

As an alternative to Spearman's rho, *Kendall's tau* measures the proportional concordant pairs minus the proportional discordant pairs in samples. A pair of observations (x_i, x_j) and (y_i, y_j) is called *concordant* when the product $(x_i - x_j)(y_i - y_j)$ is positive; and called *discordant* when the product is negative. Kendall's tau is defined as

$$\widetilde{\tau} = \frac{n_c - n_d}{n(n-1)/2} \tag{17.3}$$

where n_c is the number of concordant pairs of ranks, n_d is the number of discordant pairs of ranks, and $n(n-1)/2$ is the total number of possible pairs of ranks.

3.4   Comparison of Different Correlation Coefficients

Before we discuss the correlation coefficient for nominal variables, we first compare different correlation coefficients for ordinal and metric variables, since metric variables have the option of either using Pearson's product-moment correlation, or being converted to rank order first, and then treated as ordinal variables.

Pearson's rho measures the strength of a *linear* relationship (Figure 17.1a and Figure 17.1b), whereas Spearman's rho and Kendall's tau measure any *monotonic* relationship between two variables (Figure 17.1a, b, c and Table 17.2). If the relationship between the two variables is non-monotonic, all three correlation coefficients fail to detect the existence of a relationship (Figure 17.1e).

Both Spearman's rho and Kendell's tau are rank-based non-parametric measures of association between variable X and Y. Although they use different logic for computing the correlation coefficient, they seldom lead to markedly different conclusions (Siegel & Castellan, 1988).

The rank-based correlation coefficients are more robust against outliers. In Figure 17.1f, the data set is the same as in Figure 17.1d, except three outliers. As shown in Table 17.3, Spearman's rho and Kendall's tau are more robust against these outliers, whereas Pearson's rho is not.

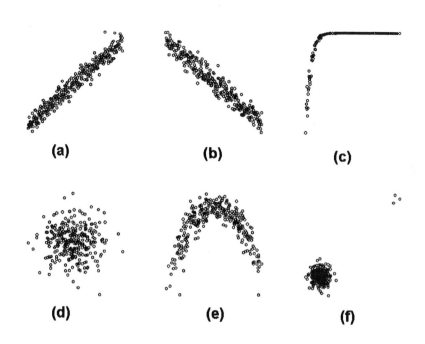

Figure 17.1. Relationships between two variables. (a) positive linear correlation; (b) negative linear correlation; (c) and (e) are non-linear relationships; (d) no relationship; and (f) is the same data set as (d) but with three outliers and a different scaling.

Table 17.3. Correlation coefficients in Figure 17.1.

Dataset	Pearson's rho	Spearman's rho	Kendall's tau
Figure 17.1(a)	0.98	0.98	0.87
Figure 17.1(b)	-0.98	-0.98	-0.87
Figure 17.1(c)	0.50	0.99	0.98
Figure 17.1(d)	-0.02	-0.03	-0.02
Figure 17.1(e)	-0.06	-0.02	-0.02
Figure 17.1(f)	0.68	0.00	0.00

3.5 Pearson's Contingency Coefficient

Pearson's contingency coefficient (c) is a measurement of association for nominal data when they are laid out in a contingency table format. c is defined as:

$$c = \sqrt{\frac{\chi^2}{\chi^2 + n}} \qquad (17.4)$$

where χ^2 is the computed chi-square value (Sheskin, 2000) that measures how far the observed frequencies deviate from the expected frequencies, and n is the total number of observations in the contingency table. The range of c is from 0 to 1, where zero indicates no association.

3.6   Statistical Tests of Correlation

After assessing the strength of association in the observations, we would like to know whether the association in the observed samples can be generalized to the population. For example, the relationship between blood pressures and angiotensin gene expression levels in the 10 mice we studied may be interesting, but are they representative of the entire mouse population? Sometimes we rephrase this question by asking whether the results from the observations could have occurred by chance alone, or whether some systematic effect produced the results.

Different statistical test procedures are designed for answering this question. Details of the test procedures of Pearson's product-moment correlation coefficient, Spearman's rho, Kendall's tau, and chi-square test can be found in (Sheskin, 2000). They are easily accessible from any statistical package such as S-Plus (Seattle, WA), SAS (Cary, NC), SPSS (Chicago, IL) or R (Ihaka & Gentleman, 1996).

4.   MINING ASSOCIATION RULES

In the previous section, we discussed how to statistically measure and test the association between two variables. That approach is only applicable when we know which two variables are of interest. In other words, we must have a hypothesis before hand. That scenario is not always true for microarray and other genomic studies, where we have a huge data set but little prior knowledge of the relationship among the variables. Thus, the methodologies from *Knowledge Discovery in Databases (KDD)*, or *data mining*, should also be discussed here. Instead of testing a specific hypothesis of association, *association rules mining* algorithms discover the intrinsic regularities in the data set.

The goal of association rules mining is to extract understandable rules (association patterns) from a given data set. A classic example of association rules discovery is market basket analysis. It determines which items go together in a shopping cart at a supermarket. If we define a *transaction* as a set of *items* purchased in a visit to a supermarket, association rules discovery can find out which products tend to sell well together. Marketing analysts can use these association rules to rearrange the store shelves or to design attractive package sales, which makes market basket analysis a major success in business applications.

4.1 Association Rules

Definition 17.1. *Association Rules*
Let $I = \{i_1, i_2, \ldots, i_m\}$ be a set of m items. Let D be a data set of n business "transactions" as defined above $T = \{t_1, t_2, \ldots, t_n\}$, where each "transaction" t_j consists of a set of items such that $t_j \subseteq I$. Note, that the items in each transaction are expressed as *yes* (present) or *no* (absent); the quantity of items bought is not considered. The rule describing an association between the presence of certain items is called an *association rule*. The general form of an association rule is $A \rightarrow B$, indicating "if A, then B", where $A \subseteq I$, $B \subseteq I$, and $A \cap B = \{\}$. A and B are item sets that can be either single items or a combination of multiple items. A is called the *rule body* (also called an *if-clause*) and B is the *rule head* (also called a *then-clause*). The breakdown of transactions according to whether A and B are true are given in Table 17.4.

Table 17.4. The contingency table of transactions according to A and B.

Number of transactions	A is present	A is absent
B is present	d	e
B is absent	f	g

The total number of transactions is $n = d + f + e + g$

Definition 17.2. *Support*
For the association rule $A \rightarrow B$, the *support* (also called *coverage*) is the percentage of transactions that contain all the items in A and B. It is the joint probability of finding all items in A and B in the same transaction.

$$Support = \frac{d}{n} \cdot 100\% = \frac{number\ of\ transactions\ involving\ all\ items\ in\ A\ and\ B}{total\ number\ of\ transactions} \cdot 100\% \qquad (17.5)$$

Definition 17.3. *Confidence*
For the association rule $A \rightarrow B$, the *confidence* (also called *accuracy*) is the number of transactions where B occurs along with A as a percentage of all transactions where A occurs, with or without B. It is the conditional probability within the same transaction of finding items in B, given the fact that the items were found in A.

$$Confidence = \frac{d}{d + f} \cdot 100\% = \frac{number\ of\ transactions\ involving\ all\ items\ in\ B\ and\ A}{number\ of\ transactions\ involving\ all\ items\ in\ A} \cdot 100\% \qquad (17.6)$$

To better understand the association rules and their measure of confidence and support, let us consider a simple supermarket data set.

$$D = \begin{bmatrix} transaction & i_1 = milk & i_2 = snack & i_3 = bread & i_4 = magazine & i_5 = butter \\ t_1: & Yes & No & \underline{\underline{Yes}} & No & \underline{Yes} \\ t_2: & No & Yes & No & Yes & No \\ t_3: & No & Yes & Yes & Yes & No \\ t_4: & \underline{Yes} & No & No & No & \underline{Yes} \\ t_5: & \underline{Yes} & No & \underline{\underline{Yes}} & No & \underline{Yes} \\ t_6: & \underline{Yes} & No & \underline{\underline{Yes}} & No & \underline{Yes} \end{bmatrix}$$

An association rule can be presented as

$$\{butter, milk\} \rightarrow \{bread\} \qquad \text{(confidence: 75\%, support: 50\%)},$$

where the rule body $A = \{butter, milk\}$ and the rule head $B = \{bread\}$. The contingency table for the rule body and the rule head in this example is given shown in Table 17.5.

Table 17.5. Contingency table for the supermarket data set.

Number of transactions	$A = \{butter, milk\}$ is present	$A = \{butter, milk\}$ is absent
$B = \{bread\}$ is present	$d = 3$	$e = 1$
$B = \{bread\}$ is absent	$f = 1$	$g = 1$

The total number of transactions is $n = 6$.

This rule means that a person who buys both butter and milk is also very likely to buy bread. As we can see from this example, support and confidence further characterize the importance of the rule in the data set.

The support of the rule is the percentage of transactions where the customers buy all three items: butter, milk and bread.

$$\text{support} = \frac{\#\{t_1, t_5, t_6\}}{\#\{t_1, t_2, t_3, t_4, t_5, t_6\}} \cdot 100\% = \frac{3}{6} \cdot 100\% = 50\%$$

(# indicates the number of elements in the set; { } indicates the set.)

A higher percentage of the support will ensure the rule applies to a significant fraction of the records in the data set. In other words, support indicates the relative frequency with which both the rule body and the rule head occur in the data set.

However, support alone is not enough to measure the usefulness of the rule. It may be the case that a considerable group of customers buy all three items, but there is also a significant amount of people who buy butter and milk but not bread. Thus, we need an additional measure of confidence in the rule.

In our example, the confidence is the percentage of transactions for which bread holds, within the group of records for which butter and milk hold. The confidence indicates to what extent the rule predicts correctly.

$$\text{confidence} = \frac{\#\{t_1, t_5, t_6\}}{\#\{t_1, t_4, t_5, t_6\}} \cdot 100\% = \frac{3}{4} \cdot 100\% = 75\%$$

Both support and confidence are represented by a number from 0% to 100%. A user-defined threshold (for example, 10% for support, and 70% for confidence) is used as a cutoff point for the mining algorithm to extract the rules from the data set (see the next section).

4.2 Machine Learning Algorithms for Mining Association Rules

Association rules can be discovered by the Apriori algorithm (Agrawal et al., 1996) that involves two main steps. The first step is to find item sets whose support exceeds a given minimum threshold, and these item sets are called frequent item sets. The second step is to generate the rules with the minimum confidence threshold. Mining association rules in large data sets can be both CPU and memory-demanding (Agrawal & Shafer, 1996). For a data set with many items, checking all the potential combinations for the items in step one is computationally expensive, since the search space can be very large (Hipp et al., 2000). There have been many attempts to efficiently search the solution space by either breadth-first or depth-first algorithms. Besides the commonly used Apriori algorithm, there are improvements such as parallel algorithms (Agrawal and Shafer, 1996), Direct Hashing and Pruning (DHP) (Park et al., 1997), Dynamic Itemset Counting (DIC) (Brin et al., 1997), FP-growth (Han et al., 2000), and Eclat (Zaki et al., 1997). For a review of different association rules mining algorithms, see Zaki (2000) and Hipp et al. (2000) For a discussion on generalizing association rules to handle interval data, see Han & Kamber (2001). Finding association rules is a data mining procedure, but not in the framework of statistical testing of associations. Silverstein et al. (1998) discussed how to utilize the chi-square test for independency in association rules mining. Association rules mining have been implemented in many data mining packages such as Clementine (SPSS, Chicago, IL), PolyAnalyst (Megaputer, Bloomington, IN), and Intelligent Miner (IBM, Armonk, NY).

5.   CAUSATION AND ASSOCIATION

It is always tempting to jump to cause-and-effect relationships when observing an association.

In a properly designed *experimental study*, we should often be able to infer causal relationships. For example, if we manipulate the expression of *gene A* in animals, and then measure their blood pressure, we might be able to establish that a high level of *gene A* increases blood pressure. In an *observational study*, on the other hand, we have no control over the values of the variables in the study. Instead we are only able to observe the two variables as they occur in a natural environment. For example, we randomly pick several mice from the population and measure their expression levels for *gene A* and blood pressure. In this case, we are not able to infer the causal relationship. It could be that *gene A* increases blood pressure, or the increased level of *gene A* is not the cause but the consequence of increased blood pressure.

Generally, observation of association from a stochastic process does not warrant causal relationships among the variables. In such cases, the hypothesis on causation may not be derived from the data itself. Thus, once we find the associated items, we must further analyze them to determine what exactly causes their nonrandom associations.

The explanations of association are illustrated in Figure 17.2.

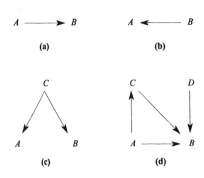

Figure 17.2. Mechanistic explanations of the association between *A* and *B* could be (a) *A* causes *B*; (b) *B* causes *A*; (c) there is a common cause *C* for both *A* and *B*; or, (d) there is a complex causality relationship among *A*, *B*, *C*, and *D*.

For further discussion of computational methodologies to discover causal relationships, the reader should refer to Glymour & Cooper (1999), and Chapter 8 of this book.

6.   APPLICATIONS IN MICROARRAY STUDIES

A number of supervised and unsupervised machine learning methodologies have been applied to microarray studies. Correlation coefficients have been the building blocks for many of these studies, including algorithms from simple clustering to the real challenge of inferring the topology of a genetic regulation networks.

6.1 Using Correlation Coefficients for Clustering

Clustering has been a commonly used exploratory tool for microarray data analysis. It allows us to recognize co-expression patterns in the data set. The goal of clustering is to put similar objects in the same group, and at the same time, separate dissimilar objects. Correlation coefficients have been used in many clustering algorithms as a similarity measure. For example, in the Eisen's Cluster program (Eisen et al., 1998), Pearson's rho, Spearman's rho, and Kendall's tau are available for hierarchical clustering.

6.2 Associating Profiles, or Associating Genes with Drug Responses

Hughes et al. (2000) utilized Pearson's product moment correlation coefficient (r) to conclude that deletion mutant of CUP5 and VMA8, both of which encode components of the vacuolar H+-ATPase complex, shared very similar microarray profiles with an $r = 0.88$; as a contrast, when the CUP5 mutant is compared with an unrelated mutant of MRT4, the correlation coefficient is $r = 0.09$.

Correlation studies have been used as a major strategy to mine the NCI-60 anticancer drug screening databases (Scherf et al., 2000). Readers can use the online tools at http://discover.nci.nih.gov/arraytools/ to correlate gene expression with anticancer drug activities.

In Figure 17.3, we demonstrate that the anticancer activity of the drug L-asparaginase is negatively correlated with the asparagine synthetase gene. It corresponds with our pharmacological knowledge that L-asparaginase is an enzyme that destroys asparagines that are required by the malignant growth of tumor cells, whereas asparagine synthetase produced more asparagines in the cell.

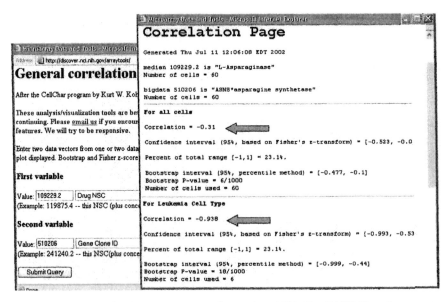

Figure 17.3. Correlation between L-asparaginase (drug NSC ID: 109229.2) and asparagine synthetase (gene clone ID: 510206). The negative correlation is more evident for leukemia cell lines.

6.3   Inferring Genetic Networks

The problem of inferring a genetic network can be formulated as following: given a data set of n genes which are the nodes in the network, and an n by m data matrix where each row is a gene (a total of n genes) and each column is a microarray experiment (a total of m experiments), find the network regulatory relationships among the genes using efficient inference procedures; i.e., finding the connecting topology of the nodes. This problem is also called *reverse-engineering of the genetic network.*

Inferring the genetic network from experimental data is a very difficult task. One simplification is to assume that, if there is an association between the two genes, then it is more likely that the two genes are connected in the genetic network. This way, the edges that connect the nodes (i.e. the regulations between the genes) could be inferred.

Lindlof and Olsson (2002) have used the correlation coefficient to build up the genetic network in yeast. Waddell and Kishino (2000) have explored the possibility of constructing the genetic network using partial correlation coefficients. All of these studies are preliminary, and they have not yet met the expectation of experimental biologists.

6.4 Mining Association Rules from Genomic Data

As opposed to using the network metaphor, the genetic regulatory machinery can also be expressed as a set of rules. For example, we can express the knowledge of genes A, B and C in the following form:

$$\{gene\ A, gene\ B\} \rightarrow \{gene\ C\}$$

to indicate if *gene A* and *gene B* are turned on, then *gene C* is also turned on.

Thus, reverse engineering the genetic network can also be formulated as deducing rule sets from data sets, where the association rules mining algorithms are applicable.

To apply association rules mining algorithms to microarray data sets, we can treat each microarray experiment as a transaction and each gene as an item. However, there are problems associated with this approach. First, the expression data have to be discretized since many mining algorithms can only efficiently interpret discretized variables. Secondly, unlike market basket analysis, for microarray data the number of items (genes) is much larger than the number of transactions (microarray experiments). This implies a huge search space when the mining algorithm tries to enumerate the possible combinations in the item sets. For example, even only selecting 4 genes as a set from a data set containing 10,000 genes, results in 4.16×10^{14} choices. Even with efficient pruning strategies, this algorithm still requires an immense amount of CPU time and memory. Thus, it is a novel computational challenge to apply association mining in microarray data.

Berrar et al. (2002) discussed a strategy for discretizing the expression values by a separability score method; as well as a strategy for feature selection to reduce the search space for the Apriori algorithm. An application to the NCI-60 anticancer drug screening data set was demonstrated.

Table 17.6. An example data set of 5 cell lines, 2 cancer classes, 3 genes, and 3 drugs.

Cell line #	Class	Genes			Drugs		
		Gene X	Gene Y	Gene Z	Drug A	Drug B	Drug C
1	CNS	H	H	H	H	H	H
2	CNS	H	H	H	H	H	H
3	BR	L	M	L	L	L	M
4	BR	H	H	H	H	H	H
5	BR	L	L	L	L	M	L

(From Berrar *et al.*, 2002. H, M, and L indicate high, medium and low in value, respectively. *CNS*: central nervous system; *BR*: breast)

As illustrated in Table 17.6, after discretizing the numerical data into high (H), medium (M) and low (L) values, a rule can be derived to indicate the intrinsic regularities in this data set:

```
if    Gene_X = H and Gene_Y = H and Gene_Z = H
   and Drug_A = H and Drug_B = H and Drug_C = H
   then   Class = CNS.
(coverage: 3/5 (60%); accuracy: 2/3 (67%))
```

Chen et al. (2001) applied the Apriori mining algorithm to find the association between transcription factors and their target genes. With an ad hoc pruning strategy before and after running the Apriori algorithm, false-positive results are reduced.

Aussem and Petit (2002) applied function dependency inference, which is closely related to the idea of association rules mining, to a reduced data set of Saccharomyces cerevisiae cell cycle data. They were able to successfully recover some of the known rules.

Berrar et al. (2001) described a case study where the general purpose Apriori algorithm failed due to the complexity of the microarray data. A new algorithm called the *maximum association algorithm*, tailored to the needs of microarray data, was presented. Its aim is to find all sets of associations that apply to 100% of the cases in a genetic risk group. By using this new algorithm, Berrar et al. found that stathmin is 100% over-expressed in the del (17p) subgroup of B-cell chronic lymphocytic leukemia, a subgroup with a lower survival rate. Stathmin is a signaling molecule that relays and integrates cellular messages of growth and proliferation. It was previously found up-regulated and associated with many high-grade leukemias (Roos et al., 1993). This rule is also of therapeutics interest, since antisense RNA inhibition of stathmin can reverse the malignant phenotype of leukemic cells (Jeha et al., 1996).

Association rules mining is often used as an exploratory method when one does not know what specific patterns to look for in a large data set. The discovered rules always deserve further analysis or experimental tests. With the increased ability to extract rules from large data sets, efficient and user-friendly post-processing of the rules are necessary, since many of the discovered rules are either trivial or irrelevant (Klemettinen et al., 1994).

6.5   Assessing the Reliability and Validity of High-Throughput Microarray Results

Correlation analysis can be used to asses the reliability and validity of microarray measurements. By reliability we mean reproducibility. If two microarray measurements conducted from the same sample on two occasions are not correlated, then an error in the measurement system is indicated. Taniguchi et al. (2001) reported that replicated DNA microarray measurements had a Pearson correlation coefficient between 0.984 and 0.995.

In addition to determining the reliability of microarray measurements, correlation analysis can be used to assess the validity of the measurements. By validity, we mean the extent to which our microarray measurements agree with another "standard" methodology. RT-PCR and Northern blots are often considered as the conventional standards for mRNA determination. For example, Zhou et al. (2002) utilized a Pearson correlation coefficient to indicate their microarray results as consistent with Northern blots with an average $r = 0.86$.

It is important to note that a set of measurements can be reliable without being valid. Microarray measurements could be highly reproducible, with a high test-retest correlation, but the measurements could turn out to be systematically biased. For example, Taniguchi et al. (2001) suggested that the sensitivity of the DNA microarrays was slightly inferior to that of Northern blot analyses.

6.6   Conclusions

Correlation coefficients provide a numerical measurement of the association between two variables. They can be used to determine the similarly between two objects when they are merged into a cluster; to assess the association between two gene expression profiles; to establish a connection between two genes in a genetic network; or to asses the agreement between two experimental methodologies.

Mining association rules for microarray data are a novel research challenge. In terms of feasibility, they might require a considerable amount of CPU time and computer memory. In reality, they potentially generate too many trivial or irrelevant rules for biological usefulness. In addition, some established algorithms for market basket analysis do not satisfy the challenge of microarray data (Berrar et al., 2001). However, as an exploratory data analysis tool, the association rules mining technique provides new insights into finding the regularities in large data sets. We expect much more research in this area.

ACKNOWLEDGEMENTS

The authors thank Jennifer Shoemaker and Patrick McConnell for valuable discussions.

REFERENCES

Agrawal R., Mannila H., Srikant R., Toivonen H., and Verkamo. I. C. (1996). Fast discovery of association rules. In "Advances in knowledge discovery and data mining" (U. M. Fayyad, Ed.), pp. 307-328, AAAI Press : MIT Press, Menlo Park, CA.

Agrawal R., and Shafer J. C. (1996). Parallel mining of association rules. IEEE Transactions on Knowledge and Data Engineering 8: 962-969.

Aussem A., and Petit J.-M. (2002). Epsilon-functional dependency inference: application to DNA microarray expression data. In Proceedings of BDA'02 (French Database Conference), Evry, France.

Berrar D., Dubitzky W., Granzow M., and Eils R. (2002). Analysis of Gene Expression and Drug Activity Data by Knowledge-based Association Mining. In Proceedings of CAMDA 02, Durham, NC, http://www.camda.duke.edu/CAMDA01/papers.asp.

Berrar D., Granzow M., Dubitzky W., Stilgenbauer S., Wilgenbus, K. D. H., Lichter P., and R. E. (2001). New Insights in Clinical Impact of Molecular Genetic Data by Knowledge-driven Data Mining. In Proc. 2nd Int'l Conference on Systems Biology, pp. 275-281, Omnipress.

Brin S., Motwani R., Ullman J. D., and Tsur S. (1997). Dynamic itemset counting and implication rules for market basket data. In "IGMOD Record (ACM Special Interest Group on Management of Data).

Chang J.-H., Hwang K.-B., and Zhang B.-T. (2002). Analysis of Gene Expression Profiles and Drug Activity Patterns by Clustering and Bayesian Network Learning. In Methods of microarray data analysis II (S. M. Lin, and K. F. Johnson, Eds.), Kluwer Academic Publishers.

Chen R., Jiang Q., Yuan H., and Gruenwald L. (2001). Mining association rules in analysis of transcription factors essential to gene expressions. In Proceedings of CBGIST 2001, Durham, NC.

Eisen M. B., Spellman P. T., Brown P. O., and Botstein D. (1998). Cluster analysis and display of genome-wide expression patterns. Proc Natl Acad Sci USA 95:14863-8.

Glymour C. N., and Cooper G. F. (1999). *Computation, causation, and discovery*. MIT Press, Cambridge, Mass.

Han J., and Kamber M. (2001). *Data mining: concepts and techniques*. Morgan Kaufmann Publishers, San Francisco.

Han J., Pei J., and Yin Y. (2000). Mining frequent patterns without candidate generation. In ACM SIGMOD Intl. Conference on Management of Data, ACM Press.

Hipp J., Guntzer U., and Nakaeizadeh G. (2000). Algorithms for Association Rule Mining - A General Survey and Comparison. In Proc. ACM SIGKDD International Conference on Knowledge Discovery and Data Mining.

Hughes T. R., Marton M. J., Jones A. R., Roberts C. J., Stoughton R., Armour C. D., Bennett H. A., Coffey E., Dai H., He Y. D., Kidd M. J., King A. M., Meyer M. R., Slade D., Lum P. Y., Stepaniants S. B., Shoemaker D. D., Gachotte D., Chakraburtty K., Simon J., Bard

M., and Friend S. H. (2000). Functional discovery via a compendium of expression profiles. Cell 102: 109-26.

Ihaka R., and Gentleman R. (1996). R: A language for data analysis and graphics. Journal of Computational and Graphical Statistics 5:299-314.

Jeha S., Luo X. N., Beran M., Kantarjian H., and Atweh G. F. (1996). Antisense RNA inhibition of phosphoprotein p18 expression abrogates the transformed phenotype of leukemic cells. Cancer Res 56:1445-50.

Klemettinen M., Mannila H., Ronkainen P., Toivonen H., and Verkamo A. I. (1994). Finding interesting rules from large sets of discovered association rules. In Third International Conference on Information and Knowledge Management (CIKM' 94), pp. 401-407, ACM Press.

Lindlof A., and Olsson B. (2002). Could correlation-based methods be used to derive genetic association networks? In Proceedings of the 6th Joint Conference on Information Sciences, pp. 1237-1242, Association for Intelligent Machinery, RTP, NC.

Park J. S., Chen M. S., and Yu P. S. (1997). Using a hash-based method with transaction trimming for mining association rules. IEEE Transactions on Knowledge and Data Engineering 9:813-825.

Roos G., Brattsand G., Landberg G., Marklund U., and Gullberg M. (1993). Expression of oncoprotein 18 in human leukemias and lymphomas. Leukemia 7:1538-46.

Scherf U., Ross D. T., Waltham M., Smith L. H., Lee J. K., Tanabe L., Kohn K. W., Reinhold W. C., Myers T. G., Andrews D. T., Scudiero D. A., Eisen M. B., Sausville E. A., Pommier Y., Botstein D., Brown P. O., and Weinstein J. N. (2000). A gene expression database for the molecular pharmacology of cancer. Nat Genet 24:236-44.

Sheskin D. (2000). *Handbook of parametric and nonparametric statistical procedures.* Chapman & Hall/CRC, Boca Raton.

Siegel S., and Castellan N. J. (1988). *Nonparametric statistics for the behavioral sciences.* McGraw-Hill, New York.

Silverstein C., Brin S., and Motwani R. (1998). Beyond market baskets: Generalizing association rules to dependence rules. Data Mining and Knowledge Discovery 2:39-68.

Taniguchi M., Miura K., Iwao H., and Yamanaka S. (2001). Quantitative assessment of DNA microarrays – comparison with Northern blot analyses. Genomics 71:34-9.

Waddell P. J., and Kishino H. (2000). Cluster inference methods and graphical models evaluated on NCI60 microarray gene expression data. Genome Inform Ser Workshop Genome Inform 11:129-40.

Zaki M. J. (2000). Scalable algorithms for association mining. IEEE Transactions on Knowledge and Data Engineering 12:372-390.

Zaki M. J., Parthasarathy S., Ogihara M., and Li W. (1997). New algorithms for fast discovery of association rules. In Proceedings of the Third International Conference on Knowledge Discovery and Data Mining (KDD-97).

Zhou Y., Gwadry F. G., Reinhold W. C., Miller L. D., Smith L. H., Scherf U., Liu E. T., Kohn K. W., Pommier Y., and Weinstein J. N. (2002). Transcriptional regulation of mitotic genes by camptothecin-induced DNA damage: microarray analysis of dose- and time-dependent effects. Cancer Res 62:1688-95.

Chapter 18

GLOBAL FUNCTIONAL PROFILING OF GENE EXPRESSION DATA

Sorin Draghici[1] and Stephen A. Krawetz[2]

[1]*Dept. of Computer Science, Karmanos Cancer Institute and the Institute for Scientific Computing, Wayne State University, 431 State Hall, Detroit, MI, 48202*
e-mail: sod@cs.wayne.edu

[2]*Dept. of Obstetrics and Gynecology, Center for Molecular Medicine and Genetics, and the Institute for Scientific Computing, Wayne State University*
e-mail: steve@compbio.med.wayne.edu

1. CHALLENGES IN TODAY'S BIOLOGICAL RESEARCH

Molecular biology and genetics are currently at the center of an informational revolution. The data gathering capabilities have greatly surpassed the data analysis techniques. If we were to imagine the Holy Grail of life sciences, we might envision a technology that would allow us to fully understand the data at the speed at which it is collected. Sequencing, localization of new genes, functional assignment, pathway elucidation, and understanding the regulatory mechanisms of the cell and organism should be seamless. Ideally, we would like knowledge manipulation to become tomorrow the way goods manufacturing is today: high automatization producing more goods, of higher quality and in a more cost effective manner than manual production. In a sense, knowledge manipulation is now reaching the pre-industrial age. Our farms of sequencing machines and legions of robotic arrayers can now produce massive amounts of data but using it to manufacture highly processed pieces of knowledge still requires skilled masters painstakingly forging through small pieces of raw data one at a time. The ultimate goal is to automate this knowledge discovery process.

Data collection is easy, data interpretation is difficult. Typical examples of high-throughput techniques able to produce data at a phenomenal rate include shotgun sequencing (Bankier, 2001; Venter et al., 2001) and gene expression microarrays (Lockhart et al., 1996; Schena et al., 1995; Shalon et al., 1996). Researchers in structural genomics have at their disposal sequencing machines able to determine the sequence of approximately 100 samples every 3 hours (see for instance the ABI 3700 DNA analyzer from Applied Biosystems). The machines can be set up to work in a continuous flow which means data can be produced at a theoretical rate of approx. 800 sequences per day per machine. Considering a typical length of a sequence segment of about 500 base pairs, it follows that one machine alone can sequence approximately 400,000 nucleotides per day. This enormous throughput enabled impressive accomplishments such as the sequencing of the human genome (Lander et al., 2001; Venter et al., 2001). Recent estimates indicate there are 306 prokaryotic and 195 eukaryotic genome projects currently being undertaken in addition to 93 published complete genomes (Bernal et al., 2001). Currently, our understanding of the role played by various genes seems to be lagging far behind their sequencing. The yeast is an illustrative example. Although the 6,600 genes of its genome have been known since 1997, only approximately 40% of them have known or inferred functions.

A second widely used high-throughput genomic technique is the DNA microarray technology (Eisen et al., 1998; Golub et al., 1999; Lockhart et al., 1996; Schena et al., 1995). In its most general form, the DNA array is a substrate (nylon membrane, glass or plastic) on which DNA is deposited in localized regions arranged in a regular, grid-like pattern. The DNA array is subsequently probed with complementary DNA (cDNA) obtained by reverse transcriptase reaction from the mRNA extracted from a tissue sample. This DNA is fluorescently labeled with a dye and a subsequent illumination with an appropriate source of light will provide an image of the array. (Alternative detection techniques include using radioactive labels.) After an image processing step is completed, the result is a large number of expression values. Typically, one DNA chip will provide expression values for thousands of genes. For instance, the recently released Affymetrix chip HGU133A contains 22,283 genes. A typical experiment will involve several chips and generate hundreds of thousands of numerical values in a few days.

The continuous use of such high-throughput data collection techniques over the years has produced a large amount of heterogeneous data. Many types of genetic data (sequence, protein, EST, etc.) are stored in many different databases. The existing data is neither perfect nor complete, but reliable information can be extracted from it. The first challenge faced by today's researchers is to *develop effective ways of analyzing the huge amount of data that has been and will continue to be collected* (Eisenberg et al., 2000; Lockhart et al., 2000; Vukmirovic et al., 2000). In other words, there is a need for global, high-throughput data analysis techniques able to keep pace with the available high throughput data collection techniques.

The second challenge focuses on the type of discoveries we should be seeking. The current frontiers of knowledge span two orthogonal directions. Vertically, there are different levels of abstractions such as genes, pathways and organisms. Horizontally, at each level of abstraction there are known, hypothesized and unknown entities. For instance, at the gene level, there are genes with a known function, genes with an inferred function, genes with an unknown function and completely unknown genes. In any given pathway, there are known interactions, inferred interactions and completely unknown interactions. However, the vertical connections between the levels are, in many cases, limited to the membership relationships of genes associated to known pathways.

Most available techniques focus on the horizontal direction, trying to expand the knowledge frontier from known entities to unknown entities or trying to individuate the specific entities involved in a given condition. For instance, there are very many approaches to identifying the genes that are differentially expressed in a specific condition. Such techniques include fold-change (DeRisi, 1997; ter Linde et al., 1999; Wellmann et al., 2000), unusual ratio (Tao et al., 1999; Schena et al., 1995; Schena et al., 1996), ANOVA (Aharoni et al., 1975; Brazma et al., 2000; Draghici et al., 2001; Draghici et al., 2002; Kerr et al., 2000; Kerr and Churchill, 2001a; Kerr and Churchill, 2001b), model based maximum likelihood (Chen et al., 1997; Lee et al., 2000; Sapir et al., 2000), hierarchical models (Newton et al., 2001), univariate statistical tests (Audic and Claverie, 1997; Claverie et al., 1999; Dudoit et al., 2000), clustering (Aach et al., 2000; Ewing et al., 1999; Heyer et al., 1999; Proteome, 2002; Tsoka et al., 2000; van Helden et al., 2000; Zhu and Zhang, 2000), principal component analysis (Eisen et al., 1998; Hilsenbeck et al., 1999; Raychaudhuri et al., 2000), singular value decomposition (Alter et al., 2000), independent component analysis (Liebermeister, 2001), and gene shaving (Hastie et al., 2000). However, the task of establishing vertical relationships, such as translating sets of differentially regulated genes into an understanding of the complex interactions that take place at pathway level, is much more difficult. Although such techniques have started to appear (e.g., inferring gene networks (DeRisi et al., 1997; D'haeseleer et al., 2000; Roberts et al., 2000; Wu et al., 2002), function prediction (Fleischmann et al., 1999; Gavin et al., 2002; Kretschmann et al., 2001; Wu et al., 2002), etc.), this approach is substantially more difficult. Thus, the second challenge is to establish advanced methods and techniques able to make such vertical inferences or at least to propose such potential inferences for human validation. In other words, the challenge is to extract system level information from component level data (Ideker et al., 2001).

2. FUNCTIONAL INTERPRETATION OF HIGH-THROUGHPUT GENE EXPRESSION EXPERIMENTS

Microarrays enable the simultaneous interrogation of thousands of genes. Using such tools, researchers often aim at constructing gene expression profiles that characterize various pathological conditions such as cancer (Golub et al., 1999; Perou et al., 2000; van't Veer et al., 2002). Various technologies, such as cDNA and oligonucleotide arrays, are now available together with a plethora of methods for analyzing the expression data produced by the chips. Independent of the platform and the analysis methods used, the result of a microarray experiment is, in most cases, a list of genes found to be differentially expressed between two or more conditions under study. The challenge faced by the researcher is to translate this list of differentially regulated genes into a better understanding of the underlying biological phenomena. The translation from a list of differentially expressed genes to a functional profile able to offer insight into the cellular mechanisms is a very tedious task if performed manually. Typically, one would take each accession number corresponding to a regulated gene, search various public databases and compile a list with, for instance, the biological processes that the gene is involved in. This task can be performed repeatedly, for each gene, in order to construct a master list of all biological processes in which at least one gene was involved. Further processing of this list can provide a list of those biological processes that are common between several of the regulated genes. It is expected that those biological processes that occur more frequently in this list would be more relevant to the condition studied. The same type of analysis could be carried out for other functional categories such as biochemical function, cellular role, etc.

3. FUNCTIONAL PROFILING WITH ONTO-EXPRESS

Onto-Express (OE) is a tool designed to facilitate this process. This is accomplished by mining known data and compiling a functional profile of the experiment under study. OE constructs a functional profile for each of the Gene Ontology (GO) categories (Ashburner et al., 2000): cellular component, biological process and molecular function as well as biochemical function and cellular role, as defined by Proteome (Proteome, 2002). The precise definitions for these categories and the other terms used in OE's output can be found in GO (Ashburner et al., 2000). As biological processes can be regulated within a local chromosomal region (e.g. imprinting), an additional profile is constructed for the chromosome location. OE uses a database with a proprietary schema implemented and maintained in our laboratory (Draghici and Khatri, 2002). We use data from GenBank, UniGene, LocusLink, PubMed, and Proteome.

The current version of Onto-Express is implemented as a typical 3-tier architecture. The back-end is a relational DB implemented in Oracle 9i and running on a SunFire V880, 4CPUs, 8 GB RAM, 200GB accessing a 500 GB RAID array and tape jukebox backup. The application performing the data mining and statistical analysis is written in Java and runs on a separate server (Dell PowerEdge). The front end is a Java applet served by a Tomcat/Apache web server running on a Sun Fire V100 web server appliance.

OE's input is a list of genes found to be regulated in a specific condition. Such a list may be constructed using any technology: microarrays, SAGE, Westerns blots (e.g., high throughput PowerBlots (Biosciences, 2002)), Northerns blots, etc. This is why the utility of this application goes well beyond the needs of microarray users. At present, our database includes the human and mouse genomes.

The input of Onto-Express is a list of genes specified by either accession number, Affymetrix probe IDs or UniGene cluster IDs. At present, the Onto-Express database contains human and mouse data. More organisms will be added, as more annotation data becomes available. A particular functional category can be assigned to a gene based on specific experimental evidence or by theoretical inference (e.g., similarity with a protein having a known function). Onto-Express explicitly shows how many genes in a category are supported by experimental evidence (labelled with "experimented") and how many are predicted ("predicted"). Those genes for which it is not known whether they were assigned to the given functional category based on a prediction or experimental evidence are reported as "non-recorded". The results are provided in graphical form and emailed to the user on request. By default, the functional categories are sorted in decreasing order of number of genes as shown in Figure 18.1. The functional categories can also be sorted by confidence (see details about the computation of the p-values below) with the exception of the results for chromosomes where the chromosomes are always displayed in their order. There is one graph for each of the biochemical function, biological process, cellular role, cellular component and molecular function categories. A specific graph can be requested by choosing the desired category from the pull-down menu and subsequently clicking the "Draw graph" button. Clicking on a category displays a hyper-linked list of the genes in that category. The list contains the UniGene cluster

IDs uniquely identifying the genes. Clicking on a specific gene provides more information about that gene.

The following example will illustrate OE's functionality. Let us consider an array containing 1,000 genes used to investigate the effect of a substance X. Using classical statistical and data analysis methods we decide that 100 of these genes are differentially regulated by substance X. Let us assume that the 100 differentially regulated genes are involved in the following biological processes: 80 of the 100 genes are involved in positive control of cell proliferation, 40 in oncogenesis, 30 in mitosis and 20 in glucose transport. These results are tremendously useful since they save the researcher the inordinate amount of effort required to go through each of the 100 genes, compile lists with all the biological processes and then cross-compare those biological processes to determine how many genes are in each process (Khatri et al., 2002). In comparison, a manual extraction of this information would literally take several weeks and would be less reliable and less rigorous.

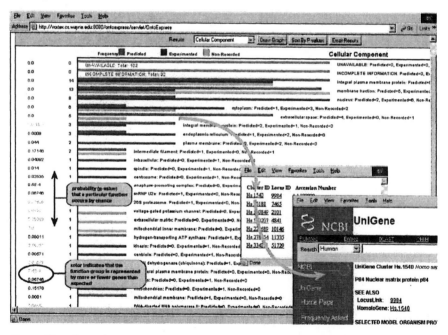

Figure 18.1. The main features of the Onto-Express output. The functional categories can be sorted by number of genes (shown) or by *p*-values (with the exception of the results for chromosomes where the chromosome are displayed in their order). Requesting a specific graph is done by choosing the desired category from the pull-down menu and subsequently clicking on "Draw graph".

The large number of genes involved in cell proliferation, oncogenesis and mitosis in the functional profile above, might suggest substance X affects a cancer pathway. However, a reasonable question is: what would happen if all genes on the array were involved in cell proliferation?

Would the presence of cell proliferation at the top of the list be significant? Clearly, the answer is no. If most or all genes on the array are involved in a certain process, then the fact that that particular process appears at the top is not significant. To correct this, the current version of the software allows the user to specify the array type used in the microarray experiment. Based on the genes present on this array, OE calculates the expected number of occurrences of a certain category.

Now, the data mining results are as in Table 18.1 and the interpretation of the functional profile appears to be completely different.

Table 18.1. Statistical significance. The number of genes involved in a given biological process can be misleading on its own. In this example, positive control of cell proliferation may appear to be the most important biological process affected since 80 of the 100 differentially regulated genes are involved in it. However, this process loses its significance when placed in the context that 800 of the 1,000 genes on the chip are involved in positive control of cell proliferation.

biological process	genes found	genes expected	
positive control of cell proliferation	80	80	not significant
oncogenesis	40	40	not significant
mitosis	30	10	significant
glucose transport	20	5	highly significant

There are indeed 80 cell proliferation genes but in spite of this being the largest number, we actually expected 80 such genes so this is not significant. The same holds true for oncogenesis. The mitosis starts to be interesting because we expected 10 genes and we observed 30, which is 3 times more than expected. However, the most interesting is the glucose transport. We expected only 5 genes and we observed 20, i.e. 4 times more than expected. The emerging picture changes radically: instead of generating the hypothesis that substance X is a potential carcinogen, we may consider the hypothesis that X is correlated with diabetes.

The problem is that an event such as observing 30 genes when we expect 10 can still occur just by chance. The next section explains how the significance of these categories is calculated based on their frequency of occurrence in the initial set of genes M, the total number of genes N, the frequency of occurrence in the list of differentially regulated genes x and the number of such differentially regulated genes K. The statistical confidence thus calculated, will allow us to distinguish between significant events and possibly random events.

3.1 Statistical Approaches

Several different statistical approaches can be used to calculate a p-value for each functional category F. Let us consider there are N genes on the chip used. Any given gene is either in category F or not. In other words, the N genes are of two categories: F and *non-F* (*NF*). This is similar to having an urn filled with N balls of two colors such as red (F) and green (not in F). M of these balls are red and $N - M$ are green. The researcher uses their choice of data analysis methods to select which genes are regulated in their experiments. Let us assume that they picked a subset of K genes. We find that x of these K genes are red and we want to determine the probability of this happening by chance.

So, our problem is: given N balls (genes) of which M are red and $N - M$ are green, we pick randomly K balls and we ask what is the probability of having picked exactly x red balls. This is sampling without replacement because once we pick a gene from the chip, we cannot pick it again.

The probability that a category occurs exactly x times just by chance in the list of differentially regulated genes is appropriately modeled by a hypergeometric distribution with parameters (N, M, K) (Casella, 2002):

$$P(X = x \mid N, M, K) = \frac{\binom{M}{x}\binom{N-M}{K-x}}{\binom{N}{K}} \tag{18.1}$$

Based on this, the probability of having x or fewer genes in F can be calculated by summing the probabilities of picking 1 or 2 or ... or $x - 1$ or x genes of category F (Tavazoie et al., 1999):

$$p_u = P(X = 1) + P(X = 2) + \ldots + P(X = x) = \sum_{i=0}^{x} \frac{\binom{M}{i}\binom{N-M}{K-i}}{\binom{N}{K}} \tag{18.2}$$

This corresponds to a one-sided test in which small *p*-values correspond to under-represented categories. The *p*-value for over-represented categories can be calculated as $p_o = 1 - p_u$ when $p_u > 0.5$.

The hypergeometric distribution is difficult to calculate when the number of genes is large (e.g., arrays such as Affymetrix HGU133A contain 22,283 genes). However, when N is large, the hypergeometric distribution tends to the binomial distribution (Casella, 2002). A similar approach was used by Cho et al. to discern whether hierarchical clusters were enriched in specific functional categories (Cho et al., 2001).

If a binomial distribution is used, the probability of having x genes in F in a set of K randomly picked genes is given by the classical formula of the binomial probability in which the probability of extracting a gene from F is estimated by the ratio of genes in F present on the chip M/N and the corresponding p-value can be respectively calculated as:

$$P(X = x \mid K, M/N) = \binom{K}{x}\left(\frac{M}{N}\right)^{x}\left(1 - \frac{M}{N}\right)^{K-x} \tag{18.3}$$

and

$$p = \sum_{i=0}^{x}\binom{K}{i}\left(\frac{M}{N}\right)^{i}\left(1 - \frac{M}{N}\right)^{K-i} \tag{18.4}$$

The main difference between the binomial and hypergeometric distributions is that the binomial models a sampling with replacement. Thus, selecting a gene involved in F should not influence the probability of selecting another gene involved in F. However, in our experiments, we do sampling without replacement since when a gene is picked, we cannot pick it again and the set of unpicked genes in F is reduced by one, thus changing the probability of future picks from F. Because of this, the two distributions behave a bit differently. For example, one cannot use the hypergeometric to calculate the probability of having $x > M$ genes since this would be equivalent to picking more F genes than there are on the microarray. However, the expression of the binomial probability density function will still provide a meaningful probability since in sampling with replacement one gene can be picked more than once and it is possible to pick more F genes than present on the microarray. However, the expression of the binomial probability density function will still provide a meaningful probability since in sampling with replacement one gene can be picked more than once and it is possible to pick more F genes than present on the microarray. Unfortunately, the computation of the hypergeometric distribution is not feasible for lists longer than $K > 150$ genes. However, for such large values, the hypergeometric distribution tends to behave like a binomial. In consequence, the binomial formula in Equation 18.3 can be used to compute the p-values.

Alternative approaches include a Chi-square test for equality of proportions (Fisher and van Belle, 1993) and Fisher's exact test (Man et al., 2000). For the purpose of applying these tests, the data can be organized as shown in Table 18.2. The dot notation for an index is used to represent the summation on that index.

In this notation, the number of genes on the microarray is $N = N_{.1}$, the number of genes in functional category F is $M = n_{11}$, the number of genes selected as differentially regulated is $K = N_{.2}$ and the number of differentially regulated genes in F is $x = n_{12}$. Using this notation, the Chi-square test involves calculating the value of the χ^2 statistics as follows:

$$\chi^2 = \frac{N_{..}\left(\left|n_{11}n_{22} - n_{12}n_{21}\right| - \frac{N_{..}}{2}\right)^2}{N_{1.}N_{2.}N_{.1}N_{.2}} \tag{18.5}$$

where $\frac{N_{..}}{2}$ in the numerator is a continuity correction term that can be omitted for large samples (Glover and Mitchell, 2002). The value thus calculated can be compared with critical values obtained from a χ^2 distribution with $df = (2 - 1)(2 - 1) = 1$ degree of freedom.

However, the χ^2 test for equality of proportion cannot be used for small samples. The rule of thumb is that all expected frequencies: $E_{ij} = (N_{i.} \cdot N_{.j})/N_{..}$ should be greater than or equal to 5 for the test to provide valid conclusions. If this is not the case, Fisher's exact test can be used instead (Fisher and van Belle, 1993; Kennedy et al., 1981). Fisher's exact test considers the row and column totals $N_{1.}$, $N_{2.}$, $N_{.1}$, $N_{.2}$ fixed and uses the hypergeometric distribution to calculate the probability of observing each individual table combination as follows:

$$P = \frac{N_{1.}! \cdot N_{2.}! \cdot N_{.1}! \cdot N_{.2}!}{N_{..}! \cdot n_{11}! \cdot n_{12}! \cdot n_{21}! \cdot n_{22}!} \tag{18.6}$$

Table 18.2. The significance of a particular functional category F can be calculated using a 2×2 contingency table and a Chi-square or Fisher's exact test for equality of proportions. The N genes on a chip can be divided into genes that are involved in the functional category of interest F ($n_{11} = M$) and genes that are not involved in F (n_{21}). The K genes found to be differentially regulated can also be divided into genes involved ($n_{21} = x$) and not involved (n_{22}) in F.

	Genes on chip	Diff. regulated genes	
having function F	n_{11}	n_{12}	$N_{1.} = \sum_{j=1}^{2} n_{1j}$
not having F	n_{21}	n_{22}	$N_{2.} = \sum_{j=1}^{2} n_{2j}$
	$N_{.1} = \sum_{i=1}^{2} n_{i1}$	$N_{.2} = \sum_{i=1}^{2} n_{i2}$	$N_{..} = \sum_{i,j} n_{ij}$

Using this formula, one can calculate a table containing all the possible combinations of $n_{11}n_{12}n_{21}n_{22}$.

The *p*-value corresponding to a particular occurrence is calculated as the sum of all probabilities in this table lower than the observed probability corresponding to the observed combination (Man et al., 2000).

Finally, Audic and Claverie (1997) have used a Poisson distribution and a Bayesian approach to calculate the probability of observing a given number of tags in SAGE data. As noted by Man et al. (2000), this approach can be used directly to calculate the probability of observing n_{12} genes of a certain functional category F in the selected subset given that there are n_{11} such genes on the microarray:

$$P(n_{12} \mid n_{11}) = \left(\frac{N_{.2}}{N_{.1}}\right)^{n_{12}} \cdot \frac{(n_{11} + n_{12})!}{n_{11}! \cdot n_{12}! \cdot \left(1 + \frac{N_{.2}}{N_{.1}}\right)^{n_{11}+n_{12}+1}} \tag{18.7}$$

The *p*-values are calculated as a cumulative probability density function (cdf) as follows (Audic and Claverie, 1997; Man et al., 2000):

$$p = \min\left\{ \sum_{k=0}^{k \le n_{12}} P(k \mid n_{11}), \sum_{k=n_{12}}^{\infty} P(k \mid n_{11}) \right\} \tag{18.8}$$

Extensive simulations performed by Man et al. compared the Chi-square test for equality of proportions with Fisher's exact test and Audic and Claverie's test and showed that the Chi-square test has the best power and robustness (Man et al., 2000).

Onto-Express provides implementations of the χ^2 test, Fisher's exact test as well as the binomial test. Fisher's exact test is required when the sample size is small and the chi-square test cannot be used. For a typical microarray experiment with $N \approx 10,000$ genes on the chip and $K \approx 100 = 1\%N$ selected genes, the binomial approximates very well the hypergeometric and is used instead. For small, custom microarrays (fewer than 200 genes), the χ^2 is used. The program calculates automatically the expected values and uses Fisher's exact test when χ^2 becomes unreliable (expected values less than 5). Thus, *the choice between the three different models is automatic, requiring no statistical knowledge from the end-user.*

We did not implement Audic and Claverie's test because: i) it has been shown that χ^2 is at least as good (Man et al., 2000), and ii) while very appropriate for the original problem involving ESTs, the use of a Poisson distribution may be questionable for our problem.

The exact biological meaning of the calculated *p*-values depends on the list of genes submitted as input. For example, if the list contains genes that are upregulated and mitosis appears more often than expected, the conclusion may be that the condition under study stimulates mitosis (or more generally, cell proliferation) in a statistically significant way. If the list contains genes that are downregulated and mitosis appears more often than expected (exactly as before), then the conclusion may be that the condition significantly inhibits mitosis.

4. DISCUSSION

Onto-Express has been applied to a number of publicly available data sets. For example, a microarray strategy was recently used by van 't Veer et al. to identify 70 genes that can be used as a predictor of clinical out- come for breast cancer (van't Veer et al., 2002). A subsequent analysis revealed that several key mechanisms such as cell cycle, cell invasion, metastasis, angiogenesis and signal transduction genes were significantly upregulated in cases of breast cancer with poor prognosis. However, as shown below, a comprehensive global analysis of the functional role associated with the differentially regulated genes has revealed novel biological mechanisms involved in breast cancer. Using the global strategy provided by Onto-Express the 231 genes significantly correlated with breast cancer prognosis were categorized into 102 different biological processes. Seventy-two of these groups had significant p-values ($p < 0:05$). Of these 72 groups, only 17 are represented by two or more members of the same biological pathway. These encompass most of the processes postulated to be indicative of poor prognosis including cell cycle, cell invasion, metastasis, and signal transduction (van't Veer et al., 2002). Interestingly, angiogenesis, cell cycle control, cell proliferation, and oncogenesis, are not significantly represented ($p > 0:05$) but a host of novel pathways were identified. These included protein phosphorylation, a common cellular response to growth factor stimulation and anti-apoptosis (apoptosis = programmed cell death). Both are believed to be intimately linked, acting to preserve homeostasis and developmental morphogenesis. Clearly, these processes can impact cancer. This data is used as a sample data set at: http://vortex.cs.wayne.edu. We invite the readers to login and use Onto-Express to analyze the data themselves in the light of the information provided in (van't Veer et al., 2002).

4.1 Utility, Need, and Impact

A tool such as Onto-Express can be used in two different ways. Many microarray users embark upon "hypotheses generating experiments" in which the goal is to find subsets of genes differentially regulated in a given condition. In this context, Onto-Express can be used to analyze and interpret the results of the experiment in a rigorous statistical manner (see section 3).

However, another major application is in experiment design. An alternative to the "hypotheses generating experiments" is the "hypothesis driven experiments" in which one first constructs a hypothesis about the phenomenon under study and then performs directed experiments to test the hypothesis. Currently, no two chips offer exactly the same set of genes. There is a natural tendency to select the chip with the most genes but this may not necessarily be the best choice and certainly, not the most cost effective. When a hypothesis of a certain mechanism does exist, we argue that one should use the chip(s) that best represent the corresponding pathways. Onto-Express can suggest the best chip or set of chips to be used to test a given hypothesis. This can be accomplished by analyzing the list of genes on all existing arrays and providing information about the pathways and biological mechanisms covered by the genes on each chip. If chip A contains 10,000 genes but only 80 are related to a given pathway and chip B contains only 400 genes but 200 of them are related to the pathway of interest, the experiment may provide more information if performed with chip B instead of A. This can also translate into significant cost savings.

An early version of Onto-Express was first made available in February 2002 (Khatri et al., 2002). In the period February-June, our user base grew to 590 valid registered users from 47 countries. The web traffic analysis shows a daily average of 74.28 page views by 18.28 unique visitors (including the weekend). Onto-Express has also been mentioned in several news articles (Janssen, 2002; Tracy, 2002; Uehling, 2002). Version 1 of Onto-Express is available free of charge at: http://vortex.cs.wayne.edu. This version constructs functional profiles for the cellular role, cellular component, biological process and molecular function as well as biochemical function and chromosome location. Version 2 of Onto-Express adds the computation of the statistical significance of the results.

4.2   Other Related Work and Resources

A tremendous amount of genetic data is available on-line from several public databases (DBs). NCBI provides sequence, protein, structure and genome DBs, as well a taxonomy and a literature DB. Of particular interest are UniGene (non-redundant set of gene-oriented clusters) and LocusLink (genetic loci). SWISS-PROT is a curated protein sequence DB that provides high-level annotation and a minimal level of redundancy (Bairoch and Apweiler, 2000). Kyoto Encyclopedia of Genes and Genomes (KEGG) contains a gene catalogue (annotated sequences), a pathway DB containing a graphical representation of cellular processes and a LIGAND DB (Kanehisa and Goto, 2000; Kanehisa et al., 2000; Ogata et al., 1999).

GenMAPP is an application that allows the user to create and store pathways in a graphic format, includes a multiple species gene database and allows a mapping of a user's expression data on existing pathways (Dahlquist et al., 2002). Other related databases and on-line tools include: PathDB (metabolic networks) (Waugh et al., 2000), GeneX (NCGR) (source independent microarray data DB; Mangalam et al., 2001), Arrayexpress (EBI, 2001a), SAGEmap (Lash et al., 2000), μArray (EBI, 2001a), ArrayDB (NHGRI, 2001), ExpressDB (Aach et al., 2000), and Stanford Microarray Database (Sherlock et al., 2001; Stanford, 2001). Two meta-sites containing information about various genomic and microarray on-line DBs are (Shi, 2001) and (CNRS, 2001).

Data format standardization is necessary in order to automate data processing (Brazma, 2001). The Microarray Gene Expression Data Group (MGED) is working to standardize the Minimum Information About a Microarray Experiment (MIAME), the format (MAGE) and ontologies and normalization procedures related to microarray data (Brazma et al., 2001; EBI, 2001b). Of particular interest is the Gene Ontology (GO) effort which aims to produce a dynamic, controlled vocabulary that can be applied to all organisms even as knowledge of gene and protein roles in cells is accumulating and changing (Ashburner et al., 2000; Ashburner et al., 2001). Expression profiles of genes across tissues can be obtained with tissue microarrays (Kononen et al., 1998; Bubendorf et al., 1999a; Bubendorf et al., 1999b; Schraml et al., 1999; Sallinen et al., 2000; Moch et al., 2001; Nocito et al., 2001a; Nocito et al., 2001b; mousses et al., 2002). Other techniques allowing a high-throughput screening includes the Serial Analysis of Gene Expression (SAGE) (Velculescu et al., 1995) and PowerBlots (Biosciences, 2002). Although such techniques have very high throughput when compared with techniques such as Northern blots or RT-PCR, they still require a considerable amount of laboratory effort. Data mining of the human dbEST has been used previously to determine tissue gene expression profiles (Bortoluzzi et al., 2000; Hishiki et al., 2000; Hwang et al., 2000; Sese et al., 2001; Vasmatzis et al., 1995).

5. CONCLUSION

In contrast to the approach of looking for key genes of known specific pathways or mechanisms, global functional profiling is a high-throughput approach that can reveal the biological mechanisms involved in a given condition. Onto-Express is a tool that translates the gene expression profiles showing how various genes are changed in specific conditions into functional profiles showing how various functional categories (e.g., cellular functions) are changed in the given conditions. Such profiles are constructed based on public data and Gene Ontology categories and terms. Furthermore, Onto-Express provides information about the statistical significance of each of the pathways and categories used in the profiles allowing the user to distinguish between cellular mechanisms significantly affected and those that could be involved by chance alone.

REFERENCES

Aach J., Rindone W., and Church G.M. Systematic management and analysis of yeast gene expression data (2000). Genome Research, 10:431-445.

Aharoni A., Keizer L.C.P., Bouwneester H.J., Sun Z., et al.(1975). Identification of the SAAT gene involved in strawberry flavor biogenesis by use of DNA microarrays. The Plant Cell, 12:647-661.

Alter O., Brown P., and Botstein D. (2000). Singular value decomposition for genome-wide expression data processing and modeling. Proc. Natl. Acad. Sci., 97(18):10101-10106.

Ashburner M., Ball C.A., Blake J.A., Botstein D. et al. (2001). Creating the gene ontology resource: Design and implementation. Genome Research, 11(8):1425-1433.

Ashburner M., Ball C.A., Blake J.A., Botstein D. et al. (2000). Gene ontology: tool for the unification of biology. Nature Genetics, 25:25-29.

Audic S. and Claverie J.-M. (1997). The significance of digital gene expression profiles. Genome Research, 10(7):986-995.

Audic S. and Claverie J.-M. (1998). Vizualizing the competitive recognition of TATA-boxes in vertebrate promoters. Trends in Genetics, 14:10-11.

Bairoch A. and Apweiler R. (2000). The SWISS-PROT protein sequence database and its supplement TrEMBL in 2000. Nucleic Acids Research, 28(1):45-48.

Bankier A. (2001). Shotgun DNA sequencing. Methods in Molecular Bilolgy, 167:89-100.

Bernal A., Ear U., and Kyrpides N. (2001). Genomes online database (GOLD): a monitor of genome projects world-wide. Nucleic Acids Research, 29(1):126-127.

Biosciences B. (2002). PowerBlot Western Array Screening Service. Technical report, BD Biosciences. Available at http://www.bdbiosciences.com.

Bortoluzzi S., d'Alessi G., Romualdi C., and Daneli G. (2000). The human adult skeletal muscle transcriptional profile reconstructed by a novel computational approach. Genome Research, 10(3):344-349.

Brazma A. and Vilo J. (2000). Gene expression data analysis. Federation of European Biochemical Societies Letters, 480(23893):17-24.

Brazma A. (2001). On the importance of standardisation in life sciences. Bioinformatics, 17(2):113-114.

Brazma A., Hingamp P., Quackenbush J., Sherlock G. et al. (2001). Minimum information about a microarray experiment (MIAME) – toward standards for microarray data. Nature Genetics, 29(4):365-371.

Bubendorf L., Kononen J., Koivisto P., Schraml P. et al. (1999). Survey of gene amplifications during prostate cancer progression by high-throughout fluorescence in situ hybridization on tissue microarrays. Cancer Research, 59(4):803-806.

Bubendorf L., Kolmer M., Kononen J., Koivisto P. et al. (1999). Hormone therapy failure in human prostate cancer: analysis by complementary DNA and tissue microarrays. Journal of the National Cancer Institute, 91(20):1758-1764.

Casella G. (2002). Statistical inference. Duxbury.

Chen Y., Dougherty E.R., and Bittner M.L. (1997). Ratio-based decisions and the quantitative analysis of cDNA microarray images. Journal of Biomedical Optics, 2(4):364-374.

Cho R., Huang M., Campbell M., Dong H. et al. (2001). Transcriptional regulation and function during the human cell cycle. Nature Genetics, 27:48-54.

Claverie J.-M. (1999). Computational methods for the identification of differential and coordinated gene expression. Human Molecular Genetics, 8(10):1821-1832.

CNRS (2001). Microarray databases. Technical report, Centre National de la Recherche Scietifique. Available at http://www.biologie.ens.fr/en/genetiqu/puces/bddeng.html.

Dahlquist K., Salomonis N., Vranizan K., Lawlor S., and Conklin B. (2002). GenMAPP, a new tool for viewing and analyzing microarray data on biological pathways. Nature Genetics, 31(1):19-20.

DeRisi J.L., Iyer V.R., and Brown P.O. (1997). Exploring the metabolic and genetic control of gene expression on a genomic scale. Science, 278:680-686, 1997.

DeRisi J.L., Penland L., Brown P.O., Bittner M.L. et al. (1996). Use of a cDNA microarray to analyse gene expression patterns in human cancer. Nature Genetics, 14(4):457-460.

D'haeseleer P., Liang S., and Somogyi R. (2000). Genetic network inference: From co-expression clustering to reverse engineering. Bioinformatics, 16(8):707-726.

Draghici S. and Khatri P. (2002). Onto-Express web site. Technical report, Wayne State University. Available at http://vortex.cs.wayne.edu.

Draghici S. (2002). Statistical intelligence: effective analysis of high-density microarray data. Drug Discovery Today, 7(11):S55-S63.

Draghici S., Kuklin A., Hoff B., and Shams S. (2001). Experimental design, analysis of variance and slide quality assessment in gene expression arrays. Current Opinion in Drug Discovery and Development, 4(3):332-337.

Dudoit S., Yang Y.H., Callow M., and Speed T. (2000). Statistical models for identifying differentially expressed genes in replicated cDNA microarray experiments. Technical Report 578, University of California, Berkeley.

EBI (2001a). ArrayExpress. Technical report, European Bioinformatics Institute. Available at http://www.ebi.ac.uk/arrayexpress/index.html.

EBI (2001b). Microarray gene expression database group. Technical report, European Bioinformatics Institute. Available at http://www.mged.org/.

Eisen M., Spellman P., Brown P., and Botstein D. (1998). Cluster analysis and display of genome-wide expression patterns. In Proc. of the Nat. Acad. of Sci., 95:14863-14868.

Eisenberg D., Marcotte E.M., Xenarios I., and Yeates T.O. (2000). Protein function in the post-genomic era. Nature, 405:823-826.

Ewing R.M., Kahla A.B., Poirot O., Lopez F., Audic S., and Claverie J.-M. (1999). Large-scale statistical analyses of rice ESTs reveal correlated patterns of gene expression. Genome Research, 9:950-959.

Fisher L.D. and van Belle G. (1993). Biostatistics: a methodology for health sciences. John Wiley and Sons, New York.

Fleischmann W., Moller S., Gateau A., and Apweiler R. (1999). A novel method for automatic functional annotation of proteins. Bioinformatics, 15(3):228-233.

Gavin A., Bosche M., Grandi K.R.P. et al. (2002). Functional organization of the yeast proteome by systematic analysis of protein complexes. Nature, 415(6868):141-147.

Glover T. and Mitchell K. (2002). An introduction to biostatistics. McGraw-Hill, New York.

Golub T.R., Slonim D.K., Tamayo P., Huard C., Gaasenbeek M., Mesirov J.P., Coller H., Loh M.L., Downing J.R., Caligiuri M.A., Bloomfield C.D., and Lander E.S. (1999) Molecular classification of cancer: class discovery and class prediction by gene expression monitoring. Science, 286(5439):531-537.

Hastie T., Tibshirani R., Eisen M.B., Alizadeh A., Levy R., Staudt L., Chan W., Botstein D., and Brown P. (2000). "Gene shaving" as a method for indentifying distinct sets of genes with similar expression patterns. Genome Biology, 1(2):1-21.

Heyer L.J., Kruglyak S., and Yooseph S. (1999). Exploring expression data: Identification and analysis of coexpressed genes. Genome Research, 9:1106-1115.

Hill A.A., Hunter C.P., Tsung B.T., Tucker-Kellogg G., and Brown E.L. (2000). Genomic analysis of gene expression in C. elegans. Science, 290:809-812.

Hilsenbeck S., Friedrichs W., Schiff R., O'Connell P., Hansen R., Osborne C., and Fuqua S.W. (1999). Statistical analysis of array expression data as applied to the problem of Tamoxifen resistance. Journal of the National Cancer Institute, 91(5):453-459.

Hishiki T., Kawamoto S., Morishita S., and BodyMap O.K. (2000). A human and mouse gene expression database. Nucleic Acids Research, 28(1):136-138.

Hwang D., Dempsy A., Lee C.-Y., and Liew C.-C. (2000). Identifcation of differentially expressed genes in cardiac hypertrophy by analysis of expressed sequence tags. Genomics, 66(1):1-14.

Ideker T., Galitski T., and Hood L. (2001). A new approach to decoding life: systems biology. Annual Review Of Genomics And Human Genetics, (2):343-372.

Janssen D. (2002). The information behind the informatics. Genomics and Proteomics. Available at http://www.genpromag.com/feats/0205gen23.asp.

Kanehisa M. and Goto S. (2000). KEGG: Kyoto encyclopedia of genes and genomes. Nucleic Acids Research, 28(1):27-30.

Kanehisa M., Goto, S., Kawashima S., and Nakaya A. (2002). The KEGG databases at GenomeNet. Nucleic Acids Research, 30(1):42-46.

Kennedy J.W., Kaiser G.W., Fisher L.D., Fritz J.K., Myers W., Mudd J., and Ryan T. (1981). Clinical and angiographic predictors of operative mortality from the collaborative study in coronary artery surgery (CASS). Circulation, 63(4):793-802.

Kerr M.K. and Churchill G.A. (2001a). Experimental design for gene expression analysis. Biostatistics, (2):183-201.
Available at http://www.jax.org/research/churchill/pubs/index.html.

Kerr M.K. and Churchill G.A. (2001b). Statistical design and the analysis of gene expression. Genetical Research, 77:123-128.
Available at http://www.jax.org/research/churchill/pubs/index.html.

Kerr M.K., Martin M., and Churchill G.A. (2000). Analysis of variance for gene expression microarray data. Journal of Computational Biology, 7:819-837.

Khatri P., Draghici S., Ostermeier C., and Krawetz S. (2002). Profiling gene expression utilizing Onto-Express. Genomics, 79(2):266-270.

Kononen J., Bubendorf L., Kallioniemi A., Barlund M. et al. (1998). Tissue microarrays for high-throughput molecular profiling of tumor specimens. Nature Medicine, 4(7):844-847.

Kretschmann E., Fleischmann W. (2001). Automatic rule generation for protein annotation with the C4.5 data mining algorithm applied on SWISS-PROT. Bioinformatics, 17(10):920-926.

Lander E., Linton L. et al. (2001). Initial sequences and analysis of the human genome. Nature, 409(6822):860-921.

Lash A.E., Tolstoshev C.M., Wagner L., Shuler G.D., Strausberg R.L., Riggins G.J., and Altschul S.F. (2000). SAGEmap: A public gene expression resource. Genome Research, 10:1051-1060.

Lee M.-L.T., Kuo F.C., Whitmore G.A., and Sklar J. (2000). Importance of replication in microarray gene expression studies: Statistical methods and evidence from repetitive cDNA hybridizations. Proc. Natl. Acad. Sci., 97(18):9834-9839.

Liebermeister W. (2001). Independent component analysis of gene expression data. In Proc. of German Conference on Bioinformatics GCB'01.
Available at http://www.bioinfo.de/isb/gcb01/poster/.

Lockhart D.J. and Winzeler E.A. (2000). Genomics, gene expression and DNA arrays. Nature, 405:827-836.

Lockhart D.J., Dong H., Byrne M., Folletie M., Gallo M.V., Chee M.S., Mittmann M., Want C., Kobayashi M., Horton H., and Brown E.L. (1996). DNA expression monitoring by hybridization of high density oligonucleotide arrays. Nature Biotechnology, 14:1675-1680.

Magrane M. and Apweiler R. (2002). Organisation and standardisation of information in SWISS-PROT and TrEMBL. Data Science Journal, 1(1):13-18.

Man M.Z., Wang Z., and Wang Y. (2000). POWER SAGE: comparing statistical tests for SAGE experiments. Bioinformatics, 16(11):953-959.

Mangalam H., Stewart J., Zhou J., Schlauch K., Waugh M., Chen G., Farmer A.D., Colello G., and Weller J.W. (2001). GeneX: An open source gene expression database and integrated tool set. IBM Systems Journal, 40(2):552-569.
Available at http://www.ncgr.org/genex/.

Moch H., Kononen T., Kallioniemi O., and Sauter G. (2001). Tissue microarrays: what will they bring to molecular and anatomic pathology? Advances in Anatomical Pathology, 8(1):14-20.

Mousses S., Bubendorf L., Wagner U., Hostetter G., Kononen J., Cornelison R., Goldberger N., Elkahloun A., Willi N., Koivisto P., Ferhle W., Rafield M., Sauter G., and Kallioniemi O. (2002). Clinical validation of candidate genes associated with prostate cancer progression in the cwr22 model system using tissue microarrays. Cancer Research, 62(5):1256-1260.

Newton M., Kendziorski C., Richmond C., Blattner F.R., and Tsui K. (2001). On differential variability of expression ratios: Improving statistical inference about gene expression changes from microarray data. Journal of Computational Biology, 8:37-52.

NHGRI (2001). ArrayDB. Technical report, National Human Genome Research Institute.
Available at http://genome.nhgri.nih.gov/arraydb/schema.html.

Nocito A., Bubendorf L., Tinner E.M., Suess K. et al. (2001a). Microarrays of bladder cancer tissue are highly representative of proliferation index and histological grade. Pathology, 194(3):349-357.

Nocito A., Kononen J., Kallioniemi O., and Sauter G. (2001b). Tissue microarrays (tmas) for high-throughput molecular pathology research. International Journal of Cancer, 94(1):1-5.

Ogata H., Goto S., Sato K., Fujibuchi W., Bono H., and Kanehisa M. (1999). KEGG: Kyoto encyclopedia of genes and genomes. Nucleic Acids Research, 27(1):29-34.

Perou C. M., Sørlie T., Eisen M.B., van de Rijn M., Jeffrey S.S., Rees C.A., Pollack J.R., Ross D.T., Johnsen H., Akslen L.A., Fluge Ø., Pergamenschikov A., Williams C., Zhu S.X., Lønning P.E., Børresen-Dale A.-L., Brown P.O., and Botstein D. (2000). Molecular portraits of human breast tumours. Nature, 406:747-752.

Pietu G., Mariage-Samson R., Fayein N.-A., Matingou C., Eveno E. et al. (1999). The genexpress IMAGE knowledge base of the human brain transcriptome: A prototype integrated resource for functional and computational genomics. Genome Research, 9:195-209.

Proteome (2002). Proteome BioKnowledge Library. Technical report, Incyte Genomics. Available at http://www.incyte.com/sequence/proteome.

Raychaudhuri S., Stuart J.M., and Altman R. (2000). Principal components analysis to summarize microarray experiments: Application to sporulation time series. In Proceedings of the Pacific Symposium on Biocomputing, volume 5, pages 452-463.

Roberts C.J., Nelson B., Marton M.J., Stoughton R., Meyer M.R., Bennett H.A., He Y.D., Dia H., Walker W.L., Hughes T.R., Tyers M., Boone C., and Friend S.H. (2000). Signaling and circuitry of multiple MAPK pathways revealed by a matrix of global gene expression profiles. Science, 287:873-880.

Sallinen S., Sallinen P., Haapasalo H., Helin H., Helen P., Schraml P., Kallioniemi O., and Kononen J. (2000). Identification of differentially expressed genes in human gliomas by DNA microarray and tissue chip techniques. Cancer Research, 60(23):6617-6622.

Sapir M. and Churchill G.A. (2000). Estimating the posterior probability of differential gene expression from microarray data. Technical Report, Jackson Labs, Bar Harbor, ME. Available at http://www.jax.org/research/churchill/pubs/.

Schena M. (2000). Microarray Biochip Technology. Eaton Publishing.

Schena M., Shalon D., Davis R., and Brown P. (1995). Quantitative monitoring of gene expression patterns with a complementary DNA microarray. Science, 270:467-470.

Schena M., Shalon D., Heller R., Chai A., Brown P., and Davis R. (1996) Parallel human genome analysis: microarray-based expression monitoring of 1000 genes. Proc. National Academy of Science USA, 93:10614-10519.

Schraml P., Kononen J., Bubendorf L., Moch H., Bissig H., Nocito A., Mihatsch M., Kallioniemi O., and Sauter G. (1999). Tissue microarrays for gene amplification surveys in many different tumor types. Clinical Cancer Research, 5(8):1966-1975.

Sese J., Nikaidou H., Kawamoto S., Minesaki Y., Morishita S., and Okubo K. (2001). BodyMap incorporated PCR-based expression proling data and a gene ranking system. Nucleic Acids Research, 29(1):156-158.

Shalon D., Smith S.J., and Brown P.O. (1996). A DNA microarray system for analyzing complex DNA samples using two-color fluorescent probe hybridization. Genome Research, 6:639-645.

Sherlock G., Hernandez-Boussard T., Kasarskis A., Binkley G. et al. (2001). The Stanford Microarray Database. Nucleic Acid Research, 29(1):152-155.

Shi L. (2001). DNA microarray – monitoring the genome on a chip. Technical report. Available at http://www.gene-chips.com/.

Stanford (2001). SMD - Stanford Microarray Database. Technical report, Stanford University. Available at http://genome-www4.Stanford.EDU/MicroArray/SMD/.

Stokes M.E., Davis C.S., and Koch G.G. Categorical Data Analysis Using the SAS System. SAS Institute, Carry, NC.

Sudarsanam P., Iyer V.R., Brown P.O., and Winston F. (2000). Whole-genome expression analysis of snf/swi mutants of Saccharomyces cerevisiae. Proc. Natl. Acad. Sci., 97(7):3364-3369.

Tamayo P., Slonim D., Mesirov J., Zhu Q., Kitareewan S., Dmitrovsky E., Lander E.S., and Golub T.R. (1999). Interpreting patterns of gene expression with self-organizing maps:

Methods and application to hematopoietic differentiation. Proc. Natl. Acad. Sci, 96:2907-2912.

Tao H., Bausch C., Richmond C., Blattner F.R., and Conway T. (1999). Functional genomics: Expression analysis of Escherichia coli growing on minimal and rich media. Journal of Bacteriology, 181(20):6425-6440.

Tavazoie S., Hughes J.D., Campbell M.J., Cho R.J., and Church G.M. (1999). Systematic determination of genetic network architecture. Nature Genetics, 22:281-285.

ter Linde J.J. M., Liang H., Davis R.W., Steensma H.Y., Dijken J.P. V., and Pronk J.T. (1999). Genome-wide transcriptional analysis of aerobic and anaerobic chemostat cultures of Saccharomyces cerevisiae. Journal of Bacteriology, 181(24):7409-7413.

Tracy S. (2002). Onto-Express – a tool for high-throughput functional analysis. Scientific Computing and Instrumentation, in press.

Tsoka S. and Ouzounis C.A. (2000). Recent developments and future directions in computational genomics. Federation of European Biochemical Societies Letters, (23897):1-7.

Uehling M. (2002). Open Channel Software Revamps Onto-Express. Technical report, BioIT World. Available at http://www.bio-itworld.com/products/050702_onto-express.html.

van Helden J., Rios A.F., and Collado-Vides J. (2000). Discovering regulatory elements in non-coding sequences by analysis of spaced dyads. Nucleic Acids Research, 28(8):1808-1818.

van't Veer L.J., Dai H., van de Vijver M.J., He Y.D. et al. (2002). Gene expression profiling predicts clinical outcome of breast cancer. Nature, 415:530-536.

Vasmatzis G., Essand M., Brinkmann U., Lee B., and Pastan I. (1995). Discovery of three genes specifically expressed in human prostate by expressed sequence tag database analysis. Proc. of the National Academy of Science USA, 95(1):300-304.

Velculescu V., Zhang L., Vogelstein B., and Kinzler K. (1995). Serial analysis of gene expression. Science, 270(5235):484-487.

Venter J.C., Adams M.D. et al. (2001). The sequence of the human genome. Science, 291(5507):1304-1351.

Vukmirovic O.G. and Tilghman S.M. (2000). Exploring genome space. Nature, 405:820-822.

Waugh M.E., Bulmore D.L., Farmer A.D., Steadman P.A. et al. (2000). PathDB: A metabolic database with sophisticated search and visualization tools. In Proc. of Plant and Animal Genome VIII Conference, San Diego, CA, January 9-12.

Wellmann A., Thieblemont C., Pittaluga S., Sakai A. et al. (2000). Detection of differentially expressed genes in lymphomas using cDNA arrays: identification of clustering as a new diagnostic marker for anaplastic large-cell lymphomas. Blood, 96(2):398-404.

White K.P., Rifkin S.A., Hurban P., and Hogness D.S. (1999). Microarray analysis of Drosophila development during metamorphosis. Science, 286:2179-2184.

Wu L., Hughes T., Davierwala A., Robinson M., Stoughton R., and Altschuler S. (2002). Large-scale prediction of saccharomyces cerevisiae gene function using overlapping transcriptional clusters. Nature Genetics, 31(3):255-265.

Zhu J. and Zhang M. (2000). Cluster, function and promoter: Analysis of yeast expression array. In Pacific Symposium on Biocomputing, pages 476-487.

Chapter 19

MICROARRAY SOFTWARE REVIEW

Yuk Fai Leung[1,2], Dennis Shun Chiu Lam[1], Chi Pui Pang[1]

[1]*Department of Ophthalmology and Visual Sciences, The Chinese University of Hong Kong,*
e-mail: {yfleung,dennislam,cppang}@cuhk.edu.hk

[2]*Genomicshome.com,*
e-mail: yfleung@genomicshome.com

1. INTRODUCTION

Microarray analysis is a burgeoning field in recent years. From the statistical inference of differential gene expression to expression profiling and clustering (Chapters 13-16), classification by various supervised and unsupervised algorithms (Chapters 7-12), and pathway reconstruction, microarray data analysis can provide a rich source of information that no one could imagine before. In the meantime, the great demand on data analysis capability has created an unprecedented challenge for life scientists to grasp the never-ending data analysis developments in time. The aim of this review is to provide an overview of various microarray software categorized by their purposes and characteristics, and aid the selection of suitable software. Readers are referred to a more comprehensive online version of the list[1] which contains further details on the particular software system requirements, features, prices, related publications and other useful information. This online software list is an ongoing project of a functional genomics Web site[2].

---

[1] http://ihome.cuhk.edu.hk/~b400559/arraysoft.html
[2] yfleung's functional genomics home:
http://ihome.cuhk.edu.hk/~b400559 or http://genomicshome.com (permanent domain)

2. PRIMER/PROBE DESIGN SOFTWARE

The focus on microarray data analysis has been the later stages in Data Mining. However, we should not neglect the early stage and should remember that probe and experimental design affect the quality of late stage analysis. Depending on the scale and throughput of the experiment, clones can be spotted from a normalized library without prior sequencing, sequence-verified cDNA clones such as *expressed sequence tags* (ESTs), specific parts of genes, or even specific oligonucleotides. A list of software for probe and the corresponding primer design is shown in Table 19.1.

Table 19.1. Examples of probe/primer design software.

Software	URL
Array designer	http://www.premierbiosoft.com/dnamicroarray/dnamicroarray.html
GAP (Genome-wide Automated Primer finder servers)	http://promoter.ics.uci.edu/Primers/
OligoArray	http://berry.engin.umich.edu/oligoarray/
Primer3	http://www-genome.wi.mit.edu/genome_software/other/primer3.html
ProbeWiz Server	http://www.cbs.dtu.dk/services/DNAarray/probewiz.html

There are a number of considerations when using EST clones (Tomiuk and Hofman, 2001). For example, rigorous informatic analysis is necessary to eliminate their redundancies, given the fact that nonidentical clones might actually correspond to the same gene, and some clones that correspond to different genes might be accidentally assigned to the same gene by chimeric EST clones generated during cDNA library construction or because of the existence of physiologically overlapping transcribed regions. There are also situations such as *alternative splicing*[3], *alternative polyadenylation*[4] and *alternative promoter usage*[5] that can produce variants of a single gene. These confounding factors should also be considered during probe selection. In these situations the same cDNA probe would probably bind many variants of the same gene and give a mixed signal. At the same time the human genome contains many gene families with only subtle sequence differences. They are quite difficult to be differentiated by cDNA probes. Therefore many

---

[3] *Alternative splicing* refers to a gene regulatory mechanism in which more than one mRNA product being produced due to distinct exon splicing events of a common mRNA.

[4] *Alternative polyadenylation* refers to a gene regulatory mechanism in which more than one mRNA product being produced due to different polyadenylation signals are used in the untranslated 3' end of a gene.

[5] *Alternative promoter usage* refers to a gene regulatory mechanism in which more than one mRNA product being produced due to different promoters are used in the 5' end of a gene.

researchers are seriously investigating on the use of oligonucleotides that can be specifically designed to hybridize to a single variant.

In our opinion, this pre-experimental informatic endeavor is very important. When we performed a microarray experiment on our *in vitro* glaucoma model, using a commercial cDNA microarray as our platform, the problems of clone redundancy and inaccuracy were only apparent at the later stage in analysis (Leung et al., 2002). For example, two clones with exactly the same sequence annotation in the curated database were included in the array due to earlier different annotations in other databases. A hybrid clone only produced in cancerous translocation was also found to be differentially expressed. However this translocation is not very likely to happen in glaucoma and created a difficulty in addressing the differential expression to the particular gene domain in the hybrid clone. We believe these problems could have been eliminated by proper probe selection and array design, in particular using oligonucleotide arrays.

3. IMAGE ANALYSIS SOFTWARE

There are three fundamental processes of image analysis: *gridding*[6], *segmentation*[7] and *information extraction*[8] (Yang et al., 2001), which are the standard functions in commonly used *image analysis software* (Table 19.2).

Table 19.2. Examples of image analysis software.

Software	URL
AIDA Array Metrix	http://www.raytest.de/products/software/aida/array/array.html
ArrayPro	http://www.mediacy.com/arraypro.htm
ArrayVision	http://www.imagingresearch.com/products/ARV.asp
Dapple	http://www.cs.wustl.edu/~jbuhler/research/dapple/
F-scan	http://abs.cit.nih.gov/fscan/
GenePix Pro	http://www.axon.com/GN_GenePixSoftware.html
ImaGene	http://www.biodiscovery.com/imagene.asp
Iconoclust	http://www.clondiag.com/products/sw/iconoclust/
Iplab	http://www.scanalytics.com/product/hts/microarray.html
Lucidea Automated Spotfinder	http://www1.amershambiosciences.com/aptrix/upp01077.nsf/Content/Products?OpenDocument&parentid=460766&moduleid=165065

[6] *Gridding* is a process to locate each spot on the slide.

[7] *Segmentation* is a process to differentiate the pixels within a spot-containing region into foreground (true signal) and background.

[8] *Information extraction* includes two parts, the spot intensity extraction and background intensity extraction. Spot intensity extraction refers to the calculation of fluorescence signal from the foreground from segmentation process, while background intensity extraction utilizes different algorithms to estimate the background signal due to the non-specific hybridization on the glass.

Software	URL
Phoretix Array	http://www.phoretix.com/products/array_products.htm
P-scan	http://abs.cit.nih.gov/pscan/index.html
QuantArray	http://www.packardbioscience.com/products/products.asp?content_item_id=521
ScanAlyze	http://rana.lbl.gov/EisenSoftware.htm
Spot	http://www.cmis.csiro.au/iap/spot.htm
TIGR Spotfinder	http://www.tigr.org/software/
UCSF Spot	http://jainlab.ucsf.edu/Projects.html

Gridding appears not to be a difficult problem for most of the image analysis software, though some manual adjustments are often necessary. The greatest challenges are performing the segmentation and background estimation efficiently, because the sizes and shapes of the spots can vary considerably. There are several segmentation algorithms including fixed circle, adaptive circle, adaptive shape and histogram, whereas algorithms for background estimation include constant background, local background and morphological opening. Every method has its strengths and weaknesses. Unfortunately no single one is perfect in all situations. *UCSF Spot* adopts a quite different experimental segmentation approach, which applies DNA counterstain DAPI on the array (Jain et al., 2002). Only the positions with DNA would be stained. The resulting counterstain image assists the segmentation process, which apparently resolves the limitations of the algorithmic approach.

Certain commercial software packages have measures to evaluate the quality of the spots in the images. For instance in one software package there are a number of parameters including diameter, spot area, footprint, circularity, signal/noise and spot, background and replicate uniformity for judging whether a spot is of sufficiently good quality to be included in later stage analysis. However, these quality measures usually idealize the spots being analyzed, i.e. consistent spot size, circularity, and signal intensity. This can be quite deviated from reality. Thus, some spots whose signal is good enough for data analysis might be rejected because they do not fulfill the defined criteria. Therefore, many researchers are still inclined to inspect the spots by naked eye instead of relying on these automated algorithms even though hours are needed for checking arrays with tens of thousands of spots. Unfortunately different researchers might have different standards on judging the borderline cases and their intuitive standards can change from time to time. As a result, manual inspection is not sufficiently objective to maintain the quality consistency among experiments and across laboratories. There is an urgent need for a rigorous definition of a good-quality spot. We believe a combination of DNA counterstain method as mentioned before and the existing quality measures can be a plausible solution. The counterstain provides information about the actual spot morphology and DNA distribution in the spot, which can be a better basis for applying quality measures to evaluate the spots.

4. DATA MINING SOFTWARE

It is always perplexing to select a Data Mining software because so many similar software packages are available in the field. Many scientists are very concerned whether the software they choose is really suitable for their experimental design and able to keep up with the fast changing analysis field. At the same time, there is the cost consideration. Some commercially available Data Mining software carry a high price tag that makes it a big commitment to use them. Perhaps this anxiety stems from the fact that the majority of scientists who wish to perform microarray experiments are relatively unfamiliar with the basis of large-scale Data Mining. We wish the earlier chapters on Data Mining fundamentals have resolved this anxiety. There are also a few microarray data analysis review papers (Brazma and Vilo, 2000; Quackenbush, 2001; Wu, 2001; Sherlock, 2001; Nadon and Shoemaker, 2002) and books (Jagota, 2001; Knudsen, 2002), which are extremely helpful for learning data analysis. Here we will complement with an overview of the advantages and limitations of various sub-categories of Data Mining software.

In general, a Data Mining software performs data preprocessing (Chapters 2-3) and normalization (Chapter 4), dimensionality reduction (Chapters 5-6), statistical inference of differential expression, clustering (Chapters 13-16), and classification (Chapters 7-12), and visualization of the analysis results. It can be available as a Web tool, as a standalone solution or as a client-server application. There are roughly 4 types of Data Mining software depending on comprehensiveness: *Turnkey system, comprehensive software, specific analysis software* and *extension/accessory software*. Two other related but unique *statistic* and *pathway reconstruction software* will be further discussed in Sections 5 and 6.

4.1 Turnkey System

A *Turnkey system* is defined as a computer system that is customized for a particular application. The term derives from the idea that the end user only needs to turn a key and the system is ready for immediate operation. A microarray turnkey Data Mining system includes everything like operating system, server software, database, client software, statistics software and even hardware customized for the whole Data Mining process. Some of the microarray turnkey Data Mining systems are listed in Table 19.3. While some of them (e.g., *Genetraffic*) are built on open source software like

Linux, *R* statistical language, *PostgreSQL*, and *Apache* Web server, some (e.g., *Rosetta Resolver*) are using proprietary server and database systems like *SunOS* and *Oracle*, respectively.

Table 19.3. Examples of turnkey Data Mining system.

Software	URL
arraySCOUT	http://www.lionbioscience.com/solutions/arrayscout
BASE (BioArray Software Environment)	http://base.thep.lu.se/
Expressionist	http://www.genedata.com/products/expressionist/
Genedirector	http://www.biodiscovery.com/genedirector.asp
Genetraffic	http://www.iobion.com/products/products.html
Rosetta Resolver	http://www.rosettabio.com/products/resolver/default.htm
Silicon Genetics Enterprise Solution	http://www.silicongenetics.com/cgi/SiG.cgi/Products/Solutions/index.smf?UID=14602

A Turnkey system aims at providing all components fine-tuned for microarray data analysis and supports the developing standard and language like *Minimum Information About a Microarray Experiment* (*MIAME*) (Brazma et al., 2001) and *MicroArray and Gene Expression Markup Language* (*MAGE-ML*)[9], respectively. The client-server setup supports multiple users which is especially beneficial for sharing data within a large research group like in pharmaceutical companies. It also allows a better control of data security by restricting different access privileges for different user groups. However this type of software is not quite suitable for small laboratories with limited budget, partly because such commercial systems can be quite expensive. Although the cost can be much lower for those using open source software, turnkey system in general requires dedicated supporting staff for routine maintenance. This makes the installation and operation of such a system in small laboratories a substantial task, particularly in those that are inexperienced in related computer science.

4.2 Comprehensive Software

A comprehensive software incorporates many different analyses at different stages of microarray analysis like data preprocessing, dimensionality reduction, normalization, clustering and visualization in a single package (Table 19.4). This type of software does not have any accompanied database although they are usually equipped with an interface for *Open DataBase Connectivity* (*ODBC*), a standard for accessing different database systems. This interface enables the users to archive their data in commonly used databases.

[9] http://www.mged.org/Workgroups/MAGE/mage-ml.html

Table 19.4. Examples of comprehensive Data Mining software

Software	URL
Acuity	http://www.axon.com/GN_Acuity.html
AMIADA (Analyzing MIcroArray DAta)	http://web.hku.hk/~xxia/software/amiada.html
ArrayStat	http://www.imagingresearch.com/products/AST.asp
BRB ArrayTools	http://linus.nci.nih.gov/BRB-ArrayTools.html
Cluster	http://rana.lbl.gov/EisenSoftware.htm
DNA-arrays analysis tools	http://bioinfo.cnio.es/dnarray/analysis/
DNA-Chip Analyzer (dChip)	http://www.dchip.org/
Expression Profiler	http://ep.ebi.ac.uk/
GeneLinker Gold	http://microarray.genelinker.com/products.html#Gene LinkerGold
GeneMaths	http://www.applied-maths.com/ge/ge.htm
GeneSight	http://www.biodiscovery.com/genesight.asp
GeneSpring	http://www.sigenetics.com/Products/GeneSpring/index.html
Genesis	http://genome.tugraz.at/Software/GenesisCenter.html
J-Express	http://www.molmine.com/index_a.html
MAExplorer	http://www.lecb.ncifcrf.gov/MAExplorer/
Partek software suites	http://www.partek.com/html/products/products.html
TIGR ArrayViewer	http://www.tigr.org/software/
TIGR Multiple Experiment Viewer (TMEV)	http://www.tigr.org/software/
Xcluster	http://genome-www.stanford.edu/~sherlock/cluster.html
Xpression NTI	http://www.informaxinc.com/solutions/xpression/index.html

The advantage of using this type of software is their comprehensiveness. Most of the current data analysis tools are available within a single package. The researchers can handle various analyses with ease once they have learned the basic operations. Therefore, the total learning curve can be shorter. There are often some brilliant analytical ideas incorporated in the software that can streamline the data analysis process. For example, in *GeneSight* the graphical set builder allows interactive data analysis by dragging different analysis modules, like log-transformation and normalization, into the workspace. This eliminates the possibility of missing or repeating data processing steps during multi-step data analysis from raw data to final output. In *GeneSpring* there are automated gene annotation and ontology construction tools that mine the public databases for the possible gene functions on behalf of the users. This can considerably save time when compared to mining the databases manually.

However, there are several potential limitations for using this type of software. Firstly, there might be data compatibility problems or conversion inconvenience. Unless the software contains filters for accepting raw data from different vendors, the users might have to re-organize the raw or preprocessed data according to the specific format themselves. Besides, the software may not be flexible enough to the latest analysis development, especially when the field is still evolving rapidly. Sometimes the users might have to wait for the next software update before the desired analytical method is being incorporated, and this can take quite long. As a result, certain comprehensive software (e.g., *GeneSpring, J-express & MAExplorer*) start to include the plugin functionality that allows users to create custom analysis functions promptly. The cost for using the commercially available software can be quite high for some laboratories with tight budgets. At the same time users might have to create their data archival system if the software does not have a "sister" database software. Another potential problem of this type of comprehensive software packages is their capability of performing various data analyses with ease. The inexperienced user might overlook certain statistical limitations of data analyses and generate inaccurate results, which they might still regard as valid. This can only be solved by a better knowledge on the statistical fundamentals of various analyses. Some of these limitations are also true for the turnkey system.

4.3   Specific Analysis Software

Specific analysis software is defined as a software which performs only one analysis or a few specific analyses (Table 19.5). The distinction between comprehensive and specific analysis software is not clear-cut, but in general a specific analysis software is more specialized in a particularly confined analytical problem, while a comprehensive software aims at providing an all-in-one package for the general user. For example, *PAM* performs sample classification from gene expression data, *CTWC* performs clustering based on a specific algorithm and *GeneGluster* performs normalization and filtering, as well as clustering using *Self-Organizing Map* (SOM). Specific analysis software is usually accompanied with a journal article that details the statistical and mathematical background of the method. This greatly helps understanding the basis of the analysis. Cutting-edge data analysis tools are often released as specific software by their authors at personal websites. If the specific analysis is embraced by the general public, it will most probably be incorporated into other comprehensive software packages at a later stage. Since this type of software is quite specialized, substantial preprocessing and re-organization of the data might be necessary before input into the software.

Table 19.5. Examples of specific analysis Data Mining software.

Software	URL
ANOVA programs for microarray data	http://www.jax.org/research/churchill/software/anova/index.html
Cleaver	http://classify.stanford.edu/
CIT (Cluster Identification Tool)	http://www.vai.org/vari/bioinformatics.htm
CTWC (Coupled Two-Way Clustering)	http://ctwc.weizmann.ac.il/
Cyber T	http://genomics.biochem.uci.edu/genex/cybert/
GeneCluster	http://www-genome.wi.mit.edu/cancer/software/software.html
GeneViz	http://www.contentsoft.de/geneviz.htm
INCLUSive	http://www.esat.kuleuven.ac.be/~dna/Biol/Software.html
MicroHelper	http://www.changbioscience.com/microhelperinfo.html
PAM (Prediction Analysis for Microarrays)	http://www-stat.stanford.edu/~tibs/PAM/
R cluster	http://genomics.biochem.uci.edu/cgi-bin/genex/rcluster/index.cgi
SAM (Significance Analysis of Microarrays)	http://www-stat.stanford.edu/~tibs/SAM/index.html
SNOMAD (Standardization and NOrmalization of MicroArray Data)	http://pevsnerlab.kennedykrieger.org/snomadinput.html
VERA & SAM	http://www.systemsbiology.org/VERAandSAM/

4.4   Extensions of Existing Data Mining Software

As mentioned in Section 4.2, some comprehensive software packages promptly incorporate the latest analysis developments by plugin modules. For example, *ArrayMiner* is a new clustering tool using proprietary clustering algorithms and is available as a standalone tool or plugin of *GeneSpring*, a comprehensive software. Besides, there are a few tools that extend the power of an analysis by providing intuitive visualization functionalities. For example, *TreeView*, *Slcview* and *Freeview* are the dendrogram and clusterogram viewers for the famous classical clustering software – *Cluster* (Eisen et al. 1998).

Table 19.6. Examples of extension/accessoary Data Mining software.

Software	URL
ArrayMiner	http://www.optimaldesign.com/ArrayMiner/ArrayMiner.htm
Freeview & FreeOView	http://magix.fri.uni-lj.si/freeview/
Slcview	http://slcview.sourceforge.net/
TreeView	http://rana.lbl.gov/EisenSoftware.htm

5. STATISTICS SOFTWARE AND TECHNICAL PROGRAMMING LANGUAGES

No matter how advanced or complicated is the analysis, it must be based on valid statistical fundamentals which should be easily handled by statistics software or technical programming languages that can perform statistical analysis. Actually, numerous microarray publications report on the use of such software to perform data analyses like clustering and statistical inference of differential expression. Table 19.7 shows some commonly used statistical software and technical programming languages. A common feature of these software packages is their flexibility that every step of the analysis can be fine-tuned by appropriate programming. As commonly-used statistical analyses are their standard components, analyses with different perspectives can be easily carried out. Most of them are command-line driven and can extend their functionality by programming new tools as extensions. This allows the incorporation of new analyses promptly.

However, the command-line driven nature of this type of software comes with a price, which is the impossibility of learning this type of software intuitively. In the meantime, users are required to have a thorough understanding of the structure of their data, and the statistical background and limitations of a particular analysis. Furthermore, they need to learn the technical programming language *per se* in order to perform even an operation, not to mention conducting the whole analysis correctly. Therefore the novice will need to spend extra efforts to get used to that analysis environment. We believe this is actually an advantage in disguise. The difficulties that the novice encounters during the learning stage can actually help to build up his/her foundations on proper analysis and in turn generate analysis results of better quality.

Table 19.7. Examples of statistics software and technical programming languages

Software	URL
Excel	http://www.microsoft.com/office/excel/default.htm
MATLAB	http://www.mathworks.com/products/matlab/
Octave	http://www.octave.org/
SAS	http://www.sas.com/products/index.html
SPSS	http://www.spss.com/spss10/
S-PLUS	http://www.insightful.com/products/default.asp
Statistica	http://www.statsoftinc.com/toc.html
R	http://www.R-project.org/

5.1 R Packages for Microarray Analysis

R is a language and environment for statistical computing and graphics. It has immense potential for microarray data analysis. It is an open source software available under *GNU*[10] *General Public Licence (GPL)*[11]. It is very similar to the *S* system and is highly compatible with *S-Plus*. *R* provides a wide variety of statistical tools, which can be very useful for microarray data analysis (Table 19.8). It is also highly extensible and allows users to add additional functionalities by defining new functions. There are numerous dedicated efforts to write *R* extensions, called *"packages"*, for microarray analysis nowadays and some of these packages are listed in Table 19.9.

Table 19.8. General R packages useful for microarray analysis

R packages	URL
cclust (Convex Clustering Methods and Clustering Indexes)	http://cran.r-project.org/src/contrib/PACKAGES.html#cclust
Cluster	http://cran.r-project.org/src/contrib/PACKAGES.html#cluster
Mclust	http://cran.r-project.org/src/contrib/PACKAGES.html#mclust
multiv (Multivariate Data Analysis Routines)	http://cran.r-project.org/src/contrib/PACKAGES.html#multiv

Table 19.9. Examples of R packages dedicated to microarray analysis

R packages	URL
affy (Methods for Affymetrix Oligonucleotide Arrays)	http://biosun01.biostat.jhsph.edu/~ririzarr/Raffy/
Bioconductor	http://www.bioconductor.org/
CyberT	http://genomics.biochem.uci.edu/genex/cybert/
GeneSOM	http://lib.stat.cmu.edu/R/CRAN/src/contrib/PACKAGES.html#GeneSOM
Mixture modelling	http://www.sph.umich.edu/~ghoshd/COMPBIO/mixture1/index.html
PAM (Prediction Analysis for Microarrays)	http://www-stat.stanford.edu/%7Etibs/PAM/Rdist/index.html
permax	http://cran.r-project.org/src/contrib/PACKAGES.html#permax
OOMAL (Object-Oriented Microarray Analysis Library) *	http://www3.mdanderson.org/depts/cancergenomics/oomal.html
SMA (Statisics for Microarray Analysis)	http://www.stat.berkeley.edu/users/terry/zarray/Software/smacode.html
SMA extension (com.braju.sma)	http://www.braju.com/R/com.braju.sma/
YASMA (Yet Another Statistical Microarray Analysis)	http://www.cryst.bbk.ac.uk/~wernisch/yasma.html

*Requires *S-plus*

For example, the *SMA* package provides a cutting-edge data intensity and spatial dependent normalization method called *print-tip grouped LOWESS*,

[10] GNU project: http://www.gnu.org
[11] GPL: http://www.R-project.org/COPYING

as well as an empirical Bayes method for replicated microarray data analysis (Yang et al., 2002; Lönnstedt et al., 2002). *Bioconductor* is an ambitious project which aims at providing a bioinformatics infrastructure to assist the development of tools for analyzing genomics data, with primary emphasis on microarray analysis. Most of the software produced by *Bioconductor* project will be in the form of *R* packages. Apparently *R* environment can provide a common platform to link together different groups with common interest in microarray data analysis. Users can have unlimited capability and flexibility to embrace new analyses under such an open source standardized environment (Stewart et al., 2001). We have witnessed the success of open source software like *Linux* in computing science and The *European Molecular Biology Open Software Suite (EMBOSS)* for molecular biology, and foresee a similar development in the microarray analysis field. Therefore, we recommend to everyone who is serious in microarray data analysis to start learning a high-level statistical language like *R*, to be prepared for the future challenges in this fast evolving data analysis era. We also recommend a few books (Selvin, 1998; Venables and Ripley, 1999; Krause and Olson, 2000) that we have found very helpful in learning *R/S-Plus* for statistical analysis.

6. PATHWAY RECONSTRUCTION SOFTWARE

A gene never acts alone in a biological system. Functional mapping by data analysis permits a better understanding of the underlying biological networks and correlation between genotype and phenotype (Horvath and Baur, 2000; Leung and Pang, 2002; Leung 2002). Currently, targeted mutation (Roberts et al., 2000; Hughes et al., 2000) or time series/cell cycle expression profiling experiments (Tavazoie et al., 1999) are the most feasible methods of network reconstruction. These experiments share the common goal of identifying cause-consequence relationships. The targeted mutation experiment, in particular, can provide direct evidence about which genes are controlled by other genes. For example, Tavazoie et al. (1999) performed clustering in their cell cycle expression dataset and identified upstream DNA sequence elements specific to each cluster by *AlignACE* (Roth et al., 1998). This might provide clues to uncover the putative cis-regulatory elements that co-regulate the genes within that cluster. There is also an increasing interest in investigating various reverse engineering approaches like *Boolean network* (D'haeseleer et al., 2000), *Bayesian network* (Chapter 8) or *Singular Value Decomposition* (Chapter 5) to identify the gene network architecture from gene activity profiles. A more thorough review on various developing network reconstruction techniques were detailed by de Jong (2002). A list of useful software for achieving this purpose is shown in Table 19.10. Some of them like *AlignACE, MEME, Sequence Logos* and *GeneSpring* are helpful for finding upstream consensus elements, while *GenMAPP, Rosetta Resolver* and *Pathway Processor* are capable of visualizing the reconstructed pathways.

Table 19.10. Examples of pathway reconstruction software.

Software	URL
AlignACE	http://atlas.med.harvard.edu/
GenMAPP (Gene MicroArray Pathway Profiler)	http://www.genmapp.org
GeneSpring	http://www.sigenetics.com/Products/GeneSpring/index.html
MEME (Multiple Expectation-maximization for Motif Elicitation)	http://meme.sdsc.edu/meme/website/
PathFinder	http://bibiserv.techfak.uni-bielefeld.de/pathfinder/
Pathway Processor	http://cgr.harvard.edu/cavalieri/pp.html
PubGene	http://www.pubgene.org/
Sequence Logos	http://www.bio.cam.ac.uk/seqlogo/
Rosetta Resolver	http://www.rosettabio.com/products/resolver/default.htm

There are two challenges for pathway reconstruction in the future. Firstly, more sophisticated statistical tests are essential to address the likelihoods of the inferred pathways. Recently, Grosu et al. (2002) developed a pathway construction software called *Pathway Processor* that not only can visualize expression data on metabolic pathways, but also evaluate which metabolic pathways are most affected by transcriptional changes in whole-genome expression experiments. Using Fisher's exact test, biochemical pathways are scored according to the probability that as many or more genes in a pathway would be significantly altered in a given experiment than by chance alone.

The second challenge is in fact a grand challenge for biologists in the post-genomic era. The focus on pathway reconstruction has been mostly qualitative and trying to obtain the most probable global interaction network from the data. However, we should not neglect that a gene interaction network, as well as the biological system, is actually a dynamic system that is controlled by numerous parameters, which in turn affect the ultimate response to a particular situation. For example, proteins might be regulated by various post-translational modifications and have different sub-cellular localization, substrates and intermediates might have different initial concentrations, enzymes might have different rate constants, reactions being catalyzed might have different equilibrium constants and protein-protein interactions between the pathway members might have different affinity/dissociation constants. Merely identifying a qualitative topological pathway network without addressing its dynamical behavior is inadequate to provide a thorough understanding of the complex nature of the biological system being studied.

Unfortunately, it is not a trivial task to obtain all these information, especially on a genomic scale. This is because many of the genes that are supposed to be involved in a functional pathway have yet to be identified from the finished genome sequences. Their basic functions have yet to be characterized, not to mention all those kinetic parameters. An integration of microarray analysis with other genomics approaches like *proteomics* (Hebestreit, 2001) and *metabolomics* (Fiehn, 2002; Phelps et al., 2002) is critical to obtain the essential information to reveal the complete picture of the system. When there is enough information, a relatively complex quantitative pathway model can be constructed which aids the understanding of complex dynamical behavior of biological pathways and systems (Leung et al., 2001). For instance, Schoeberl et al. (2002) constructed a model based on ordinary differential equations to describe the dynamics of the epidermal growth factor (EGF) signal transduction pathway in which the changes in concentration over time of 94 compounds after EGF stimulation were calculated. The model provides insight into EGF signal transduction and is useful for generating new experimental hypotheses.

7. DATABASE AND LABORATORY INFORMATION MANAGEMENT SYSTEM (LIMS) SOFTWARE & PUBLIC MICROARRAY DATABASES

Database and *Laboratory Information Management System* (LIMS) software has a unique position in microarray software. There are different data being generated at various stages of a microarray experiment, for example, clone and plate identities, microarray configurations, arraying conditions, raw signal intensities and analyzed data. Although LIMS tends to concentrate on the earlier stage of array manufacturing information management while database focuses more on the later stage - archival of analyzed data -, they share the common function to record and organize a large collection of data for rapid search and retrieval. The comparison among different sets of experiments is the basis of microarray analysis and faithful archival of information from various stages is critical for making this comparison meaningful and efficient.

Table 19.11 lists some of the common microarray database/LIMS software packages available nowadays. Most of them are based on relational database architecture that is either proprietary (*Genedirector*), using open source database like *PostgreSQL* (*Genetraffic*) or interfacing with commercially available databases like *Oracle* (*SMD*). A detailed comparison of microarray database software has been detailed elsewhere (Gardiner-Garden and Littlejohn, 2001). The future microarray database/LIMS development and even the experimental design should consider following a common standard like MIAME so that researchers can compare the results from different sources readily.

Table 19.11. Examples of database/LIMS software

Software	URL
Acuity	http://www.axon.com/GN_Acuity.html
AMAD	http://www.microarrays.org/software.html
ARGUS	http://vessels.bwh.Harvard.edu/software/argus/default.htm
ArrayDB	http://genome.nhgri.nih.gov/arraydb/
ArrayInformatics	http://www.packardbioscience.com/products/819.asp
BASE (BioArray Software Environment)	http://base.thep.lu.se/
CloneTracker	http://www.biodiscovery.com/clonetracker.asp
Genedirector	http://www.biodiscovery.com/genedirector.asp
GeNet	http://www.silicongenetics.com/cgi/SiG.cgi/Products/GeNet/index.smf?UID=14602
Genetraffic	http://www.iobion.com/
GeneX	http://genex.ncgr.org/
maxd (Manchester ArrayExpress Database)	http://www.bioinf.man.ac.uk/microarray/maxd/
NOMAD	http://ucsf-nomad.sourceforge.net/
Partisan ArrayLIMS	http://www.clondiag.com/products/sw/partisan/
Phoretix Array Professional	http://www.phoretix.com/products/array_professional.htm
Stanford Microarray Database (SMD) package	http://genome-www5.stanford.edu/MicroArray/SMD/download/

A closely related species of database software is *public microarray database*, a public repository of microarray data (Table 19.12). Having public databases are essential for the research community because they can provide raw data to validate published array results and aid the development of new data analysis tools. They also permit further understanding of your own data by comparing with the data from other groups or even *meta-mining*[12]. We believe the future public microarray databases should segregate into different specialties. There should be a generic database to allow everyone to deposit his data, a species-specific database that concentrates on particular species (e.g., yeast, Arabidopsis and Drosophila), and a disease specific database (e.g., cancer and cardiovascular). Such differentiation would greatly help the advance of research in different fields. However there

---

[12] Meta-mining refers to higher order data mining from the results of previous data mining results.

are many concerns raised on the effective ways for data dissemination, quality of the deposited data and protection of intellectual property (IP) rights for submitting data to public databases (Miles, 2001; Becker, 2001; Geschwind, 2001). These concerns have exemplified the urgent need to develop a practical microarray standard for the sake of data comparison and thoroughly discuss the IP issues among the research communities.

Table 19.12. Examples of public microarray databases

Database	URL
ArrayExpress	http://www.ebi.ac.uk/arrayexpress/
ChipDB	http://staffa.wi.mit.edu/chipdb/public/
ExpressDB	http://twod.med.harvard.edu/ExpressDB/
Gene Expression Omnibus (GEO)	http://www.ncbi.nlm.nih.gov/geo/
GeneX	http://genomics.biochem.uci.edu/genex/
Human Gene Expression Index (HuGE Index)	http://www.hugeindex.org/
READ (RIKEN cDNA Expression Array Database)	http://read.gsc.riken.go.jp/
RNA Abundance Database	http://www.cbil.upenn.edu/RAD2/
Saccharomyces Genome Database (SGD): Expression Connection	http://genome-www4.Stanford.EDU/cgi-bin/SGD/expression/expressionConnection.pl
Standford Microarray Database	http://genome-www.stanford.edu/microarray
Yale Microarray Database	http://info.med.yale.edu/microarray/
yeast Microarray Global Viewer (yMGV)	http://www.transcriptome.ens.fr/ymgv/

8. ANNOTATION SOFTWARE

Annotation software annotates the genes on the array by cross-referencing to public databases (Table 19.13). This type of software as well as a lot of turnkey systems and comprehensive software provide different levels of annotation functions. *Gene Ontology* (*GO*) database[13] is perhaps the most important database that every annotation software should refer to. GO provides a dynamic and well-controlled vocabulary for the description of the molecular functions, biological processes and cellular components of gene products that can be applied to all organisms even when the knowledge of gene and protein functions in cells is accumulating and changing. This is very useful for comprehending the functions of differentially regulated genes and pathways in which the clusters of genes are being co-regulated.

---

[13] http://www.geneontology.org

Table 19.13. Examples of annotation software.

Software	URL
DRAGON (Database Referencing of Array Genes ONline)	http://207.123.190.10/dragon.htm
Resourcerer	http://pga.tigr.org/tigr-scripts/magic/r1.pl

9. CONCLUSION

Information is the king and knowledge is the queen in this post-genomics era. He who can mine the useful information from the messy raw data and turn it into knowledge is God. The software that helps the mining process is Michael the Archangel who leads the victorious battle against the complex biological problems. If you want to win the battle, start working with your angel now!

ACKNOWLEDGEMENTS

We would like to thank Dr. Duccio Cavalieri from Bauer Center of Genomics Research, Harvard University, for his insightful discussion on Pathway Processor, and Ms. Alice Y.M. Lee for her assistance in proofreading this manuscript.

REFERENCES

Becker KG. (2001). The sharing of cDNA microarray data. Nat Rev Neurosci. 2:438-440.

Brazma A., Vilo J. (2000) Gene expression data analysis. FEBS Lett. 480:17-24.

Brazma A., Hingamp P., Quackenbush J., Sherlock G., Spellman P., Stoeckert C., Aach J., Ansorge W., Ball C.A., Causton H.C., Gaasterland T., Glenisson P., Holstege F.C., Kim I.F., Markowitz V., Matese J.C., Parkinson H., Robinson A., Sarkans U., Schulze-Kremer S., Stewart J., Taylor R., Vilo J., Vingron M. (2001) Minimum information about a microarray experiment (MIAME)-toward standards for microarray data. Nat Genet. 29:365-371.

D'haeseleer P., Liang S., Somogyi R. (2000). Genetic network inference: from co-expression clustering to reverse engineering. Bioinformatics. 16:707-726.

de Jong H. (2002). Modeling and simulation of genetic regulatory systems: a literature review. J Comput Biol. 9:67–103.

Eisen M.B., Spellman P.T., Brown P.O., Botstein D. (1998). Cluster analysis and display of genome-wide expression patterns. Proc Natl Acad Sci USA. 95:14863-14868.

Fiehn O. (2002). Metabolomics--the link between genotypes and phenotypes. Mol Biol. 48:155-171.

Gardiner-Garden M., Littlejohn T.G. (2001). A comparison of microarray databases. Brief Bioinform. 2:143-158.

Geschwind D.H. (2001). Sharing gene expression data: an array of options. Nat Rev Neurosci. 2:435-438.

Grosu P., Townsend J.P., Hartl D.L., Cavalieri D. (2002). Pathway Processor: a tool for integrating whole-genome expression results into metabolic networks. Genome Res. 12:1121-1126.

Hebestreit H.F. (2001). Proteomics: an holistic analysis of nature's proteins. Curr Opin Pharmacol. 2001; 1:513-520.

Horvath S., Baur M.P. (2000). Future directions of research in statistical genetics. Stat Med. 19:3337-3343.

Hughes T.R., Marton M.J., Jones A.R., Roberts C.J., Stoughton R., Armour C.D., Bennett H.A., Coffey E., Dai H., He Y.D., Kidd M.J., King A.M., Meyer M.R., Slade D., Lum P.Y., Stepaniants S.B., Shoemaker D.D., Gachotte D., Chakraburtty K., Simon J., Bard M., Friend S.H. (2000). Functional discovery via a compendium of expression profiles. Cell. 102:109-26.

Jain A.N., Tokuyasu T.A., Snijders A.M., Segraves R., Albertson D.G., Pinkel D. (2002). Fully automatic quantification of microarray image data. Genome Res. 12:325-332.

Jagota A. (2001). *Microarray Data Analysis and Visualization*. Bioinformatics by the Bay Press.

Knuden S. (2002). *A Biologist's Guide to Analysis of DNA Microarray Data*. New York: John Wiley & Sons, 2002.

Krause A., Olson M. (2000). *The Basics of S and S-Plus*. New York: Springer Verlag.

Leung Y.F., Lam D.S.C., Pang C.P. (2001). In silico biology: observation, modeling, hypothesis and verification. Trends Genet. 17:622-623.

Leung Y.F. (2002). Microarray data analysis for dummies... and experts too? Trends Biochem Sci. in press.

Leung Y.F., Pang C.P. (2002). Eye on bioinformatics – dissecting complex disease trait in silico. Applied Bioinform., in press.

Leung Y.F., Tam P.O.S., Lee W.S., Yam G.H.F., Chua J.K.H., Lam D.S.C., Pang C.P. (2002). The dual role of dexamethasone on anti-inflammation and outflow resistance in human trabecular meshwork. Submitted.

Lönnstedt I., Speed T.P. (2002). Replicated Microarray Data. Stat Sinica 12:31-46.

Miles M.F. (2001). Microarrays: lost in a storm of data? Nat Rev Neurosci. 2:441-443.

Nadon R., Shoemaker J. (2002). Statistical issues with microarrays: processing and analysis. Trends Genet. 18:265-271.

Phelps T.J., Palumbo A.V., Beliaev A.S. (2002). Metabolomics and microarrays for improved understanding of phenotypic characteristics controlled by both genomics and environmental constraints. Curr Opin Biotechnol. 13:20-24.

Quackenbush J. (2001). Computational genetics computational analysis of microarray data. Nat Rev Genet. 2:418-427.

Roberts C.J., Nelson B., Marton M.J., Stoughton R., Meyer M.R., Bennett H.A., He Y.D., Dai H., Walker W.L., Hughes T.R., Tyers M., Boone C., Friend S.H. (2000). Signaling and circuitry of multiple MAPK pathways revealed by a matrix of global gene expression profiles. Science 287:873-880.

Roth F.P., Hughes J.D., Estep P.W., Church G.M. (1998). Finding DNA regulatory motifs within unaligned noncoding sequences clustered by whole-genome mRNA quantitation. Nat Biotechnol. 16:939-945.

Schoeberl B., Eichler-Jonsson C., Gilles E.D., Muller G. (2002). Computational modeling of the dynamics of the MAP kinase cascade activated by surface and internalized EGF receptors. Nat Biotechnol. 20:370-375.

Selvin S. (1998). *Modern Applied Biostatistical Methods: Using S-Plus.* New York: Oxford University Press.

Sherlock G. (2001). Analysis of large-scale gene expression data. Brief Bioinform. 2:350-362.

Stewart J.E., Mangalam H., Zhou J. (2001). Open Source Software meets gene expression. Brief Bioinform. 2:319-328.

Tavazoie S., Hughes J.D., Campbell M.J., Cho R.J., Church G.M. (1999). Systematic determination of genetic network architecture. Nat Genet. 22:281–285.

Tomiuk S., Hofmann K. (2001). Microarray probe selection strategies. Brief Bioinform. 2:329-340.

Venables WN, Ripley BD. (1999). *Modern Applied Statistics With S-Plus.* New York: Springer Verlag.

Wu T.D. (2001). Analysing gene expression data from DNA microarrays to identify candidate genes. J Pathol 195:53-65.

Yang Y.H., Buckley M.J., Speed T.P. (2001). Analysis of cDNA microarray images. Brief Bioinform 2:341-349.

Yang Y.H., Dudoit S., Luu P., Lin D.M., Peng V., Ngai J., Speed T.P. (2002). Normalization for cDNA microarray data: a robust composite method addressing single and multiple slide systematic variation. Nucleic Acids Res. 30:E15.

Chapter 20

MICROARRAY ANALYSIS AS A PROCESS

Susan Jensen

SPSS (UK) Ltd,St. Andrew's House, West Street, Woking GU21 6EB, UK,
e-mail: sjensen@spss.com

1. INTRODUCTION

Throughout this book, various pieces of the microarray data analysis puzzle are presented. Bringing the pieces together in such a way that the overall picture can be seen, interpreted and replicated is important, particularly where auditability of the process is important. In addition to auditability, results of the analysis are more likely to be satisfactory if there is a method to the apparent madness of the analysis. The term "Data Mining" is generally used for pattern discovery in large data sets, and Data Mining methodologies have sprung up as formalised versions of common sense that can be applied when approaching a large or complex analytical project.

Microarray data presents a particular brand of difficulty when manipulating and modelling the data as a result of the typical data structures therein. In most industries, the data can be extremely "long" but fairly "slim" – millions of rows but perhaps tens or a few hundreds of columns. In most cases (with the usual number of exceptions), raw microarray data has tens of rows, where rows represent experiments, but thousands to tens of thousands of columns, each column representing a gene in the experiments.

In spite of the structural difference, the underlying approach to analysis of microarray retains similarities to the process of analysis of most datasets. This chapter sets out an illustrative example of a microarray Data Mining process, using examples and stepping through one of the methodologies.

2. DATA MINING METHODOLOGIES

As academic and industrial analysts began working with large data sets, different groups have created their own templates for the exploration and modelling process. A template ensures that the results of the experience can be replicated, as well as providing a framework for an analyst just beginning to approach a domain or a data set. It leaves an audit trail that may prevent repetition of fruitless quests, or allows new technology to intelligently explore results that were left as dead ends.

Published Data Mining methodologies have been produced by academic departments, software vendors and by independent special interest groups. There is general agreement on the basic steps comprising such a methodology, beginning with an understanding of the problem to be addressed, through to deploying the results and refining the question for the next round of analysis. Some methodologies concentrate solely on the work with the data, while others reinforce the importance of the context in which the project is undertaken, particularly in assessing modelling results.

Transparency of the Data Mining process is important for subsequent auditing and replication of the results, particularly in applications that face regulatory bodies. A very successful combination in Data Mining is a self-documenting analytical tool along with an established Data Mining methodology or process. There are several Data Mining tools (reviewed in Chapter 19) that implement a "visual programming" concept, where the steps taken in each phase of a methodology are documented and annotated by the creator, in the course of the project development. Examples of "visual programming" will be presented in later sections in this chapter. Importantly, though, any given methodology will generally be supported by whatever tool or set of tools is used for analysis, as the methodology is more a function of the user than the software.

An EU-funded project resulted in the Cross-Industry Standard Process for Data Mining (CRISP-DM; see http://www.crisp-dm.org). A consortium of some 200 representatives from industry, academia and software vendors set out to develop a common framework based upon the occasionally painful experiences that they had in early Data Mining activities. The idea was to improve both their own project management and to assist others tackling what could be an intimidating task, and the CRISP-DM document was published by the consortium in 1999.

Web searches and anecdotal information indicate that CRISP-DM is currently the single most widely-referenced Data Mining methodology, probably because the documentation is detailed, complete and is public access. CRISP-DM is visible as a project-management framework for consulting firms, and is actively supported by various Data Mining software vendors, such as NCR Teradata and SPSS, who were among the consortium participants. An example of extensions to CRISP-DM is in RAMSYS, (Moyle and Jorge, 1999), a methodology to improve remote collaboration on Data Mining projects. RAMSYS was (and continues to be) developed by members of SolEUNet (Solomon European Virtual Enterprise).

Because of this visibility, I will use CRISP-DM as the framework for an example of working through the analysis or mining of microarray data in the subsequent section.

3.   APPLICATION TO MICROARRAY ANALYSIS

Whatever the origin of data, Data Mining tasks have unsurprising similarities, beginning with "what am I trying to do here?" and ending with "what have I done and what can I do with it?" In the following sections I will lay out the more formal version of those questions in a microarray data analysis project, in CRISP-DM steps (Figure 20.1). At the beginning of each section will be a general flow diagram outlining the sub-steps within each of the steps, to give a gentle introduction to that part of the process, to be fleshed out by application to microarray data.

For a detailed, step-by-step, industry-neutral guide on steps and tips in using CRISP-DM, the booklet can be downloaded from the Web site.

The approach of a Data Mining process will be familiar, roughly following the structure of a scientific paper. The process begins with the introduction of objectives and goals, to the methods and materials of data cleaning, preparation and modelling, to evaluation of results, concluding with discussion and deployment – where the results lead, what benefit gained, next steps to take.

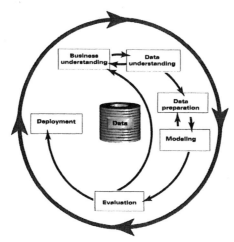

Figure 20.1. CRISP-DM Data Mining process flow, emphasising the feedbacks and iterative nature. From the CRISP-DM Web site (http://www.crisp-dm.org).

3.1   Business Understanding

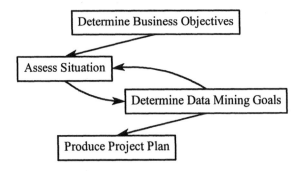

Figure 20.2. Steps within the CRISP-DM stage of Business Understanding; arrows indicate usual direction of work flow.

Analysis of microarray data is generally done with a specific objective or goal in mind, whether mapping active regions of a genome, classification of tissue types by response to a drug treatment, looking for gene expression patterns that might provide early indicators of a disease, and so on. This differs from a more classic Data Mining problem where someone is faced with a database and must think of appropriate, testable questions to ask of the data, in order to structure the information gained.

Because of the (likely) existing purpose in microarray data collection, the objectives (both the "business objective" and "Data Mining goal"; Figure 20.2.) of the analysis would likely already be determined. The fate of output of the project will obviously impact the goal: is it for scientific publication, is it part of an early drug discovery project where the results are to be fed to the cheminformatics department, or is it testing internal processes or equipment?

The public access leukemia gene expression data set described by Golub et al., (1999) can be used as a classification example throughout this process. The simplified and illustrative Data Mining process diagrams in this chapter were produced using the Clementine® Data Mining workbench.

Golub et al., (1999) describe one of the first attempts to classify cancer and tumour classes based solely on gene expression data, which is the clear business objective. The Data Mining goal was more specific: to use an initial collection of samples belonging to known classes to create a class predictor that could classify new, unknown samples (Golub et al., 1999, p531). The next level of detail, algorithm selection and evaluation, is addressed in the modelling phase of the process (below), after developing an understanding of the data involved.

Within the business understanding phase, once the business objectives and Data Mining goal are stated, the next step would be to inventory the resources. How much time might be required, and will the people doing the analysis be available to complete the project? If not, are there others that could step in? What computing resources will be required, appropriate hardware and software for completing the analysis to satisfy the objective?

An idea of how long the analysis will take, including all the data manipulation required, is probably an unknown until you have been through it once. During that time, the analyst(s) may be interrupted to attend conferences, or follow up on other projects. Has time or contingency plans been built in to deal with the interruption?

If CRISP-DM were being used to help maintain an audit trail of the analysis, a project plan would be produced at this stage, with the inventory of resources, duration, inputs, outputs and dependencies of the project.

3.2   Data Understanding

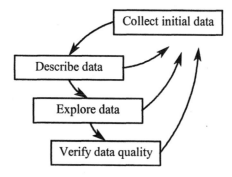

Figure 20.3. Steps within the CRISP-DM stage of Data Understanding; arrows indicate usual direction of work flow

Data from microarray readers can be stored in text format, databases or Excel® spreadsheets, and may resemble that depicted in Table 20.1. The collation of data from several experiments may be done manually or via database appending processes. If the analysis involves combining several formats, such as expression data with clinical data, then at this stage the formats would be checked and any potential problems documented. For example, if the clinical data were originally collected for a different purpose, unique identifiers for individuals may not match between the files, and have to be reconciled in order to merge the data later.

Table 20.1. Example of microarray data (Golub et al., 1999), experiments in columns, genes in rows; calls removed, and 6 of 33 experiments are visible here.

	Gene Description	Gene Accession Number	1	2	3	4	5	6
1	AFFX-BioB-5_at (endogenous control)	AFFX-BioB-5_at	-214	-139	-76	-135	-106	-138
2	AFFX-BioB-M_at (endogenous control)	AFFX-BioB-M_at	-153	-73	-49	-114	-125	-85
3	AFFX-BioB-3_at (endogenous control)	AFFX-BioB-3_at	-58	-1	-307	265	-76	215
4	AFFX-BioC-5_at (endogenous control)	AFFX-BioC-5_at	88	283	309	12	168	71
5	AFFX-BioC-3_at (endogenous control)	AFFX-BioC-3_at	-295	-264	-376	-419	-230	-272
6	AFFX-BioDn-5_at (endogenous control)	AFFX-BioDn-5_at	-558	-400	-650	-585	-284	-558
7	AFFX-BioDn-3_at (endogenous control)	AFFX-BioDn-3_at	199	-330	33	158	4	67
8	AFFX-CreX-5_at (endogenous control)	AFFX-CreX-5_at	-176	-168	-367	-253	-122	-186
9	AFFX-CreX-3_at (endogenous control)	AFFX-CreX-3_at	252	101	206	49	70	87
10	AFFX-BioB-5_st (endogenous control)	AFFX-BioB-5_st	206	74	-215	31	252	193
11	AFFX-BioB-M_st (endogenous control)	AFFX-BioB-M_st	-41	19	19	363	155	325
12	AFFX-BioB-3_st (endogenous control)	AFFX-BioB-3_st	-831	-743	-1135	-934	-471	-631
13	AFFX-BioC-5_st (endogenous control)	AFFX-BioC-5_st	-853	-239	-962	-577	-490	-625

The audit trail would, in this section, document any problems encountered when locating, storing or accessing the data, for future reference, particularly if data quality becomes a question when results are returned from the analysis.

It is during this data understanding phase that the analyst would begin exploring the data, graphically analysing attributes, looking for possible relationships between variables, and checking the distribution of classes that might be predicted (c.f. Figure 20.3., depicted in Figure 20.4.). As with most data analyses, if there is an interesting pattern that comes out immediately, it is probably data contamination.

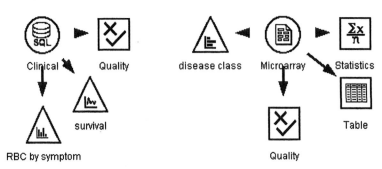

Figure 20.4. Simplified example of process of examining and getting acquainted with data during the data understanding phase of a project. RBC indicates red blood cell count in the clinical data.

Time spent on examination of the data at this phase will pay off, as that is when data import or export errors are detected, and coding or naming problems are found and corrected (and documented). By the end of this phase, the analyst should have confidence in the quality and integrity of the data, and have a good understanding of the structure relative to the desired goal of the analysis.

3.3 Data Preparation

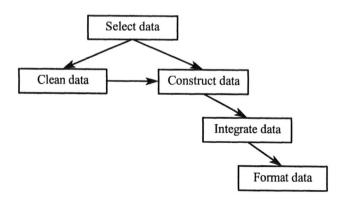

Figure 20.5. Steps within the CRISP-DM stage of Data Preparation; arrows indicate usual direction of work flow.

The objective of this step, as indicated in Figure 20.5., is to get all of the data ready to do the modelling. The data understanding and data preparation phases of any project usually represent about 80% of the work and about 90% of the tedium and frustration in dealing with data from any industry. Since techniques on data cleaning and imputation appropriate to the microarray analysis field are still evolving, much time may be spent in preparation in order to test, for example, the effects of different types of data normalisation or missing value imputation. There are several chapters in this book that detail various aspects of preparing microarray data for analysis, so I will speak in more general terms of selection, cleaning, and enriching the data.

Transposing the data, so that the experiments/subjects are in rows and genes in columns, may sound trivial but is one of the early and major hurdles for large data sets. Some steps in cleaning the data are easiest with genes as rows, others easiest with genes as columns. It is relatively simple to do the transposition in Excel provided that the end result will have only 255 genes or columns, which is unlikely with modern microarray data sets, which can hold up to 30,000 genes. Otherwise, it can be easily accomplished in a couple of steps in some Data Mining software, or through construction and implementation of repeated Perl scripts.

The preparation process generally begins with selection of the data to be used from the set collated. If Affymetrix chips are used, then removal of the calls and control fields is necessary for a clean analysis set. Expression levels may have registered at very low and very high levels, and the extremes are generally not meaningful, so deciding on upper and lower thresholds, and setting the outlier values to those thresholds will remove unnecessary noise. For example, it may be decided that values less than 20 and greater than 1,600 are not biologically meaningful; rather than simply exclude those values, the values less than 20 are replaced with 20, and those greater than 1,600 are replaced with the upper threshold of 1,600. Then the next step of finding and removing the invariant genes (i.e. those genes that are not differentially expressed) from the data will help to reduce the data set to a more meaningful set of genes (c.f. Figure 20.6.).

Decisions about standardising, centralising and/or normalising of the expression data, and imputation of missing values are discussed in previous chapters. If there are choices to be made among these data manipulation stages, it may be that several of the options will be implemented to test the results. The result may be that where there was one raw data file, there is now a developing library of several files with different permutations of the manipulation: e.g. data with overall-mean missing value imputation, data with group-mean imputation, data with k-nearest neighbour imputation, each of the imputed data sets with mean-centering, each of the imputed data sets with 0-1 centring, and so on.

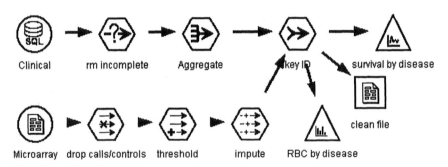

Figure 20.6. Simplified example of data preparation. This step involves cleaning, selecting, aggregating, dealing with missing values, and a lot of time and energy on the part of the analyst.

Documentation of the library resulting from these decisions is a good idea, for the sanity of the analyst and to ensure that all the sets are treated similarly when the modelling phase is reached, as well as for assistance when dealing with queries about the results. Further manipulation, such as creation of new variables, tags indicating tissue class, categorisation of a target field into an active/inactive dichotomy or discretization of numeric fields, should also be documented for future reference and reproducibility.

If the microarray data are to be used in conjunction with clinical data, then the clinical data will be cleaned during this phase as well, so that the resulting formats will be compatible or at least comprehensible. Decisions may have to be made on whether to aggregate, for example, lab exam results of the subjects by averaging, summing or taking the variance of the data for merging with the single line of gene expression data for each patient.

3.4   Modelling

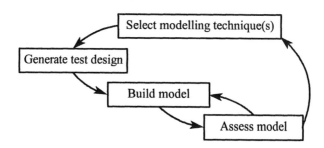

Figure 20.7. Steps within the CRISP-DM stage of Modelling; arrows indicate usual direction of work flow.

As mentioned in the last section, a single raw data file may, after the data preparation phase, have turned into a series of data sets that describe a matrix of normalisation and missing value replacement techniques, each of which to be analysed similarly for results comparison. It is often a good idea to be reminded of the goal of the project at this point to keep the analyses focused as you decide which modelling technique or techniques will be applied before embarking on the appropriate steps shown in Figure 20.7. For a given project, the list of algorithms should be narrowed to those appropriate and possible given the time, software and money allotted to the project.

There is an evolving variety of ways to reduce, classify, segment, discriminate, and cluster microarray data, as covered in Chapters 5 through 18 of this book. By the time you finish your project, more techniques, and tests on existing techniques, will have been published or presented at conferences. Because of these changes, it is very important to keep in mind (and preferably on paper) the assumptions, expected types of outcomes and limitations of each technique chosen to be used in the analysis, as well as how the outcome of each technique to be used relates to the Data Mining goal.

Documentation of why a particular modelling technique is used is also advisable because behaviours of techniques only become apparent after considerable use in different environments. As such, recommendations of when it is appropriate to use a given technique will likely change over time, and it is useful to have a reference. For example, if a project decided that *boosting* (where the case weights in a data set are adjusted by the modelling algorithm to accentuate the hard-to-classify cases, which should improve the modelling of those cases) was ideal in a 5-class classification problem, and a year later a paper is published that points out shortcomings in such a use of boosting, then the project will have on record that it used the best information available at the time. The record should also make it more straightforward to repeat the analysis with a change in the boosting effect.

Going back to the example using the Golub et al. (1999) data set, the data will be thresholded, invariant genes removed, data normalised and clean at this point, and so modelling can begin. In an ideal world, the data could be run and the model(s) and results admired immediately, but in reality, once modelling begins, more data manipulation usually becomes necessary. It may be that the thresholding levels were too lax; for example instead of a lower limit of 20 and an upper of 1,600, 40 and 1,500 would be more appropriate. Again, keep track of the changes made and run the models again.

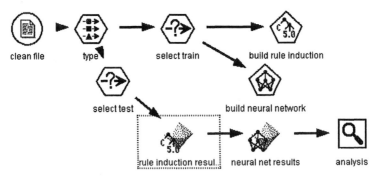

Figure 20.8. Simplified example of building and comparing modelling techniques on a single data set. Automation can be used to set up and run a series of data sets through the same analysis, reducing human error and tedium. Rule induction and neural networks (ANNs) are two commonly-used Data Mining algorithms.

When results come back, interpretability of the results is very algorithm- and software-dependent – some output is aesthetically pleasing, some takes a bit more effort. Examples of output represented in Figure 20.8. are presented in Figure 20.9. Rule Induction (Decision Tree) will present the gene(s) most significantly contributing to the discrimination between the classes (in this case two classes) and a Neural Network will present a list of relative importance to the discrimination. At this point, algorithms may be discarded or modified because they are not producing the detail of output that is required to address the business objective. If the analyst prefers the Decision Tree output format and interpretability, but the Decision Tree insists that just one gene is doing all the discrimination, then perhaps a *bagging* approach to the decision tree would be more appropriate than relying on a single tree: bagging consists of select a series of subsets of genes to run against the class, and combine the resulting trees for an overall predictor (c.f. Chawla et al., 2001).

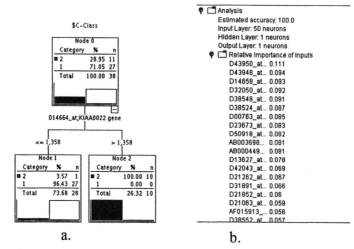

a.   b.

Figure 20.9. Examples of model output: a) rule induction (decision tree) classification, and b) neural network (ANN) classification.

Nor is it necessary to just use one modelling technique in isolation; if two techniques are good classifiers but pick up different idiosyncrasies of the data, use the results of both to predict new cases. For example, accept the prediction of a particular class only in those cases where both modelling techniques agree, or accept the prediction of a class where at least one modelling technique accurately classifies (Figure 20.10.). The analyst can be creative with the use and combination of modelling techniques when searching for novel patterns.

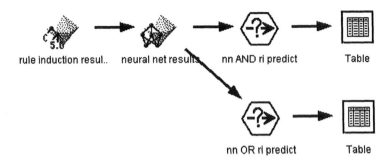

Figure 20.10. Simplified example of post-model processing to select and examine results of models, including a portion of Figure 20.8. and continuing the stream. The "nn" indicates neural networks (ANNs) and "ri" indicates rule induction. Stream illustrates the selection of cases where both algorithms agree on a classification or prediction, in the AND select node; and where either one classifies a case into a particular group in the OR select node.

Once the modelling techniques have been decided, a test design should accompany the selections, which describes the process by which the models will be tested for generalisability beyond the model-building data set and for success of the modelling exercise. Perhaps the most important aspect of the test design is deciding how the data are to be split into train and test samples, or use of cross-validation. In non-microarray data sets with many cases (more cases than variables) a train:test ratio can comfortably be up to 80:20. In microarray data a classification problem might be trying to discriminate among 4 classes spread across 40 cases. If 80% of the data are used to build the model, only 2 cases of each class will be left for testing the model's generalisability. If it fails, it will be difficult to see where and why it failed, and how to improve the model. Whatever the ratio chosen, training and testing data are also often selected at random from the raw data set. In such small data sets, if 80% of the data are randomly selected as training, it is possible that only half of the classes would show up in the testing data set at all. Because of these dangers, cross-validation is rapidly picking up in popularity, and most modelling software supports cross-validation. Golub et al. (1999) describe the use of both types of validation in the classification analysis.

What will be the success criteria, or what sort of error rates will be used to determine the quality of model result? Answers will vary with data and modelling technique chosen. In a supervised modelling task, will classification rate be sufficient or will there be more weight put on the control of false positives? Using unsupervised techniques for class discovery, it may be more difficult to put a success or failure label on the results; looking for clusters of similarly expressing genes may be a good start, but will that give you the answer that will fulfil project objectives – will the cheminformatics department be able to work with the cluster results or do they want further labelling of the clusters? Do the survival probabilities of the clusters make sense? Clustering techniques most often require human intervention, and an analysis of whether or not the clusters make sense will determine success. Again, as part of the documentation process, this subjective stage and possible ranges of results should be noted so that if an aberrant result comes up it can be rectified quickly and productivity is not lost.

If there are more than a few data sets to be analysed, generate a tick-list where completion of modelling steps for each data set can be noted, with space to document where the results can then be found. Having made it this far through the process, it is a good idea to ward against project collapse should the analyst be offered a better job elsewhere, hit the lecture circuit or decide not to return from a conference.

Algorithms, whether statistical or machine learning, generally have many parameters to tweak, and unless the analyst is trying to replicate a particular protocol, the parameters should be sensibly tweaked to see what happens. Some of the results will be rubbish, some will be useful, but all should be assessed as to whether or not they make sense. Iterate model building and assessment, throwing away models that are not useful or sensible, but keep track of what has been tried or it will inevitably be scrolled through again.

When the tick-list of models has been run through, there may be a series of models to rank if only the most successful of them are to be kept for use on future projects. This is where the previously-determined success criteria are used to greatest value, as it is easy to start hedging about the definition of success if the pet algorithm or freeware did not make into the top 5 hitlist. Having success criteria in place will also be a signal of when the parameter tweaking can be stopped and the process evaluated. Many projects simply stop when time or money runs out, while the actual analytical objectives had been long-satisfied and time has been lost in the drug discovery or publication process.

3.5   Evaluation of Project Results

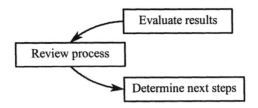

Figure 20.11. Steps within the CRISP-DM stage of Evaluation; arrows indicate usual direction of work flow.

In the modelling phase, the individual models' results are assessed for error and sensibility, and in this evaluation phase, the whole process is similarly assessed as outlined in Figure 20.11. Was there a bottleneck in the project – was it a problem with computers, people, data access? How can that be removed or decreased? How did the modelling perform relative to the Data Mining goal? Was a single model found that had the best classification rate? Or will a combination of models be used as the classification algorithm, with some pre- and post-model processing? What about the business objective? If the Data Mining goal has been achieved, then classification of new cases based on expression levels should be a trivial leap, as the data manipulation and cleaning required of the new cases, as well as the modelling techniques to be used, will be documented and possibly even automated by this point.

Did the project uncover results not directly related to the project objective that perhaps should redefine the objective or even define the next project? Results are a combination of models and findings, and peripheral findings can lead to both unexpected profits and attractive dead ducks. As such, it may be prudent to keep track of them, but perhaps save the pursuit of them for the next project.

3.6 Deployment

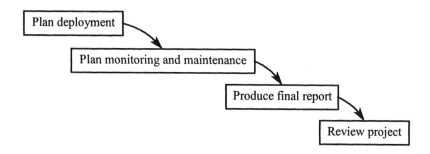

Figure 20.12. Steps within the CRISP-DM stage of Deployment; arrows indicate usual direction of work flow.

Deployment is a very broad term. Essentially "doing something with the results", it can cover anything from writing a one-off report to describe the analytical process, the writing of a thesis or publication, to informing an ongoing laboratory protocol, to the building of knowledge that will help take the field forward. Figure 20.12 shows the general steps that are taken when approaching the deployment phase of a project.

If the results or process are to be part of a laboratory protocol, once the desirable report (visualisation, modelling protocol or text/Powerpoint® output) is established, will it be a regular report? If so, the final process that was established through the above steps can be automated for re-use with similar data sets. The complexities of the data storage, cleaning and manipulation, model creation and comparison, and output-report production could be built into the automation to run with a minimum of human intervention, to save human patience for interpretation of the results. A production and maintenance schedule could be built to support the ongoing exertions, possibly with modelling-accuracy thresholds put in place to check the ongoing stability and validity of the model. In a production mode, alerts could be set up to warn when the models are no longer adequate for the data and should be replaced.

From the deployment phase of the Data Mining process, a summary of findings and how they relate to the original objectives – essentially a discussion section – would be a sensible addition to audit documentation. If you are the analyst that is offered a better job, it would be much nicer of you to leave a trail that your replacement could pick up on, and you would hope for the same from your new position.

The stream-programming concept of Data Mining shows its strengths at the evaluation and deployment stages, in a production environment as well as in cases of one-off analysis projects. The steps are self-documenting and visual, and the models quickly rebuilt using the same pre- and post-modelling processing steps as well as the same model parameters, as found with the traditional code programming. Visual programming permits a new analyst to pick up or repair an existing project with relatively small loss of productivity.

4.   EPILOGUE

An analyst in the field of microarray data will be suffering the Red Queen syndrome for many years to come, as new algorithms and data manipulations are evolved, tested and discarded. If chips become less expensive, as usually happens with technology, then in the near future more experiments on a given project may be run and larger data sets generated, leading to yet another set of algorithms being appropriate for the peculiarities of microarray data. As such, establishing a framework for Data Mining that includes flexible boundaries and change, planning for the next steps to include testing of new algorithms or to tighten or loosen model success criteria, will prevent getting caught in the tulgey wood.

REFERENCES

Chawla N., Moore T., Bowyer K., Hall L., Springer C., Kegelmeyer P. (2001). Investigation of bagging-like effects and decision trees versus neural nets in protein secondary structure prediction. Workshop on Data Mining in Bioinformatics, KDD.

CRISP-DM: Cross-Industry Standard Process for Data Mining. Available at http://www.crisp-dm.org.

Golub T.R., Slonim D.K., Tamayo P., Huard C., Gaasenbeek M., Mesirov J.P., Coller H., Loh M., Downing J.R., Caligiuri M.A., Bloomfield C.D., Lander E.S. (1999). Molecular classification of cancer: class discovery and class prediction by gene expression monitoring. Science 286(5439):531-537.
(Data set available at: http://www-genome.wi.mit.edu/mpr/data_set_ALL_AML.html)

Moyle S., Jorge A. (2001) RAMSYS – A methodology for supporting rapid remote collaborative data mining projects. In Christophe Giraud-Carrier, Nada Lavrac, Steve Moyle, and Branko Kavsek, editors, *Integrating Aspects of Data Mining, Decision Support and Meta-Learning: Internal SolEuNet Session*, pp. 20-31. ECML/PKDD'01 workshop notes.

Index